VOLUME FIVE

ADVANCES IN STEM CELLS AND THEIR NICHES

The Cancer Stem Cell Niche

VOLUME FIVE

ADVANCES IN STEM CELLS AND THEIR NICHES

The Cancer Stem Cell Niche

Edited by

SUSIE NILSSON
Biomedical Manufacturing Commonwealth Scientific and Industrial Research Organisation (CSIRO), Melbourne, Australia;
Australian Regenerative Medicine Institute, Monash University, Melbourne, Australia

ELSEVIER

ACADEMIC PRESS
An imprint of Elsevier

Academic Press is an imprint of Elsevier
50 Hampshire Street, 5th Floor, Cambridge, MA 02139, United States
525 B Street, Suite 1650, San Diego, CA 92101, United States
The Boulevard, Langford Lane, Kidlington, Oxford OX5 1GB, United Kingdom
125 London Wall, London EC2Y 5AS, United Kingdom

First edition 2021

ISBN: 978-0-323-85325-5
ISSN: 2468-5097

For information on all Academic Press publications
visit our website at https://www.elsevier.com/books-and-journals

Publisher: Zoe Kruze
Editorial Project Manager: Leticia M. Lima
Production Project Manager: James Selvam
Cover Designer: Miles Hitchen

Typeset by SPi Global, India

Contents

Contributors

Juan Aréchaga
Department of Cell Biology and Histology, Faculty of Medicine and Nursery; Analytical & High Resolution Biomedical Microscopy Core Facility (SGIker), University of the Basque Country (UPV/EHU), Leioa, Spain

Ivan Bertoncello
Lung Health Research Centre, Department of Pharmacology and Therapeutics, University of Melbourne, Melbourne, VIC, Australia

Giacomo Canesin
Department of Surgery, Cancer Research Institute, Beth Israel Deaconess Medical Center and Harvard Medical School, Boston, MA, United States

Benjamin Cao
Biomedical Manufacturing, Commonwealth Scientific and Industrial Research Organization; Australian Regenerative Medicine Institute, Monash University, Clayton, VIC, Australia

Jessica De Angelis
Tissue and Tumor Microenvironments Group, The Kennedy Institute of Rheumatology, University of Oxford, Oxford, United Kingdom

Alejandro Díez-Torre
Analytical & High Resolution Biomedical Microscopy Core Facility (SGIker), University of the Basque Country (UPV/EHU), Leioa, Spain

Yan Gu
Department of Medicine, McMaster University; The Research Institute of St Joe's Hamilton; Urological Cancer Center for Research and Innovation (HCCRI), St. Joseph's Hospital, Hamilton, ON, Canada

Shen Y. Heazlewood
Biomedical Manufacturing, Commonwealth Scientific and Industrial Research Organization; Australian Regenerative Medicine Institute, Monash University, Clayton, VIC, Australia

Ingunn Holen
University of Sheffield, Medical School, Sheffield, United Kingdom

Lucía Jimenez-Rojo
Department of Cell Biology and Histology, Faculty of Medicine and Nursery, University of the Basque Country (UPV/EHU), Leioa, Spain

Anil Kapoor
Department of Medicine, McMaster University; The Research Institute of St Joe's Hamilton; Urological Cancer Center for Research and Innovation (HCCRI), St. Joseph's Hospital, Hamilton, ON, Canada

Anjali P. Kusumbe
Tissue and Tumor Microenvironments Group, The Kennedy Institute of Rheumatology, University of Oxford, Oxford, United Kingdom

Taosha Li
Life-Tech Industry Alliance, Shenzhen, China

Xiaozeng Lin
Department of Medicine, McMaster University; The Research Institute of St Joe's Hamilton; Urological Cancer Center for Research and Innovation (HCCRI), St. Joseph's Hospital, Hamilton, ON, Canada

Jingyi Peng
Department of Medicine, McMaster University; The Research Institute of St Joe's Hamilton; Urological Cancer Center for Research and Innovation (HCCRI), St. Joseph's Hospital, Hamilton, ON, Canada

Claire L. Pritchard
Biomedical Manufacturing, Commonwealth Scientific and Industrial Research Organization; Australian Regenerative Medicine Institute, Monash University, Clayton, VIC, Australia

Lewis Quayle
University of Sheffield, Medical School, Sheffield, United Kingdom

Saravana K. Ramasamy
Institute of Clinical Sciences; MRC London Institute of Medical Sciences, Imperial College London, London, United Kingdom

Unai Silván
BCMaterials, Basque Center for Materials, Applications and Nanostructures, UPV/EHU Science Park, Leioa; IKERBASQUE, Basque Foundation for Science, Bilbao, Spain

Unnikrishnan Sivan
Tissue and Tumor Microenvironments Group, The Kennedy Institute of Rheumatology, University of Oxford, Oxford, United Kingdom

Damu Tang
Department of Medicine, McMaster University; The Research Institute of St Joe's Hamilton; Urological Cancer Center for Research and Innovation (HCCRI), St. Joseph's Hospital, Hamilton, ON, Canada

CHAPTER ONE

Acute lymphoblastic leukemia and the bone marrow microenvironment

Claire L. Pritchard[a,b], Shen Y. Heazlewood[a,b], and Benjamin Cao[a,b,*]
[a]Biomedical Manufacturing, Commonwealth Scientific and Industrial Research Organization, Clayton, VIC, Australia
[b]Australian Regenerative Medicine Institute, Monash University, Clayton, VIC, Australia
*Corresponding author: e-mail address: ben.cao@csiro.au

Contents

1. Introduction

Hematopoietic stem cells (HSC) are the rare, multipotent cells that reside in the bone marrow (BM) and are responsible for the life-long production of all types of mature blood cells (reviewed in (Kiel & Morrison, 2008). The signals that induce HSC differentiation and self-renewal at the population level is a tightly regulated process, carefully controlled through both cell intrinsic mechanisms as well as cell extrinsic interactions

Advances in Stem Cells and their Niches, Volume 5
ISSN 2468-5097
https://doi.org/10.1016/bs.asn.2020.11.001

with their BM microenvironment to ensure sustainable hematopoiesis without depletion of the stem cell pool (reviewed in Domingues, Cao, Heazlewood, Cao, & Nilsson, 2017; Mendelson & Frenette, 2014). Nevertheless, dysregulation of steady state hematopoiesis both due to extra- and intra-cellular changes drive the development of malignant hematopoietic cells, such as leukemia (reviewed in Mullighan, 2012; Tabe & Konopleva, 2014). These diseased cells can re-appropriate the BM microenvironment and inhibit the production of functional blood cells (reviewed in Barwe, Quagliano, & Gopalakrishnapillai, 2017; Lim et al., 2016).

One predominant subtype of leukemia, acute lymphoblastic leukemia (ALL), is characterized by the developmental arrest of immature lymphoblasts, resulting in the rapid onset and progression of excess blasts within the BM and blood. ALL is the most common childhood cancer and despite the unmistakable improvements to cure rates, drug resistance continues to be a substantial problem in the 10–20% of patients that relapse (Einsiedel et al., 2005; Hunger & Mullighan, 2015; Pui & Evans, 2006; Stary et al., 2014). Two major phenotypes of ALL have been defined; B-cell ALL (B-ALL) and T-cell ALL (T-ALL). Although B-ALL is the most common form of ALL, approximately 15% and 25% of newly diagnosed ALL cases in children and adults are T-ALL, respectively, and are associated with worse prognosis compared to B-ALL (Aifantis, Raetz, & Buonamici, 2008).

Unlike acute myeloid leukemia (AML), where leukemic stem cells (LSC) have been identified as a phenotypically distinct stem cell population that drive disease progression, (Bonnet & Dick, 1997; Jordan et al., 2000), no clear LSC population has been identified in ALL due to the numerous clones that drive this heterogenic disease, suggesting that a subset of ALL cells maintain stem like properties, allowing for disease progression and expansion (Cobaleda et al., 2000; le Viseur et al., 2008; Rehe et al., 2013). In addition, the findings of Fries et al. demonstrate that genetic heterogeneity and clonal dominance differs between the BM and peripheral blood, highlighting the potential impact of tissue site selection on the detection of minor subclones in B-ALL (Fries et al., 2020) that may increase the risk of minimal residual disease (MRD) and relapse (Berry et al., 2017; Borowitz et al., 2008; Fries et al., 2020). Furthermore, chemoprotection is also offered *via* interactions with the tissue microenvironment, altering ALL metabolic pathways, survival, proliferative and metastatic responses and shielding leukemic cells from clinical treatment. Interestingly, the role of the BM microenvironment in leukemic cell propagation and drug resistance differs for B- and T-ALL. T-cell polarized lymphoid progenitors receive signals for survival and differentiation from the thymus (reviewed in Muro, Takayanagi, & Nitta, 2019) and T-ALL cells do not preferentially localize to specialized niches within the BM, but instead are dynamic and randomly distributed (Hawkins et al., 2016). A thorough exploration of how T-ALL cells interact with their BM microenvironment, driving drug resistance is beyond this examination, and reviewed elsewhere (Vadillo, Dorantes-Acosta, Pelayo, & Schnoor, 2018; Witkowski, Kousteni, & Aifantis, 2020). In contrast, normal B cell development relies on both cell autonomous mechanisms as well as factors secreted by osteoblasts and stromal cells within BM (reviewed in Carsetti, 2000; Hagman, 2015; Nagasawa, 2006), with B-ALL development and progression also highly dependent on interactions within their microenvironment. Notably, B-ALL cells hijack the BM microenvironment inducing structural and cellular changes, that protect and support disease progression allowing these cells to outcompete healthy resident hematopoietic cells, indicating the need for eradication of B-ALL in BM *via* inhibition of multiple mechanisms of chemoresistance.

This chapter will review how B-ALL are regulated by, and alter both the cellular and extracellular components of the BM microenvironment (Fig. 1) and will explore the mechanisms exploited by leukemic cells to abrogate normal hematopoietic function and induce resistance to clinical therapies.

2. Cellular components of the ALL niche

2.1 Osteoblasts and osteoclasts

Reduced bone mass and loss of bony trabeculae is characteristic of B-ALL patients (Leeuw, Koudstaal, Wiersema-Buist, Kamps, & Timens, 2003; Nguyen et al., 2015), with bone density further impaired by clinical treatment, resulting in additional bone demineralization and skeletal toxicity (reviewed in Mostoufi-Moab & Halton, 2014). Bone homeostasis requires balanced function of both osteoblasts and osteoclasts, key osteogenic lineages of mesenchymal stem cells (reviewed in Naji et al., 2019). Lining the bone, osteoblasts are responsible for forming new bone but are also key regulators of HSC proliferation/quiescence, differentiation and survival (reviewed in Taichman & Emerson, 1998). Specifically, osteoblasts secrete macrophage colony-stimulating factor (M-CSF) and receptor activator of NF-kB ligand (RANKL) (Suda et al., 1999; Tanaka et al., 1993), that activate osteoclast maturation and bone reabsorption, resulting in bone degradation through the release of proteolytic enzymes including matrix metalloproteinase-9 (MMP-9) (Kusano et al., 1998) and cathepsin K (reviewed in Goto,

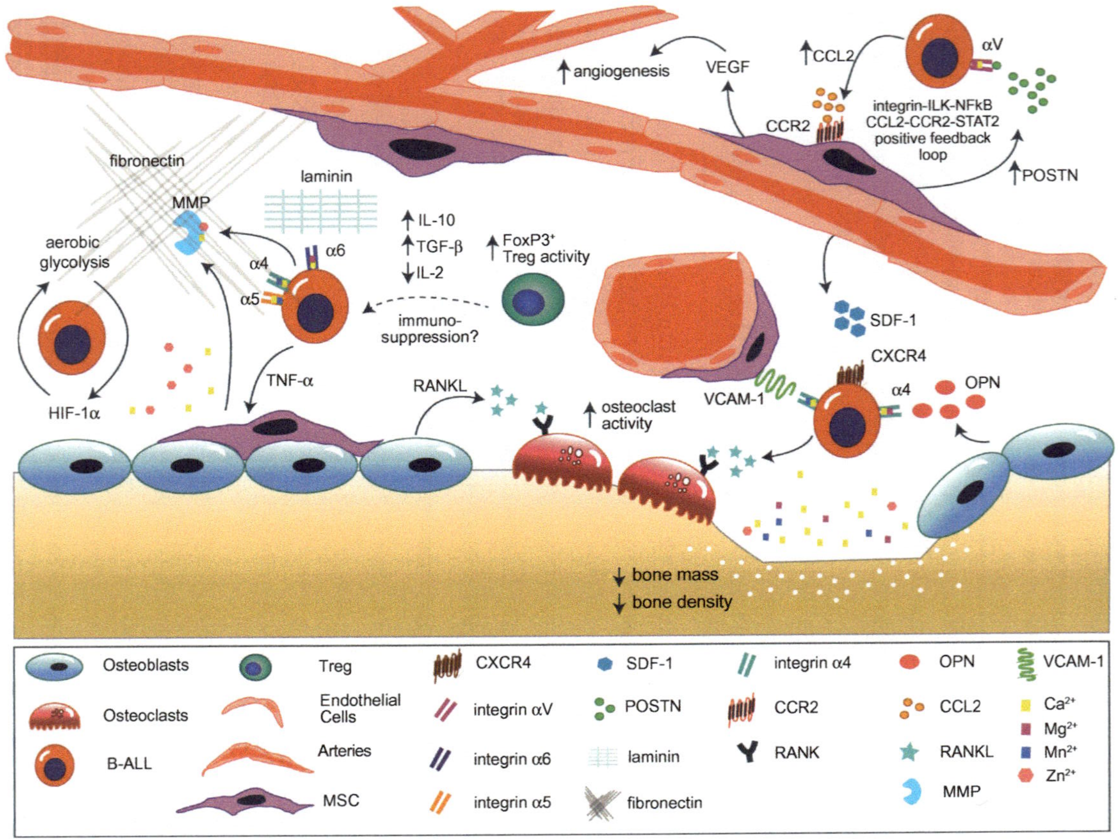

Fig. 1 See figure legend on opposite page.

Yamaza, & Tanaka, 2003). These secreted enzymes degrade type I and IV collagen, stromal derived factor 1 (SDF-1), osteopontin (OPN) and stem cell factor (SCF) (Kollet et al., 2006), which regulate hematopoietic function. While osteoblasts are key sources of RANKL, B-ALL cells have recently been shown to secrete RANKL, advancing bone degradation by manipulating RANK/RANKL homeostatic signaling (Cheung et al., 2018; Rajakumar et al., 2016, 2020).

By remodeling their microenvironment, leukemic cells suppress healthy hematopoiesis to support their own growth and transition into chemoresistant states (reviewed in Barwe et al., 2017; Sison & Brown, 2011). As essential sources of extracellular matrix (ECM) and soluble factors that support leukemiagenesis, osteoblastic cell lineages are key targets for the leukemic disruption of hematopoiesis, though their role remains multifaceted. ALL cells have been shown to prevent osteogenic differentiation of mesenchymal stromal/stem cells (MSC) *in vitro* by withholding NOTCH activation (Yang, Xu, Chen, & Wang, 2015), which may be a survival response, as increased osteoblasts *in vivo* suppressed ALL survival and metastasis in leukemic mice (Krevvata et al., 2014). Both B- and T-ALL are responsible for niche remodeling through destruction of osteoblasts

Fig. 1 Schematic depiction of B-ALL cell regulation in the BM microenvironment. B-ALL engraftment contributes to remodeling of the BM microenvironment through several mechanisms. Leukemic cells and MSC both activate osteoclasts via RANK/RANKL signaling, resulting in increased bone resorption and decreased bone mass/density. In addition, ALL can induce expression of MMPs, either directly, or by inducing MMP secretion from MSC *via* TNF-α signaling. VEGF secreted by MSC also increases BM vascularization and it has been recently postulated that ALL disease progression is associated with intense OxPhos in B-ALL, which leads to the hypoxic leukemic microenvironment despite sufficient O_2 supply. Alternatively, B-ALL have been associated with altered metabolism dependent on aerobic glycolysis, characterized by increased lactate production in BM. Lactate has been shown to activate HIF-1α in the absence of hypoxia, which stimulate genes that promote glucose uptake (glucose transporters) and vascularization (e.g. VEGF). Several adhesive interactions exist between ALL and the microenvironment to mediate adhesion (α4/VCAM-1; α4/tcOPN; α5/FN; α6/laminin; CXCR4/SDF-1), homing (α4/VCAM-1; CXCR4/SDF-1) and chemoresistance (α4/VCAM-1; α4/tcOPN; CXCR4/SDF-1; α6/laminin; α4 and α5/FN). In addition, POSTN secreted by MSC has been shown to activate integrin signaling in ALL via binding to αV, stimulating CCL2 secretion and activating CCL2-CCR2-STAT3 signaling in MSC, promoting further POSTN release that ultimately results in increased ALL proliferation/burden and infiltration in peripheral organs. Other cells in BM that may play a role in B-ALL resistance include Treg, which adopt an enhanced and activated immunosuppressive phenotype that may be involved in B-ALL immune evasion and endothelial cells, which secrete SDF-1 and mediate homing and engraftment of ALL through binding to CXCR4.

in vivo (Cheung et al., 2018; Hawkins et al., 2016), directly suppressing hematopoietic stem and progenitor cells (HSPC) and lymphocytic homeostasis (Wang et al., 2016). To further deplete the osteoblast contribution to the ECM, ALL cells induce collagen, fibronectin (FN) and laminin degradation *via* MMP, either directly secreted by ALL cells (Feng et al., 2011; Hendrix et al., 1992; Ivanoff, Ivanoff, Hultenby, & Sundqvist, 1999) or through tumor necrosis factor-α (TNF-α) induced MSC (Verma et al., 2020). Despite this targeted reduction of osteoblasts, cell-cell/matrix interactions between ALL and osteoblasts also afford chemoprotection (Quagliano, Gopalakrishnapillai, & Barwe, 2017). Specifically, p11 is upregulated in relapsed B-ALL, which binds to Annexin-2 expressed by osteoblasts, mediating B-ALL homing and chemoresistance to vincristine and dexamethasone (Gopalakrishnapillai et al., 2015). Similarly, greater concentrations of OPN at the endosteum of ALL xenografts was functionally associated with anchoring leukemic cells in distinct niches through binding to integrin α_4, inducing quiescence, with OPN neutralization pre-treatment significantly reducing ALL burden post chemotherapy (Boyerinas et al., 2013).

In contrast to osteoblasts, reducing osteoclast numbers in a leukemic mouse model did not alter disease progression or survival (Krevvata et al., 2014), suggesting osteoclasts do not regulate leukemiagenesis, reflecting previous findings that these cells are superfluous for HSC function (Miyamoto et al., 2011). However, B-ALL cells hijack and enhance osteoclast function to aid in destruction of bone, with early treatment of ALL murine models with the osteoclast inhibitor zoledronic acid restoring bone loss with reduced leukemic burden in BM and extramedullary sites including the spleen and peripheral blood (Cheung et al., 2018). Thus, therapeutic restoration of bone homeostasis may be an effective strategy for the treatment of B-ALL.

2.2 Mesenchymal stem/stromal cells (MSC)

Mesenchymal stromal/stem cells have been described as both cytoprotective of ALL cells as well as anti-leukemic, and this dichotomy may lie in their heterogeneity and the fact that there is still no universally agreed method for their identification and isolation. It has been shown that MSC produces several factors including vascular endothelial growth factor (VEGF), platelet derived growth factor (PDGF) and SDF-1 to positively regulate pre-B-ALL cells, and inhibition of p38 mitogen activated protein kinases (MAPK) significantly reduces these factors, and subsequently, the proliferation of ALL cells (Gaundar, Bradstock, & Bendall, 2009). The production of these factors could be an important role for MSC in ALL survival as p38 MAPK has been found to be up-regulated in pre-B-ALL samples (Ogasawara, Yasuyama, & Kawauchi, 2003). Of note, reactive oxygen species (ROS) induced by chemotherapeutics such as cytarabine or daunorubicin stimulate MSC to adopt an activated, cancer-associated fibroblast phenotype that allows them to transfer mitochondria to B-ALL cells *via* tunneling nanotubes and prevent oxidative stress-induced apoptosis (Burt et al., 2019). This was demonstrated using primary MSC isolated from ALL patient samples as well as in a murine xenograft model of ALL, and this MSC cytoprotective process could be interrupted *in vitro* and *in vivo* by agents that disrupt the formation of microtubules (Burt et al., 2019). Other examples of MSC being cytoprotective of ALL include: direct contact activating Notch-3 and -4 or Jagged-1/-2 and Delta-like-1 (DLL-1) signaling attributed to the anti-apoptotic effect on B-ALL cells (Nwabo Kamdje et al., 2011) and secretion of MSC-derived prostaglandin E_2 (PGE_2), which activates ALL cyclic adenosine monophosphate-protein kinase A (cAMP-PKA) signaling, protecting primary B-cell precursor (BCP)-ALL cells from p53 accumulation, and thus preventing cell death (Naderi et al., 2015). Conversely, others have found MSCs inhibit the proliferation of mouse ALL cells (P388), causing cell cycle arrest and increasing the proportion of apoptotic cells (Song et al., 2015). Similarly, co-culture of human BM MSC could decrease the proliferation of the pre-B-ALL cell line BV173 (Ramasamy et al., 2007; Sarmadi et al., 2010).

Like HSC, ALL also express the integrin α_4, with high expression levels in childhood B-ALL associated with poor response to treatment, reduced event-free and overall survival (Shalapour et al., 2011), which is mediated through binding to vascular cell adhesion molecule-1 (VCAM-1) expressed by MSC (reviewed in Härzschel, Zucchetto, Gattei, & Hartmann, 2020). Indeed, the α_4/VCAM-1 axis is the predominant interaction that promotes B-cell precursor adhesion in the BM microenvironment (Ryan, Nuccie, Abboud, & Winslow, 1991), B-ALL homing (Spiegel et al., 2004) and induces chemoresistance through activation of NF-kB (Jacamo et al., 2014).

While MSC can offer protection as well as cause growth inhibition and apoptosis in ALL cells *via* the release of soluble factors or through direct contact, it has also been recently demonstrated that ALL cells can alter MSC. Co-culture of ALL cell lines REH and SUP-B15 with MSC isolated from

BM aspirates of healthy pediatric donors showed that a "leukemic niche" could affect MSC in several ways: MSC showed increased senescence-associated β-galactosidase activity and p53 gene expression, produced elevated levels of cytoplasmic ROS and entered into G2/M cell cycle arrest, up-regulated adhesion molecules such as VCAM-1, intracellular cell adhesion molecule-1 (ICAM-1) and integrin α_5 as well as decreased osteogenic potential, with the authors concluding that leukemic cells may induce a senescence process in MSC because it favors leukemic growth (Bonilla, Vanegas, & Vernot, 2019).

2.3 Endothelial cells

Bone marrow vasculature and endothelial cells are important regulators of both normal and aberrant hematopoiesis. Notably, the development of ALL is associated with increased neovascularization (Aguayo et al., 2000; Perez-Atayde et al., 1997; Pulè et al., 2002), suggesting leukemia development may be dependent on angiogenesis and thus anti-angiogenic therapies may be effective for the treatment of ALL. Furthermore, soluble factors present in the BM of B-ALL patients were found to promote vascularization *ex vivo* and BM endothelium promotes leukemic cell survival through the anti-apoptotic molecule B-cell lymphoma-2 (BCL-2) (Veiga, Costa, Sallan, Nadler, & Cardoso, 2006). Of note, VEGF and basic fibroblast growth factor (bFGF) are key proangiogenic factors that promote vascularization, although in the context of B-ALL, there are discrepancies in the literature to the association of protein levels of VEGF and bFGF and disease outcome (reviewed in Schneider et al., 2011). Nevertheless, VEGF within the microenvironment has been shown to have a direct effect on B-ALL *via* activation of VEGF receptor-1 (VEGFR-1), which plays a role in ALL distribution within the BM and exit into the peripheral circulation (Fragoso et al., 2006). Furthermore, VEGF has also been shown to induce secretion of MMP-2 and MMP-9 (Poyer et al., 2009), that promotes B-ALL invasiveness (Verma et al., 2020), including migration to the central nervous system (CNS) (Münch et al., 2017).

Studies in murine models of B-ALL show vasculature comprised of specialized *E*-selectin$^+$ endothelium that secrete SDF-1 represent distinct anatomical locations within BM that are important for homing and engraftment of leukemic cells (Sipkins et al., 2005) through binding to C-X-C chemokine receptor 4 (CXCR4) expressed on B-ALL cells (Juarez et al., 2009). Indeed, administration of CXCR4 inhibitors in mouse models has been shown to mobilize ALL cells into the peripheral circulation (Juarez, Bradstock, Gottlieb, & Bendall, 2003; Juarez et al., 2007), indicating the CXCR4/SDF-1 axis in B-ALL is an important regulator of leukemic cell migration and extramedullary infiltration (reviewed in Burger & Bürkle, 2007). In addition, ALL cells that preferentially engraft within regions of $CD34^+$ HSPC homing also disrupt the vascular microenvironment by downregulating SDF-1 and upregulating SCF, thereby impairing normal progenitor function in a leukemic microenvironment (Colmone et al., 2008), indicating BM vasculature can both influence B-ALL development but can also be concomitantly disrupted by leukemic engraftment.

2.4 Regulatory T-cells

Regulatory T-cells (Treg) are a subpopulation of $CD4^+$ T-cells that co-express the transcription factor Forkhead box P3 (Foxp3) and are important regulators of immune homeostasis and self-tolerance, providing protection against autoimmune diseases (reviewed in Sakaguchi, Yamaguchi, Nomura, & Ono, 2008). Treg are well known to play important roles in ameliorating the detrimental effects of graft vs host disease (GvHD) in allogeneic transplantation, although its role in regulating normal and aberrant hematopoiesis in BM is less clear. Nevertheless, *in vivo* imaging in mice demonstrated BM resident $FoxP3^+$ Treg co-localize with allogeneic HSC following transplantation near the endosteal region, promoting HSC survival and improving engraftment in an interleukin-10 (IL-10) dependent manner (Fujisaki et al., 2011), while also maintaining HSC quiescence through adenosine production and protecting HSC from oxidative stress (Hirata et al., 2018). Depletion of BM $FoxP3^+$ Treg resulted in expansion of the HSC pool and concomitant reduction of B-cell progenitors and mature B-cells (Pierini et al., 2017), attributed to abrogation of perivascular niche cells in BM by persisting activated T cells following removal of immunosuppressive BM Treg. These studies suggest BM Treg provides an immune privileged site for HSC, regulates oxidative stress within the BM and also modulates perivascular niche cells and thus may have similar regulatory roles in maintaining B-ALL. Indeed, increased BM Treg has been associated with B-ALL (Idris et al., 2015) where an increase in the proportion of functionally active $FoxP3^+$ Treg with greater suppressive activity, characterized by increased capacity for secretion of IL-10 and transforming growth factor-β (TGF-β) as well as reduced IL-2 was observed in Treg from B-ALL patients (Bhattacharya, Chandra, & Mandal, 2014; Wu, Qing, Wu,

Zhu, & Zhou, 2012). Interestingly, the proportion of FoxP3$^+$ cells within the CD4$^+$CD25$^+$ compartment was reduced following complete remission in B-ALL patients to levels similar to normal individuals (Bhattacharya et al., 2014). Furthermore, it was demonstrated that the frequency of Treg is a predictor of patient outcome following CD19-targeted immunotherapy with lower levels of Treg corresponding to more favorable outcomes following blinatumomab treatment (Duell et al., 2017). Together, these data suggest Treg support an immunosuppressive and protective environment for B-ALL similarly to BM HSC. Thus, reduction of Treg or inhibition of Treg function in the leukemic microenvironment may be a promising strategy for improving leukemia treatment.

2.5 Central nervous system (CNS)

Treatment for childhood ALL now often involves cranial irradiation, intrathecal chemotherapy (delivery into the cerebral spinal fluid) or systemic application of chemotherapeutics with penetration into the CNS since before these measures, as much as 50% of children with ALL suffered from relapse originating from the CNS (Evans, Gilbert, & Zandstra, 1970). Interestingly, the majority of relapses involving the CNS occur in patients who were initially negative for ALL in the CNS, and while such treatments have reduced ALL relapse from the CNS to $< 5\%$, research in this area continues and cytokines such as IL-15 have been shown to be upregulated in childhood ALL with CNS infiltration, making IL-15 a possible prognostic tool for relapse involving the CNS (Cario et al., 2007). IL-15 is important in both pre-B ALL homing to the CNS *via* P-selectin glycoprotein-1 (PSGL-1), CXCR3 and serpin family E member 1 (SERPINE-1) upregulation as well as pre-B ALL proliferation in the CNS *via* Raf/Ras/ERK signaling (Williams et al., 2014) and although IL-15 can activate natural killer (NK) cells and decrease ALL growth in the periphery, IL-15 induced NK activity cannot prevent CNS relapse of ALL due to NK cells not being present in the brain region (Frishman-Levy et al., 2015).

IL-15 is released by cells such as monocytes and dendritic cells (reviewed in Bulfone-Paus, Bulanova, Budagian, & Paus, 2006; Cario et al., 2007), and both monocytes co-cultured with BCP-ALL as well as monocytes isolated from BCP-ALL patients show an upregulation in the release of CXCL10, that can increase BCP-ALL migration and invasive capacity through direct binding with its receptor CXCR3, or through the up-regulation of matrix degrading proteins such as MMP-9 (Lee et al., 2012). The role integrins play in BCP-ALL in the CNS, such as increased invasion and adhesion as well as increased resistance to chemotherapy, has been recently reviewed elsewhere (Scharff, Modvig, Marquart, & Christensen, 2020).

Using cell lines and BCP-ALL/NOD/SCID xenografts, VEGF was also shown to promote CNS leukemia: it is highly expressed in CNS derived leukemia cells, it is required for trans-endothelial migration of ALL cells and when an anti-VEGF treatment (bevacizumab) was administered *in vivo*, leukemic burden was reduced in the CNS but not in the BM, spleen or blood (Münch et al., 2017, 2015). And following analysis on cell lines and xenographs (engraftment of NALM-6 and Jurkat cells into NSG mice), it has also been suggested that other cells in the CNS, such as astrocytes, choroid plexus epithelial cells and meningeal cells can prevent chemotherapy-induced apoptosis and increased quiescence in ALL cells through both soluble factors and physical contact (Akers, Rellick, Fortney, & Gibson, 2011; Gaynes et al., 2017; Jonart et al., 2020), thus promoting residual disease within the CNS, leading to patient relapse. This complication highlights that greater understanding of both the BM microenvironment, as well as other microenvironments that can hide residual disease must be understood, and all niches harboring ALL cells must be targeted to ensure all sources of possible relapse are destroyed in the initial treatment of ALL.

3. Extracellular components of the ALL niche

3.1 Cytokines

While cytokines such as IL-12, -23 and -27 have been shown to be useful in the treatment of ALL either directly or indirectly by activating the immune responses (reviewed in Cocco, Pistoia, & Airoldi, 2012), other cytokines are upregulated in the ALL microenvironment and lead to ALL progression. It has been shown that IL-7 not only induces the proliferation of multiple pre-B-ALL cell lines *in vitro* (Eμ-ret transgenic mouse-derived cell lines 83, T309, 289, and 420.2), but IL-7 is also able to counteract the B-ALL growth inhibition of rapamycin, a mTOR inhibitor, on these cell lines (Brown et al., 2003). Because it has been reported that IL-7 levels are often elevated in the plasma of pre-B-ALL patients (Sasson et al., 2010) and rapamycin is currently being tested for the treatment of ALL (Gowda et al., 2019; Shi, Xu, Lin, Weng, & Fang, 2016), patient IL-7 levels may be an important consideration required, but possibly overlooked, to fully understand the therapeutic benefit of rapamycin in the treatment of ALL. Furthermore, the main receptor for IL-7, IL-7R, has been associated with BCP-ALL CNS infiltration and subsequent relapse (Alsadeq et al., 2018).

High CXCR4 expression has also been associated with CNS infiltration in childhood ALL (Crazzolara et al., 2001) and in addition to the CXCR4/SDF-1 axis being a critical BM niche regulator, it has now been shown that phosphorylated CXCR4 may be a good prognostic tool for the disease in adult B-ALL patients (Konoplev et al., 2011). CXCR4 is often upregulated in cases of BCP-ALL (Bradstock et al., 2000), altering their migration and localization *in vitro*, and therefore the possible proliferation and survival of ALL cells *in vivo*. Furthermore, high CXCR4 expression in childhood B-ALL has been associated with increased incidences of relapse and shortened disease-free survival (Schneider, Vasse, Al Bayati, Lenormand, & Vannier, 2002; van den Berk et al., 2014). It has been shown that stromal cells can protect mixed lineage leukemia-rearranged (MLL-R) ALL cells from Lestaurtinib-mediated cell death in culture, but when the CXCR4 inhibitor AMD3100 (plerixafor) was administered with G-CSF into a MLL-R ALL xenograft model, the cytotoxic activity of Lestaurtinib is restored, demonstrating that at least in part, leukemia-stroma interactions mediated through CXCR4 aid in the survival of ALL cells (Sison et al., 2013). Furthermore, prolonged treatment of AMD3100 decreased human BCP-ALL engraftment and expansion in NOD/SCID mice (Juarez et al., 2007) and enhances the effect of chemotherapy when AMD3100 is co-administered in xenograft models of high-risk pediatric ALL (Sison et al., 2014).

Previously shown to be cytoprotective against exogenous TNF, endogenous (intracellular) TNF expression has now been found to be inversely correlated to doxorubicin sensitivity in six ALL samples (Kobayashi et al., 1997). Furthermore, the authors reversed resistance to doxorubicin by retroviral transfection of TNF antisense cDNA into ALL cells isolated from three of the patients (Kobayashi et al., 1997). The role endogenous TNF plays as a pro-survival factor allowing the ALL cell line EU-1 to resist doxorubicin- and radiation-induced apoptosis was attributed to activation of the PI3K/Akt pathway (Gu, Findley, Zhu, & Zhou, 2006). In addition, transmembrane-TNF was found to be highly expressed by ALL cells and this correlated with poor clinical outcomes (Zhou et al., 2015), while plasma TNF levels of childhood ALL patients are increased, no changes in early treatment response or the 3-year overall survival was found (Potapnev, Petyovka, Belevtsev, Savitskiy, & Migal, 2003).

3.2 Extracellular matrix (ECM) molecules

Within the BM microenvironment, an intricate network of proteins including proteoglycans, glycoproteins and collagens provide a mesh like membrane which regulate steady state hematopoiesis. Stromal cells including osteoblasts, adipocytes, MSC as well as megakaryocytes are major sources of ECM molecules including OPN (Grassinger et al., 2009; Nilsson et al., 2005), hyaluronan (Nilsson et al., 2003), FN (reviewed in Klein, 1995; Malara et al., 2014; Nilsson et al., 1998; van der Loo et al., 1998) and tenascin-C (Ellis et al., 2013; Nakamura-Ishizu et al., 2012). Notably, leukemic cells share many cell adhesion molecules (CAM) with HSC and bind to components of the ECM, facilitating survival and chemoprotection (reviewed in Barwe et al., 2017). For example, overexpression of integrins including α_4, α_5 and α_6 (DiGiuseppe, Fuller, & Borowitz, 2009; Messinger, Chelstrom, Gunther, & Uckun, 1996; Scharff et al., 2020; Shalapour et al., 2011) have been associated with poor prognosis and contributors to ALL chemoresistance and extravasation into secondary organs.

Laminin enriched in the vasculature of the healthy BM (Malara et al., 2014; Nilsson et al., 1998) is important in the regulation of HSC proliferation and migration (reviewed in Domingues et al., 2017). Similarly, laminin$^+$ ECM of emissary bridging vessels facilitate ALL migration into the CNS through integrin α_6 binding (Yao et al., 2018), demonstrating the importance of this ECM factor in the regulation of normal HSC and leukemic maintenance. In addition, integrin α_6/laminin interactions mediate B-ALL chemoresistance by maintaining cell quiescence *in vitro* (Gang et al., 2020; Włodek & Pituch-Noworolska, 2018), with α_6 deletion resulting in forced induction of G2/M phase and corresponding caspase activation and apoptosis (Gang et al., 2020). By targeting α_6 *in vitro* and *in vivo* with a neutralizing antibody in combination with chemotherapy, leukemic cells were shown to exhibit greater chemosensitivity with modest mobilization into the peripheral blood, where they could no longer remain hidden from treatment (Gang et al., 2020).

While FN is abundantly distributed throughout the BM cavity (Nilsson et al., 1998), the role it plays in leukemiagenesis remains unclear. It has been demonstrated that ALL cells bind FN *via* integrins α_4 and α_5 *in vitro* (Harima et al., 2008; Makrynikola & Bradstock, 1993; Włodek & Pituch-Noworolska, 2018), resulting in MAPK phosphorylation in the NALM-6 cell line driving proliferation that was completely arrested following co-administration of anti-α_4/α_5 antibodies (Harima et al., 2008). Similarly, binding of Jurkat cells to FN *via* integrins α_4 and α_5 resulted in chemoprotection through Akt signaling (Liu, Leclair, Yap, & Lim, 2013). Nevertheless, the effectiveness of this chemoprotection remains to be recapitulated *in vivo*, as both *in vitro* and *in vivo* evidence suggests that integrin α_4 primarily regulates ALL adhesion through VCAM-1 (reviewed in Härzschel et al., 2020). However, B-ALL have been shown to induce MSC to secrete MMPs that degrade ECM including FN, collagen and laminin, with reduced levels of these ECM molecules being associated with B-ALL progression (Verma et al., 2020). Furthermore, pharmacological inhibition of MMP-9 and restoration of FN, collagen and laminin levels in BM was found to be associated with improved survival of B-ALL mice (Verma et al., 2020), suggesting native ECM in the BM microenvironment may be required to hinder B-ALL development. As B-ALL can be positively and negatively influenced by ECM as well as induce their degradation, the multi-faceted roles of ECM in ALL regulation could be responsible for the contradictory evidence in the literature.

Additional ECM molecules that regulate ALL progression include OPN, periostin (POSTN) and connective tissue growth factor (CTGF). B-ALL has been shown to induce secretion of POSTN by BM MSC, that binds leukemic cells *via* integrin α_V, promoting ALL proliferation and dissemination in peripheral organs (Ma et al., 2017, 2019). Furthermore, it was demonstrated that POSTN treatment induces chemokine (C—C motif) ligand 2 (CCL2) expression in B-ALL cells *via* activation of the integrin-integrin-linked kinase (ILK)-NF-kB pathway, that in turn activates the CCL2-CCR2-STAT3 signaling pathway in MSC, resulting in greater POSTN secretion and increased ALL burden in a positive feedback loop (Ma et al., 2019). Notably, this positive correlation between CCL2 and POSTN levels was also confirmed in ALL patient BM biopsies and together provides further evidence that therapeutic targeting of cooperative B-ALL/MSC crosstalk mechanisms may be an effective strategy for treating B-ALL.

Secreted by osteoblasts, OPN is concentrated to the endosteum in both hematopoietic and leukemic settings *in vivo*. In healthy BM, thrombin-cleaved OPN (tcOPN) binds $\alpha_4\beta_1$ and $\alpha_9\beta_1$ expressed on HSC to induce cell quiescence and retain HSC in the BM microenvironment (Grassinger et al., 2009). Similarly, ALL cells anchors to both full length OPN (flOPN) and tcOPN *via* integrin $\alpha_4\beta_1$ in the endosteal BM, inducing cell dormancy and promoting further secretion of OPN by leukemic cells in the BM microenvironment (Boyerinas et al., 2013). Intravital microscopy demonstrated dormant ALL cells were highly associated with areas of greatest OPN expression and OPN neutralization *in vivo* induced cell proliferation and increased leukemia burden (Boyerinas et al., 2013). However, co-administration of anti-OPN antibodies with chemotherapy significantly reduced leukemia burden *in vivo* demonstrating the targeting the OPN/integrin signaling axis effectively chemosensitizes B-ALL (Boyerinas et al., 2013).

Overexpression of CTGF has only been observed in pediatric and adult B-ALL, not T-ALL, and is associated with poor patient outcome and relapse (Kang et al., 2010; Sala-Torra et al., 2007; Vorwerk et al., 2000). Indeed, silencing or knocking down CTGF expression reduced ALL proliferation *in vitro*, arresting G1/S phase transition while increasing levels of the pro-apoptotic BCL-2 family protein BIM and sensitivity to vincristine and methotrexate (Lu et al., 2014). Incubation of ALL cells with human stromal cells (HS-5) and recombinant CTGF resulted in greater stromal proliferation and anchoring of leukemic cells, upregulating lysyl oxidase (LOX) (Wells et al., 2016), responsible for crosslinking fibrous proteins to stabilize the ECM (reviewed in Muncie & Weaver, 2018). Additionally, *in vivo* studies demonstrated that greater CTGF secretion in the BM was associated with an early increase in stromal LOX expression followed by increased collagen type 3/reticulin secretion as the disease progressed, indicating the BM was remodeled and primed by ALL cells to support subsequent leukemic expansion (Wells et al., 2016). These findings are consistent with evidence that ALL cells aid maturation of MSC and enrichment of reticulin fiber density (RFD) post chemotherapy (Duan et al., 2014). Greater RFD has previously been observed in B-ALL patients relative to T-ALL, and is associated with greater MRD (Norén-Nyström et al., 2008), further supporting the role of ALL in BM remodeling to ensure continued leukemiagenesis and chemoprotection. Non-protein metabolites add to the complexity of the leukemic microenvironment, with Lu *et al* proposing that hypoxic activation of CTGF observed in fibroblasts (Hong et al., 2006) and solid tumor cells (Kondo et al., 2006; Recchia et al., 2011) could be reflected in the hypoxic BM microenvironment observed in B-ALL models (Benito et al., 2011).

3.3 Oxygen/hypoxia/reactive oxygen species (ROS)

To prevent oxidative stress and exhaustion, it has been postulated that HSC reside in hypoxic microenvironments (Parmar, Mauch, Vergilio, Sackstein, & Down, 2007; Winkler et al., 2010), stabilizing the expression of hypoxia-inducible factor-1α (HIF-1α) that regulates HSC quiescence and cellular respiration (Jang & Sharkis, 2007; Suda, Takubo, & Semenza, 2011; Takubo et al., 2010). However, more recent evidence suggests HIF-1α may be dispensable for cell autonomous HSC maintenance despite the hypoxic nature of the BM microenvironment (Vukovic et al., 2016). Nevertheless, analysis of patient biopsy samples revealed that HIF-1α is overexpressed in ALL cells (Wellmann et al., 2004) and is negatively associated with patient prognosis (Frolova et al., 2012; Lugthart et al., 2005). In addition, transplanted B-ALL patient samples co-localized to regions of hypoxia in xenograft models, resulting in expansion of hypoxic pockets (Benito et al., 2011). Furthermore, hypoxic conditions have been shown to upregulate HIF-1α expression in ALL as well as anti-apoptotic proteins that drive chemoresistance (Petit et al., 2016). Upon activation, the HIF-1α transcription factor regulates many gene products, including expression of VEGF, which is associated with ALL angiogenesis (Wellmann et al., 2004) and glucose transporters (Frolova et al., 2012; Lugthart et al., 2005) to support metabolism. Of note, B-ALL cells rely heavily on aerobic glycolysis (Warburg effect) (Boag et al., 2006) and recent data suggests BM stromal cells enhance HIF-1α expression *via* AKT/mTOR signaling, resulting in upregulation of glucose transporters and a metabolic switch to glycolytic metabolism in ALL, promoting disease progression and chemoresistance (Frolova et al., 2012). Nevertheless, it has also been shown B-ALL disease progression in BM being is associated with intense oxidative phosphorylation, suggesting the high oxygen demand of B-ALL is the predominant contributor to both intracellular and extracellular hypoxia and that targeting mitochondrial respiration may be an effective therapeutic approach (Rytelewski et al., 2020). Interestingly, ALL dependent secretion of extracellular vesicles containing cytoskeletal structures (Johnson et al., 2017) are horizontally transferred to MSC, resulting in the metabolic reprogramming of MSC, shifting from oxidative phosphorylation to aerobic glycolysis and increasing the secretion of lactate in order to meet energy demands in a leukemic BM microenvironment (Johnson et al., 2016). Indeed, lactate and pyruvate have been shown to stimulate accumulation of HIF-1α in the absence of hypoxia (Lu, Forbes, & Verma, 2002), which may be implicated in a shift of ALL metabolism towards aerobic glycolysis to support growth. Together, the data suggests a complex interplay exists between hypoxia and metabolism of both B-ALL and their associated microenvironment cells, that act collectively to regulate disease progression.

A key consequence of greater metabolism is the abnormal production of ROS. While ROS production is counteracted by antioxidant activities in HSC, expression is lost in ALL cells (Battisti et al., 2008) resulting in oxidative stress. It has been shown that children undergoing chemotherapy who were administered vitamin E, total carotenoid, β-carotene and vitamin A displayed benefits such as fewer therapy delays, decreased toxicity, decreased hospital stays as well as a lower incidence of infection (Kennedy et al., 2004). The authors noted that a large percentage of children with ALL may have inadequate intakes of antioxidants and vitamin A, leading to increases in the adverse side effects of chemotherapy, which often increases ROS. Others have also reported that dietary antioxidants are associated with lower rates of infection and mucositis, but no association with decreased relapse or increased disease-free survival was observed (Ladas et al., 2020). Remarkably, oxidative stress is both beneficial and destructive to leukemiagenesis (Chen, Liang, Luo, & Hu, 2020; Udensi & Tchounwou, 2014). Notably, chemotherapeutic agents such as cytosine arabinoside, doxorubicin and vincristine increase intracellular ROS, disrupting mitochondrial function resulting in cell death (reviewed in Chen et al., 2020; Udensi & Tchounwou, 2014). In contrast, chemoprotection arises from targeted ALL ROS production inducing a protective feedback loop with the microenvironment, *via* adipocyte (Sheng et al., 2016) and stromal cell (Boutter et al., 2014; Jing et al., 2018) protection. Conversely, reduced ROS levels dynamically regulated by MSC *in vitro* has been shown to moderately increase ALL chemoresistance (Liu et al., 2015). Thus, metabolic plasticity in both leukemic cells and the microenvironment contributes to leukemiagenesis.

3.4 Metal ions

Within bone, essential metal ions are stored to aid in bone remodeling, enzyme function and as modulators of cell-cell/matrix interactions. However, mineral deficiency is characteristic to ALL, which associated with reduced bone density. Specifically, pediatric and adult ALL disease progression is associated with reduced serum levels of selenium and zinc corresponding with increased copper concentrations, relative to healthy subjects (Canbolat, Kavutcu, & Durak, 1994; Demir et al., 2011; Valadbeigi, Javadian, Ebrahimi-Rad, Khatami, & Saghiri, 2019; Zuo, Chen, Zhou, Li, & Mei, 2006). Both zinc and selenium are sequestered for antioxidant enzymes that protect against ROS imbalance and oxidative stress (reviewed in Mocchegiani & Malavolta, 2019), which is dysregulated in ALL (Battisti et al., 2008). Furthermore, MMPs that require zinc are upregulated in ALL and are responsible for degrading ECM proteins (Verma et al., 2020), while also disrupting tight junctions between endothelial cells, resulting in degradation of the blood brain barrier and invasion of the CNS (Feng et al., 2011; Si et al., 2018). Depletion of zinc availability *via* the chelator PTEN in Jurkat cells resulted in apoptosis, through oxidative stress and ROS activation of NF-kB signaling (Mendivil-Perez, Velez-Pardo, & Jimenez-Del-Rio, 2012), highlighting zinc sequestration as a potential therapeutic modality. However, given zinc supplementation in children being treated for ALL has been shown to decrease infection rates while increasing weight gain, improving patient outcome (Consolo, Melnikov, Cônsolo, Nascimento, & Pontes, 2013), the utility of zinc sequestration as a general therapeutic approach in ALL requires further interrogation.

During homeostasis, extracellular calcium (Ca^{2+}) is released during osteoclastic bone reabsorption (Silver, Murrills, & Etherington, 1988), producing a calcium gradient (Cao et al., 2016) that facilitates HSC homing and engraftment to the endosteal region *via* the calcium sensing receptor (CaR) (Adams et al., 2006). Expression of CaR on ALL cells has not been investigated, though B cell dependent sensing of increased [Ca^{2+}] does not occur *via* CaR (Hammond, White, Tomic, Shi, & Spaner, 2007), suggesting extracellular calcium may play a role in alternate mechanisms in leukemiagenesis. Notably, hypercalcemia resulting from elevated calcium serum concentrations have been reported in childhood and adult ALL (Inukai et al., 2007; McKay & Furman, 1993). Two traditional mechanisms have been reported to regulate hypercalcemia; osteolytic lesion and humoral activation *via* multiple mediators including RANK/RANKL signaling, parathormon (PTH), parathyroid hormone related protein (PTHrP), calcitriol (biologically active vitamin D) and inflammatory cytokines (IL-1, IL-6, TNF-α and TGF-β (Niizuma et al., 2007) and reviewed in Sargent & Smith, 2010). Although ALL blasts have been implicated in the independent secretion of PTH and PTHrP (Inukai et al., 2007; Lankisch, Kramm, Hermsen, & Wessalowski, 2004), direct functional mechanisms remain to be determined, limited partially because of rarity of patient presentation and partially due to the multitude of factors responsible for pathogenesis.

In addition to calcium, greater concentrations of magnesium and manganese localized at the endosteum (Cao et al., 2016), which has previously been shown to facilitate integrin activation of $\alpha_4\beta_1$ and $\alpha_9\beta_1$ integrins (Cao et al., 2014; Pepinsky et al., 2002). Notably, these integrins are expressed on both HSC and B-ALL and facilitate binding to tcOPN and VCAM-1 in the BM microenvironment (Boyerinas et al., 2013; Cao et al., 2016; Grassinger et al., 2009; Jacamo et al., 2014; Schreiber et al., 2009; Simmons et al., 1992). Additionally, the distribution of divalent cations within the BM could also be dictated through the binding of ECM components including OPN (Kläning, Christensen, Sørensen, Vorup-Jensen, & Jensen, 2014) and reviewed in Domingues et al. (2017). Integrins such as $\alpha_4\beta_1$ that mediate chemoresistance and migration have multiple binding partners throughout the BM (reviewed in Barwe et al., 2017), indicating that interactions with different components of the ECM may be influenced by distinct regions of divalent metal cations that have yet to be described in the leukemic microenvironment.

4. Summary

Although clinical remission and restoration of hematopoiesis is often observed following treatment of B-ALL, drug resistant leukemic cells persist in the BM, resulting in MRD and driving relapse. Progressive ALL development coincides with remodeling of the BM microenvironment and altered function of BM cell populations including MSC, osteoblasts, osteoclasts, nerve cells and Tregs, transforming the ECM components including metal ion and oxygen metabolites, cytokines and proteins, that act *via* both overlapping and unique mechanisms to either promote or hinder leukemiagenesis and/or mediate ALL immune evasion and chemoprotection.

Multifaceted targeting of chemoresistance remains critical to eradicating ALL, as evidence continues to show that MRD and relapse is driven by complex and dynamic interactions between ALL, BM cell types and the ECM. Only by merging our understanding of regulatory mechanisms during steady state hematopoiesis and ALL disease progression and chemoprotection, can therapies evolve to prevent and/or combat multifactorial relapse. Indeed, despite the success of CD19-targeted chimeric antigen receptor T-cell (CAR-T) therapy in relapsed/refractory B-ALL, 30–60% of patients still relapse after CAR treatment (Xu et al., 2019). Whether remodeling of the BM microenvironment plays a role in post-CAR relapse in B-ALL is not fully understood due to the scant data available.

Nevertheless, as CAR T-cell technologies and other novel therapies continue to be developed, patients will be provided with more curative options that is driven by a greater understanding of the complex interplay in both cell autonomous and cell extrinsic mechanisms of ALL resistance and regulation in BM.

Acknowledgments

The authors thank Elena Ortiz Perez for helpful discussions. The Australian Regenerative Medicine Institute is supported by grants from the State Government of Victoria and the Australian Government.

Conflicts of interest

The authors have no conflicts to declare.

References

Adams, G. B., Chabner, K. T., Alley, I. R., Olson, D. P., Szczepiorkowski, Z. M., Poznansky, M. C., et al. (2006). Stem cell engraftment at the endosteal niche is specified by the calcium-sensing receptor. *Nature*, *439*(7076), 599–603. https://doi.org/10.1038/nature04247.

Aguayo, A., Kantarjian, H., Manshouri, T., Gidel, C., Estey, E., Thomas, D., et al. (2000). Angiogenesis in acute and chronic leukemias and myelodysplastic syndromes. *Blood*, *96*(6), 2240–2245. https://doi.org/10.1182/blood.V96.6.2240.

Aifantis, I., Raetz, E., & Buonamici, S. (2008). Molecular pathogenesis of T-cell leukemia and lymphoma. *Nature Reviews Immunology*, *8*(5), 380–390. https://doi.org/10.1038/nri2304.

Akers, S. M., Rellick, S. L., Fortney, J. E., & Gibson, L. F. (2011). Cellular elements of the subarachnoid space promote ALL survival during chemotherapy. *Leukemia Research*, *35*(6), 705–711. https://doi.org/10.1016/j.leukres.2010.12.031.

Alsadeq, A., Lenk, L., Vadakumchery, A., Cousins, A., Vokuhl, C., Khadour, A., et al. (2018). IL7R is associated with CNS infiltration and relapse in pediatric B-cell precursor acute lymphoblastic leukemia. *Blood*, *132*(15), 1614–1617. https://doi.org/10.1182/blood-2018-04-844209.

Barwe, S. P., Quagliano, A., & Gopalakrishnapillai, A. (2017). Eviction from the sanctuary: Development of targeted therapy against cell adhesion molecules in acute lymphoblastic leukemia. *Seminars in Oncology*, *44*(2), 101–112. https://doi.org/10.1053/j.seminoncol.2017.06.005.

Battisti, V., Maders, L. D., Bagatini, M. D., Santos, K. F., Spanevello, R. M., Maldonado, P. A., et al. (2008). Measurement of oxidative stress and antioxidant status in acute lymphoblastic leukemia patients. *Clinical Biochemistry*, *41*(7–8), 511–518. https://doi.org/10.1016/j.clinbiochem.2008.01.027.

Benito, J., Shi, Y., Szymanska, B., Carol, H., Boehm, I., Lu, H., et al. (2011). Pronounced hypoxia in models of murine and human leukemia: high efficacy of hypoxia-activated prodrug PR-104. *PLoS One*, *6*(8), e23108. https://doi.org/10.1371/journal.pone.0023108.

Berry, D. A., Zhou, S., Higley, H., Mukundan, L., Fu, S., Reaman, G. H., et al. (2017). Association of minimal residual disease with clinical outcome in pediatric and adult acute lymphoblastic leukemia: A meta-analysis. *JAMA Oncology*, *3*(7), e170580. https://doi.org/10.1001/jamaoncol.2017.0580.

Bhattacharya, K., Chandra, S., & Mandal, C. (2014). Critical stoichiometric ratio of CD4(+) CD25(+) FoxP3(+) regulatory T cells and CD4(+) CD25(−) responder T cells influence immunosuppression in patients with B-cell acute lymphoblastic leukemia. *Immunology*, *142*(1), 124–139. https://doi.org/10.1111/imm.12237.

Boag, J. M., Beesley, A. H., Firth, M. J., Freitas, J. R., Ford, J., Hoffmann, K., et al. (2006). Altered glucose metabolism in childhood pre-B acute lymphoblastic leukemia. *Leukemia*, *20*(10), 1731–1737. https://doi.org/10.1038/sj.leu.2404365.

Bonilla, X., Vanegas, N.-D. P., & Vernot, J. P. (2019). Acute leukemia induces senescence and impaired osteogenic differentiation in mesenchymal stem cells endowing leukemic cells with functional advantages. *Stem Cells International*, *2019*. 3864948 https://doi.org/10.1155/2019/3864948.

Bonnet, D., & Dick, J. E. (1997). Human acute myeloid leukemia is organized as a hierarchy that originates from a primitive hematopoietic cell. *Nature Medicine*, *3*, 730. https://doi.org/10.1038/nm0797-730.

Borowitz, M. J., Devidas, M., Hunger, S. P., Bowman, W. P., Carroll, A. J., Carroll, W. L., et al. (2008). Clinical significance of minimal residual disease in childhood acute lymphoblastic leukemia and its relationship to other prognostic factors: A Children's Oncology Group study. *Blood*, *111*(12), 5477–5485. https://doi.org/10.1182/blood-2008-01-132837.

Boutter, J., Huang, Y., Marovca, B., Vonderheit, A., Grotzer, M. A., Eckert, C., et al. (2014). Image-based RNA interference screening reveals an individual dependence of acute lymphoblastic leukemia on stromal cysteine support. *Oncotarget*, *5*(22), 11501–11512. https://doi.org/10.18632/oncotarget.2572.

Boyerinas, B., Zafrir, M., Yesilkanal, A. E., Price, T. T., Hyjek, E. M., & Sipkins, D. A. (2013). Adhesion to osteopontin in the bone marrow niche regulates lymphoblastic leukemia cell dormancy. *Blood*, *121*(24), 4821–4831. https://doi.org/10.1182/blood-2012-12-475483.

Bradstock, K. F., Makrynikola, V., Bianchi, A., Shen, W., Hewson, J., & Gottlieb, D. J. (2000). Effects of the chemokine stromal cell-derived factor-1 on the migration and localization of precursor-B acute lymphoblastic leukemia cells within bone marrow stromal layers. *Leukemia*, *14*(5), 882–888. https://doi.org/10.1038/sj.leu.2401729.

Brown, V. I., Fang, J., Alcorn, K., Barr, R., Kim, J. M., Wasserman, R., et al. (2003). Rapamycin is active against B-precursor leukemia *in vitro* and *in vivo*, an effect that is modulated by IL-7-mediated signaling. *Proceedings of the National Academy of Sciences of the United States of America*, *100*(25), 15113–15118. https://doi.org/10.1073/pnas.2436348100.

Bulfone-Paus, S., Bulanova, E., Budagian, V., & Paus, R. (2006). The interleukin-15/interleukin-15 receptor system as a model for juxtacrine and reverse signaling. *Bioessays*, *28*(4), 362–377. https://doi.org/10.1002/bies.20380.

Burger, J. A., & Bürkle, A. (2007). The CXCR4 chemokine receptor in acute and chronic leukemia: A marrow homing receptor and potential therapeutic target. *British Journal of Hematology*, *137*(4), 288–296. https://doi.org/10.1111/j.1365-2141.2007.06590.x.

Burt, R., Dey, A., Aref, S., Aguiar, M., Akarca, A., Bailey, K., et al. (2019). Activated stromal cells transfer mitochondria to rescue acute lymphoblastic leukemia cells from oxidative stress. *Blood*, *134*(17), 1415–1429. https://doi.org/10.1182/blood.2019001398.

Canbolat, O., Kavutcu, M., & Durak, I. (1994). Magnesium contents of leukemic lymphocytes. *Biometals*, 7(4), 313–315. https://doi.org/10.1007/bf00144127.

Cao, B., Hutt, O. E., Zhang, Z., Li, S. H., Heazlewood, S. Y., Williams, B., et al. (2014). Design, synthesis and binding properties of a fluorescent alpha(9)beta(1)/alpha(4)beta(1) integrin antagonist and its application as an *in vivo* probe for bone marrow haemopoietic stem cells. *Organic & Biomolecular Chemistry, 12*(6), 965–978. https://doi.org/10.1039/C3ob42332h.

Cao, B., Zhang, Z., Grassinger, J., Williams, B., Heazlewood, C. K., Churches, Q. I., et al. (2016). Therapeutic targeting and rapid mobilization of endosteal HSC using a small molecule integrin antagonist. *Nature Communications*, 7(1), 11007. https://doi.org/10.1038/ncomms11007.

Cario, G., Izraeli, S., Teichert, A., Rhein, P., Skokowa, J., Moricke, A., et al. (2007). High interleukin-15 expression characterizes childhood acute lymphoblastic leukemia with involvement of the CNS. *Journal of Clinical Oncology, 25*(30), 4813–4820. https://doi.org/10.1200/JCO.2007.11.8166.

Carsetti, R. (2000). The development of B cells in the bone marrow is controlled by the balance between cell-autonomous mechanisms and signals from the microenvironment. *Journal of Experimental Medicine, 191*(1), 5–8. https://doi.org/10.1084/jem.191.1.5.

Chen, Y., Liang, Y., Luo, X., & Hu, Q. (2020). Oxidative resistance of leukemic stem cells and oxidative damage to hematopoietic stem cells under pro-oxidative therapy. *Cell Death & Disease, 11*(4), 291. https://doi.org/10.1038/s41419-020-2488-y.

Cheung, L. C., Tickner, J., Hughes, A. M., Skut, P., Howlett, M., Foley, B., et al. (2018). New therapeutic opportunities from dissecting the pre-B leukemia bone marrow microenvironment. *Leukemia, 32*(11), 2326–2338. https://doi.org/10.1038/s41375-018-0144-7.

Cobaleda, C., Gutierrez-Cianca, N., Perez-Losada, J., Flores, T., Garcia-Sanz, R., Gonzalez, M., et al. (2000). A primitive hematopoietic cell is the target for the leukemic transformation in human philadelphia-positive acute lymphoblastic leukemia. *Blood, 95*(3), 1007–1013. https://doi.org/10.1182/blood.V95.3.1007.003k35_1007_1013.

Cocco, C., Pistoia, V., & Airoldi, I. (2012). Anti-leukemic properties of IL-12, IL-23 and IL-27: Differences and similarities in the control of pediatric B acute lymphoblastic leukemia. *Critical Reviews in Oncology/Hematology, 83*(3), 310–318. https://doi.org/10.1016/j.critrevonc.2011.11.006.

Colmone, A., Amorim, M., Pontier, A. L., Wang, S., Jablonski, E., & Sipkins, D. A. (2008). Leukemic cells create bone marrow niches that disrupt the behavior of normal hematopoietic progenitor cells. *Science, 322*(5909), 1861–1865. https://doi.org/10.1126/science.1164390.

Consolo, L. Z. Z., Melnikov, P., Cônsolo, F. Z., Nascimento, V. A., & Pontes, J. C. D. V. (2013). Zinc supplementation in children and adolescents with acute leukemia. *European Journal of Clinical Nutrition, 67*(10), 1056–1059. https://doi.org/10.1038/ejcn.2013.146.

Crazzolara, R., Kreczy, A., Mann, G., Heitger, A., Eibl, G., Fink, F. M., et al. (2001). High expression of the chemokine receptor CXCR4 predicts extramedullary organ infiltration in childhood acute lymphoblastic leukemia. *British Journal of Hematology, 115*(3), 545–553. https://doi.org/10.1046/j.1365-2141.2001.03164.x.

Demir, C., Demir, H., Esen, R., Sehitogullari, A., Atmaca, M., & Alay, M. (2011). Altered serum levels of elements in acute leukemia cases in Turkey. *Asian Pacific Journal of Cancer Prevention, 12*(12), 3471–3474. Retrieved from http://journal.waocp.org/?sid=Entrez:PubMed&id=pmid:22471499&key=2011.12.12.3471.

DiGiuseppe, J. A., Fuller, S. G., & Borowitz, M. J. (2009). Overexpression of CD49f in precursor B-cell acute lymphoblastic leukemia: Potential usefulness in minimal residual disease detection. *Cytometry Part B: Clinical Cytometry, 76*(2), 150–155. https://doi.org/10.1002/cyto.b.20440.

Domingues, M. J., Cao, H. M., Heazlewood, S. Y., Cao, B., & Nilsson, S. K. (2017). Niche extracellular matrix components and their influence on HSC. *Journal of Cellular Biochemistry, 118*(8), 1984–1993. https://doi.org/10.1002/jcb.25905.

Duan, C.-W., Shi, J., Chen, J., Wang, B., Yu, Y.-H., Qin, X., et al. (2014). Leukemia propagating cells rebuild an evolving niche in response to therapy. *Cancer Cell, 25*(6), 778–793. https://doi.org/10.1016/j.ccr.2014.04.015.

Duell, J., Dittrich, M., Bedke, T., Mueller, T., Eisele, F., Rosenwald, A., et al. (2017). Frequency of regulatory T cells determines the outcome of the T-cell-engaging antibody blinatumomab in patients with B-precursor ALL. *Leukemia, 31*(10), 2181–2190. https://doi.org/10.1038/leu.2017.41.

Einsiedel, H. G., von Stackelberg, A., Hartmann, R., Fengler, R., Schrappe, M., Janka-Schaub, G., et al. (2005). Long-term outcome in children with relapsed ALL by risk-stratified salvage therapy: Results of trial acute lymphoblastic leukemia-relapse study of the Berlin-Frankfurt-Münster Group 87. *Journal of Clinical Oncology, 23*(31), 7942–7950. https://doi.org/10.1200/jco.2005.01.1031.

Ellis, S. L., Heazlewood, S. Y., Williams, B., Reitsma, A. J., Grassinger, J., Borg, J., et al. (2013). The role of Tenascin C in the lymphoid progenitor cell niche. *Experimental Hematology, 41*(12), 1050–1061. https://doi.org/10.1016/j.exphem.2013.09.009.

Evans, A. E., Gilbert, E. S., & Zandstra, R. (1970). The increasing incidence of central nervous system leukemia in children. (Children's Cancer Study Group A). *Cancer, 26*(2), 404–409. https://doi.org/10.1002/1097-0142(197008)26:2<404::aid-cncr2820260222->3.0.co;2-i.

Feng, S., Cen, J., Huang, Y., Shen, H., Yao, L., Wang, Y., et al. (2011). Matrix metalloproteinase-2 and -9 secreted by leukemic cells increase the permeability of blood-brain barrier by disrupting tight junction proteins. *PLoS One, 6*(8), e20599. https://doi.org/10.1371/journal.pone.0020599.

Fragoso, R., Pereira, T., Wu, Y., Zhu, Z., Cabeçadas, J., & Dias, S.r. (2006). VEGFR-1 (FLT-1) activation modulates acute lymphoblastic leukemia localization and survival within the bone marrow, determining the onset of extramedullary disease. *Blood, 107*(4), 1608–1616. https://doi.org/10.1182/blood-2005-06-2530.

Fries, C., Adlowitz, D. G., Spence, J. M., Spence, J. P., Rock, P. J., & Burack, W. R. (2020). Acute lymphoblastic leukemia clonal distribution between bone marrow and peripheral blood. *Pediatric Blood & Cancer, 67*(6), e28280. https://doi.org/10.1002/pbc.28280.

Frishman-Levy, L., Shemesh, A., Bar-Sinai, A., Ma, C., Ni, Z., Frenkel, S., et al. (2015). Central nervous system acute lymphoblastic leukemia: role of natural killer cells. *Blood, 125*(22), 3420–3431. https://doi.org/10.1182/blood-2014-08-595108.

Frolova, O., Samudio, I., Benito, J. M., Jacamo, R., Kornblau, S. M., Markovic, A., et al. (2012). Regulation of HIF-1α signaling and chemoresistance in acute lymphocytic leukemia under hypoxic conditions of the bone marrow microenvironment. *Cancer Biology & Therapy, 13*(10), 858–870. https://doi.org/10.4161/cbt.20838.

Fujisaki, J., Wu, J., Carlson, A. L., Silberstein, L., Putheti, P., Larocca, R., et al. (2011). *In vivo* imaging of T-reg cells providing immune privilege to the hematopoietic stem-cell niche. *Nature, 474*(7350), 216–U256. https://doi.org/10.1038/nature10160.

Gang, E. J., Kim, H. N., Hsieh, Y. T., Ruan, Y., Ogana, H. A., Lee, S., et al. (2020). Integrin α6 mediates the drug resistance of acute lymphoblastic B-cell leukemia. *Blood, 136*(2), 210–223. https://doi.org/10.1182/blood.2019001417.

Gaundar, S. S., Bradstock, K. F., & Bendall, L. J. (2009). p38MAPK inhibitors attenuate cytokine production by bone marrow stromal cells and reduce stroma-mediated proliferation of acute lymphoblastic leukemia cells. *Cell Cycle, 8*(18), 2975–2983. https://doi.org/10.4161/cc.8.18.9545.

Gaynes, J. S., Jonart, L. M., Zamora, E. A., Naumann, J. A., Gossai, N. P., & Gordon, P. M. (2017). The central nervous system microenvironment influences the leukemia transcriptome and enhances leukemia chemo-resistance. *Haematologica, 102*(4), e136–e139. https://doi.org/10.3324/haematol.2016.152926.

Gopalakrishnapillai, A., Kolb, E. A., Dhanan, P., Mason, R. W., Napper, A., & Barwe, S. P. (2015). Disruption of annexin II/p11 interaction suppresses leukemia cell binding, homing and engraftment, and sensitizes the leukemia cells to chemotherapy. *PLoS One, 10*(10), e0140564. https://doi.org/10.1371/journal.pone.0140564.

Goto, T., Yamaza, T., & Tanaka, T. (2003). Cathepsins in the osteoclast. *Journal of Electron Microscopy, 52*(6), 551–558. https://doi.org/10.1093/jmicro/52.6.551.

Gowda, C., Song, C., Ding, Y., Dhanyamraju, P.k., Iyer, S., Payne, J., et al. (2019). CK2 inhibitor CX4945 shows strong *in vivo* synergistic anti-leukemia activity with rapamycin in B-ALL due to enhanced ikaros mediated repression of mtor pathway. *Blood, 134*, 2634. https://doi.org/10.1182/blood-2019-131790.

Grassinger, J., Haylock, D. N., Storan, M. J., Haines, G. O., Williams, B., Whitty, G. A., et al. (2009). Thrombin-cleaved osteopontin regulates hemopoietic stem and progenitor cell functions through interactions with alpha9beta1 and alpha4beta1 integrins. *Blood, 114*(1), 49–59. https://doi.org/10.1182/blood-2009-01-197988.

Gu, L., Findley, H. W., Zhu, N., & Zhou, M. (2006). Endogenous TNFα mediates cell survival and chemotherapy resistance by activating the PI3K/Akt pathway in acute lymphoblastic leukemia cells. *Leukemia, 20*(5), 900–904. https://doi.org/10.1038/sj.leu.2404160.

Hagman, J. (2015). Chapter 3—Transcriptional regulation of early B cell development. In F. W. Alt, T. Honjo, A. Radbruch, & M. Reth (Eds.), *Molecular biology of B cells* (2nd ed., pp. 35–53). London: Academic Press.

Hammond, C. M., White, D., Tomic, J., Shi, Y., & Spaner, D. E. (2007). Extracellular calcium sensing promotes human B-cell activation and function. *Blood, 110*(12), 3985–3995. https://doi.org/10.1182/blood-2007-05-088468.

Harima, A., Nakaseko, C., Yokota, A., Kitagawa, M., Morimoto, C., Harigaya, K., et al. (2008). Fibronectin promotes cell proliferation of human pre-B cell line *via* its interactions with VLA-4 and VLA-5. *Hematology, 13*(4), 236–243. https://doi.org/10.1179/102453308x348315.

Härzschel, A., Zucchetto, A., Gattei, V., & Hartmann, T. N. (2020). VLA-4 expression and activation in B cell malignancies: Functional and clinical aspects. *International Journal of Molecular Sciences, 21*(6). https://doi.org/10.3390/ijms21062206.

Hawkins, E. D., Duarte, D., Akinduro, O., Khorshed, R. A., Passaro, D., Nowicka, M., et al. (2016). T-cell acute leukemia exhibits dynamic interactions with bone marrow microenvironments. *Nature, 538*(7626), 518–522. https://doi.org/10.1038/nature19801.

Hendrix, M. J. C., Seftor, E. A., Grogan, T. M., Seftor, R. E. B., Hersh, E. M., Boyse, E. A., et al. (1992). Expression of type IV collagenase correlates with the invasion of human lymphoblastoid cell lines and pathogenesis in SCID mice. *Molecular and Cellular Probes, 6*(1), 59–65. https://doi.org/10.1016/0890-8508(92)90072-6.

Hirata, Y., Furuhashi, K., Ishii, H., Li, H. W., Pinho, S., Ding, L., et al. (2018). CD150(high) bone marrow tregs maintain hematopoietic stem cell quiescence and immune privilege *via* adenosine. *Cell Stem Cell, 22*(3), 445–453.e445. https://doi.org/10.1016/j.stem.2018.01.017.

Hong, K. H., Yoo, S. A., Kang, S. S., Choi, J. J., Kim, W. U., & Cho, C. S. (2006). Hypoxia induces expression of connective tissue growth factor in scleroderma skin fibroblasts. *Clinical & Experimental Immunology, 146*(2), 362–370. https://doi.org/10.1111/j.1365-2249.2006.03199.x.

Hunger, S. P., & Mullighan, C. G. (2015). Acute lymphoblastic Leukemia in children. *New England Journal of Medicine, 373*(16), 1541–1552. https://doi.org/10.1056/NEJMra1400972.

Idris, S.-Z., Hassan, N., Lee, L.-J., Md Noor, S., Osman, R., Abdul-Jalil, M., et al. (2015). Increased regulatory T cells in acute lymphoblastic leukemia patients. *Hematology, 20*(9), 523–529. https://doi.org/10.1179/1607845415Y.0000000025.

Inukai, T., Hirose, K., Inaba, T., Kurosawa, H., Hama, A., Inada, H., et al. (2007). Hypercalcemia in childhood acute lymphoblastic leukemia: frequent implication of parathyroid hormone-related peptide and E2A-HLF from translocation 17;19. *Leukemia, 21*(2), 288–296. https://doi.org/10.1038/sj.leu.2404496.

Ivanoff, A., Ivanoff, J., Hultenby, K., & Sundqvist, K. G. (1999). Infiltrative capacity of T leukemia cell lines: a distinct functional property coupled to expression of matrix metalloproteinase-9 (MMP-9) and tissue inhibitor of metalloproteinases-1 (TIMP-1). *Clinical & Experimental Metastasis, 17*(8), 695–711. https://doi.org/10.1023/a:1006749304315.

Jacamo, R., Chen, Y., Wang, Z., Ma, W., Zhang, M., Spaeth, E. L., et al. (2014). Reciprocal leukemia-stroma VCAM-1/VLA-4-dependent activation of NF-κB mediates chemoresistance. *Blood, 123*(17), 2691–2702. https://doi.org/10.1182/blood-2013-06-511527.

Jang, Y. Y., & Sharkis, S. J. (2007). A low level of reactive oxygen species selects for primitive hematopoietic stem cells that may reside in the low-oxygenic niche. *Blood, 110*(8), 3056–3063. https://doi.org/10.1182/blood-2007-05-087759.

Jing, B., Jin, J., Xiang, R., Liu, M., Yang, L., Tong, Y., et al. (2018). Vorinostat and quinacrine have synergistic effects in T-cell acute lymphoblastic leukemia through reactive oxygen species increase and mitophagy inhibition. *Cell Death & Disease, 9*(6), 589. https://doi.org/10.1038/s41419-018-0679-6.

Johnson, S. M., Dempsey, C., Chadwick, A., Harrison, S., Liu, J., Di, Y., et al. (2016). Metabolic reprogramming of bone marrow stromal cells by leukemic extracellular vesicles in acute lymphoblastic leukemia. *Blood, 128*(3), 453–456. https://doi.org/10.1182/blood-2015-12-688051.

Johnson, S. M., Dempsey, C., Parker, C., Mironov, A., Bradley, H., & Saha, V. (2017). Acute lymphoblastic leukemia cells produce large extracellular vesicles containing organelles and an active cytoskeleton. *Journal of Extracellular Vesicles, 6*(1), 1294339. https://doi.org/10.1080/20013078.2017.1294339.

Jonart, L. M., Ebadi, M., Basile, P., Johnson, K., Makori, J., & Gordon, P. M. (2020). Disrupting the leukemia niche in the central nervous system attenuates leukemia chemoresistance. *Haematologica, 105*(8), 2130–2140. https://doi.org/10.3324/haematol.2019.230334.

Jordan, C. T., Upchurch, D., Szilvassy, S. J., Guzman, M. L., Howard, D. S., Pettigrew, A. L., et al. (2000). The interleukin-3 receptor alpha chain is a unique marker for human acute myelogenous leukemia stem cells. *Leukemia, 14*(10), 1777–1784.

Juarez, J., Bradstock, K. F., Gottlieb, D. J., & Bendall, L. J. (2003). Effects of inhibitors of the chemokine receptor CXCR4 on acute lymphoblastic leukemia cells *in vitro*. *Leukemia, 17*(7), 1294–1300. https://doi.org/10.1038/sj.leu.2402998.

Juarez, J., Dela Pena, A., Baraz, R., Hewson, J., Khoo, M., Cisterne, A., et al. (2007). CXCR4 antagonists mobilize childhood acute lymphoblastic leukemia cells into the peripheral blood and inhibit engraftment. *Leukemia, 21*(6), 1249–1257. https://doi.org/10.1038/sj.leu.2404684.

Juarez, J. G., Thien, M., Dela Pena, A., Baraz, R., Bradstock, K. F., & Bendall, L. J. (2009). CXCR4 mediates the homing of B cell progenitor acute lymphoblastic leukemia cells to the bone marrow *via* activation of p38MAPK. *British Journal of Hematology, 145*(4), 491–499. https://doi.org/10.1111/j.1365-2141.2009.07648.x.

Kang, H., Chen, I. M., Wilson, C. S., Bedrick, E. J., Harvey, R. C., Atlas, S. R., et al. (2010). Gene expression classifiers for relapse-free survival and minimal residual disease improve risk classification and outcome prediction in pediatric B-precursor acute lymphoblastic leukemia. *Blood, 115*(7), 1394–1405. https://doi.org/10.1182/blood-2009-05-218560.

Kennedy, D. D., Tucker, K. L., Ladas, E. D., Rheingold, S. R., Blumberg, J., & Kelly, K. M. (2004). Low antioxidant vitamin intakes are associated with increases in adverse effects of

chemotherapy in children with acute lymphoblastic leukemia. *American Journal of Clinical Nutrition, 79*(6), 1029–1036. https://doi.org/10.1093/ajcn/79.6.1029.
Kiel, M. J., & Morrison, S. J. (2008). Uncertainty in the niches that maintain hematopoietic stem cells. *Nature Reviews Immunology, 8*(4), 290–301. https://doi.org/10.1038/nri2279.
Kläning, E., Christensen, B., Sørensen, E. S., Vorup-Jensen, T., & Jensen, J. K. (2014). Osteopontin binds multiple calcium ions with high affinity and independently of phosphorylation status. *Bone, 66*, 90–95. https://doi.org/10.1016/j.bone.2014.05.020.
Klein, G. (1995). The extracellular matrix of the hematopoietic microenvironment. *Experientia, 51*(9–10), 914–926. https://doi.org/10.1007/BF01921741.
Kobayashi, D., Watanabe, N., Yamauchi, N., Tsuji, N., Sato, T., & Niitsu, Y. (1997). Endogenous tumor necrosis factor as a predictor of doxorubicin sensitivity in leukemic patients. *Blood, 89*(7), 2472–2479. https://doi.org/10.1182/blood.V89.7.2472.
Kollet, O., Dar, A., Shivtiel, S., Kalinkovich, A., Lapid, K., Sztainberg, Y., et al. (2006). Osteoclasts degrade endosteal components and promote mobilization of hematopoietic progenitor cells. *Nature Medicine, 12*(6), 657–664. https://doi.org/10.1038/nm1417.
Kondo, S., Kubota, S., Mukudai, Y., Moritani, N., Nishida, T., Matsushita, H., et al. (2006). Hypoxic regulation of stability of connective tissue growth factor/CCN2 mRNA by 3′-untranslated region interacting with a cellular protein in human chondrosarcoma cells. *Oncogene, 25*(7), 1099–1110. https://doi.org/10.1038/sj.onc.1209129.
Konoplev, S., Jorgensen, J. L., Thomas, D. A., Lin, E., Burger, J., Kantarjian, H. M., et al. (2011). Phosphorylated CXCR4 is associated with poor survival in adults with B-acute lymphoblastic leukemia. *Cancer, 117*(20), 4689–4695. https://doi.org/10.1002/cncr.26113.
Krevvata, M., Silva, B. C., Manavalan, J. S., Galan-Diez, M., Kode, A., Matthews, B. G., et al. (2014). Inhibition of leukemia cell engraftment and disease progression in mice by osteoblasts. *Blood, 124*(18), 2834–2846. https://doi.org/10.1182/blood-2013-07-517219.
Kusano, K., Miyaura, C., Inada, M., Tamura, T., Ito, A., Nagase, H., et al. (1998). Regulation of matrix metalloproteinases (MMP-2, -3, -9, and -13) by interleukin-1 and interleukin-6 in mouse calvaria: association of MMP induction with bone resorption. *Endocrinology, 139*(3), 1338–1345. https://doi.org/10.1210/endo.139.3.5818.
Ladas, E. J., Blonquist, T. M., Puligandla, M., Orjuela, M., Stevenson, K., Cole, P. D., et al. (2020). Protective effects of dietary intake of antioxidants and treatment-related toxicity in childhood leukemia: A report from the DALLT cohort. *Journal of Clinical Oncology, 38*(19), 2151–2159. https://doi.org/10.1200/JCO.19.02555.
Lankisch, P., Kramm, C. M., Hermsen, D., & Wessalowski, R. (2004). Hypercalcemia with nephrocalcinosis and impaired renal function due to increased parathyroid hormone secretion at onset of childhood acute lymphoblastic leukemia. *Leukemia & Lymphoma, 45*(8), 1695–1697. https://doi.org/10.1080/10428190410001683804.
le Viseur, C., Hotfilder, M., Bomken, S., Wilson, K., Rottgers, S., Schrauder, A., et al. (2008). In childhood acute lymphoblastic leukemia, blasts at different stages of immunophenotypic maturation have stem cell properties. *Cancer Cell, 14*(1), 47–58. https://doi.org/10.1016/j.ccr.2008.05.015.
Lee, Y., Chittezhath, M., André, V., Zhao, H., Poidinger, M., Biondi, A., et al. (2012). Protumoral role of monocytes in human B-cell precursor acute lymphoblastic leukemia: involvement of the chemokine CXCL10. *Blood, 119*(1), 227–237. https://doi.org/10.1182/blood-2011-06-357442.
Leeuw, J. A., Koudstaal, J., Wiersema-Buist, J., Kamps, W. A., & Timens, W. (2003). Bone histomorphometry in children with newly diagnosed acute lymphoblastic leukemia. *Pediatric Research, 54*(6), 814–818. https://doi.org/10.1203/01.PDR.0000090929.16450.DE.

Lim, M., Pang, Y., Ma, S., Hao, S., Shi, H., Zheng, Y., et al. (2016). Altered mesenchymal niche cells impede generation of normal hematopoietic progenitor cells in leukemic bone marrow. *Leukemia*, *30*(1), 154–162. https://doi.org/10.1038/leu.2015.210.

Liu, C. C., Leclair, P., Yap, S. Q., & Lim, C. J. (2013). The membrane-proximal KXGFFKR motif of α-integrin mediates chemoresistance. *Molecular and Cell Biology*, *33*(21), 4334–4345. https://doi.org/10.1128/mcb.00580-13.

Liu, J., Masurekar, A., Johnson, S., Chakraborty, S., Griffiths, J., Smith, D., et al. (2015). Stromal cell-mediated mitochondrial redox adaptation regulates drug resistance in childhood acute lymphoblastic leukemia. *Oncotarget*, *6*(40), 43048–43064. https://doi.org/10.18632/oncotarget.5528.

Lu, H., Forbes, R. A., & Verma, A. (2002). Hypoxia-inducible factor 1 activation by aerobic glycolysis implicates the Warburg effect in carcinogenesis. *Journal of Biological Chemistry*, *277*(26), 23111–23115. https://doi.org/10.1074/jbc.M202487200.

Lu, H., Kojima, K., Battula, V. L., Korchin, B., Shi, Y., Chen, Y., et al. (2014). Targeting connective tissue growth factor (CTGF) in acute lymphoblastic leukemia preclinical models: Anti-CTGF monoclonal antibody attenuates leukemia growth. *Annals of Hematology*, *93*(3), 485–492. https://doi.org/10.1007/s00277-013-1939-2.

Lugthart, S., Cheok, M. H., den Boer, M. L., Yang, W., Holleman, A., Cheng, C., et al. (2005). Identification of genes associated with chemotherapy crossresistance and treatment response in childhood acute lymphoblastic leukemia. *Cancer Cell*, 7(4), 375–386. https://doi.org/10.1016/j.ccr.2005.03.002.

Ma, Z., Zhao, X., Deng, M., Huang, Z., Wang, J., Wu, Y., et al. (2019). Bone marrow mesenchymal stromal cell-derived periostin promotes B-ALL progression by modulating CCL2 in leukemia cells. *Cell Reports*, *26*(6), 1533–1543.e1534. https://doi.org/10.1016/j.celrep.2019.01.034.

Ma, Z., Zhao, X., Huang, J., Jia, X., Deng, M., Cui, D., et al. (2017). A critical role of periostin in B-cell acute lymphoblastic leukemia. *Leukemia*, *31*(8), 1835–1837. https://doi.org/10.1038/leu.2017.149.

Makrynikola, V., & Bradstock, K. F. (1993). Adhesion of precursor-B acute lymphoblastic leukemia cells to bone marrow stromal proteins. *Leukemia*, 7(1), 86–92.

Malara, A., Currao, M., Gruppi, C., Celesti, G., Viarengo, G., Buracchi, C., et al. (2014). Megakaryocytes contribute to the bone marrow-matrix environment by expressing fibronectin, type IV collagen, and laminin. *Stem Cells*, *32*(4), 926–937. https://doi.org/10.1002/stem.1626.

McKay, C., & Furman, W. L. (1993). Hypercalcemia complicating childhood malignancies. *Cancer*, 72(1), 256–260. https://doi.org/10.1002/1097-0142(19930701)72:1<256::AID-CNCR2820720145>3.0.CO;2-D.

Mendelson, A., & Frenette, P. S. (2014). Hematopoietic stem cell niche maintenance during homeostasis and regeneration. *Nature Medicine*, *20*(8), 833–846. https://doi.org/10.1038/nm.3647.

Mendivil-Perez, M., Velez-Pardo, C., & Jimenez-Del-Rio, M. (2012). TPEN induces apoptosis independently of zinc chelator activity in a model of acute lymphoblastic leukemia and *ex vivo* acute leukemia cells through oxidative stress and mitochondria caspase-3- and AIF-dependent pathways. *Oxidative Medicine and Cellular Longevity*, *2012*, 313275. https://doi.org/10.1155/2012/313275.

Messinger, Y., Chelstrom, L., Gunther, R., & Uckun, F. M. (1996). Selective homing of human leukemic B-cell precursors to specific lymphohematopoietic microenvironments in SCID mice: A role for the β 1 integrin family surface adhesion molecules VLA-4 and VLA-5. *Leukemia & Lymphoma*, *23*(1–2), 61–69. https://doi.org/10.3109/10428199609054803.

Miyamoto, K., Yoshida, S., Kawasumi, M., Hashimoto, K., Kimura, T., Sato, Y., et al. (2011). Osteoclasts are dispensable for hematopoietic stem cell maintenance and mobilization. *Journal of Experimental Medicine, 208*(11), 2175–2181. https://doi.org/10.1084/jem.20101890.

Mocchegiani, E., & Malavolta, M. (2019). Role of zinc and selenium in oxidative stress and immunosenescence: implications for healthy aging and longevity. *Handbook of Immunosenescence: Basic Understanding and Clinical Implications*, 2539–2573. https://doi.org/10.1007/978-3-319-99375-1_66.

Mostoufi-Moab, S., & Halton, J. (2014). Bone morbidity in childhood leukemia: Epidemiology, mechanisms, diagnosis, and treatment. *Current Osteoporosis Reports, 12*(3), 300–312. https://doi.org/10.1007/s11914-014-0222-3.

Mullighan, C. G. (2012). The molecular genetic makeup of acute lymphoblastic leukemia. *Hematology, 2012*(1), 389–396. https://doi.org/10.1182/asheducation.V2012.1.389.3798360.

Münch, V., Trentin, L., Herzig, J., Demir, S., Seyfried, F., Kraus, J. M., et al. (2017). Central nervous system involvement in acute lymphoblastic leukemia is mediated by vascular endothelial growth factor. *Blood, 130*(5), 643–654. https://doi.org/10.1182/blood-2017-03-769315.

Münch, V., Trentin, L., Herzig, J., Krauss, J., Kestler, H. A., te Kronnie, G., et al. (2015). Migration of acute lymphoblastic leukemia cells into the central nervous system is regulated By VEGF. *Blood, 126*(23), 2634. https://doi.org/10.1182/blood.V126.23.2634.2634.

Muncie, J. M., & Weaver, V. M. (2018). The physical and biochemical properties of the extracellular matrix regulate cell fate. *Current Topics in Developmental Biology, 130*, 1–37. https://doi.org/10.1016/bs.ctdb.2018.02.002.

Muro, R., Takayanagi, H., & Nitta, T. (2019). T cell receptor signaling for γδT cell development. *Inflammation and Regeneration, 39*(1), 6. https://doi.org/10.1186/s41232-019-0095-z.

Naderi, E. H., Skah, S., Ugland, H., Myklebost, O., Sandnes, D. L., Torgersen, M. L., et al. (2015). Bone marrow stroma-derived PGE2 protects BCP-ALL cells from DNA damage-induced p53 accumulation and cell death. *Molecular Cancer, 14*(1), 14. https://doi.org/10.1186/s12943-014-0278-9.

Nagasawa, T. (2006). Microenvironmental niches in the bone marrow required for B-cell development. *Nature Reviews Immunology, 6*(2), 107–116. https://doi.org/10.1038/nri1780.

Naji, A., Eitoku, M., Favier, B., Deschaseaux, F., Rouas-Freiss, N., & Suganuma, N. (2019). Biological functions of mesenchymal stem cells and clinical implications. *Cellular and Molecular Life Sciences, 76*(17), 3323–3348. https://doi.org/10.1007/s00018-019-03125-1.

Nakamura-Ishizu, A., Okuno, Y., Omatsu, Y., Okabe, K., Morimoto, J., Uede, T., et al. (2012). Extracellular matrix protein tenascin-C is required in the bone marrow microenvironment primed for hematopoietic regeneration. *Blood, 119*(23), 5429–5437. https://doi.org/10.1182/blood-2011-11-393645.

Nguyen, T.-V., Melville, A., Nath, S., Story, C., Howell, S., Sutton, R., et al. (2015). Bone marrow recovery by morphometry during induction chemotherapy for acute lymphoblastic leukemia in children. *PLoS One, 10*(5), e0126233. https://doi.org/10.1371/journal.pone.0126233.

Niizuma, H., Fujii, K., Sato, A., Fujiwara, I., Takeyama, J., & Imaizumi, M. (2007). PTHrP-independent hypercalcemia with increased proinflammatory cytokines and bone resorption in two children with CD19-negative precursor B acute lymphoblastic leukemia. *Pediatric Blood & Cancer, 49*(7), 990–993. https://doi.org/10.1002/pbc.20782.

Nilsson, S. K., Debatis, M. E., Dooner, M. S., Madri, J. A., Quesenberry, P. J., & Becker, P. S. (1998). Immunofluorescence characterization of key extracellular matrix proteins in murine bone marrow *in situ*. *The Journal of Histochemistry & Cytochemistry*, *46*(3), 371–377. https://doi.org/10.1177/002215549804600311.

Nilsson, S. K., Haylock, D. N., Johnston, H. M., Occhiodoro, T., Brown, T. J., & Simmons, P. J. (2003). Hyaluronan is synthesized by primitive hemopoietic cells, participates in their lodgment at the endosteum following transplantation, and is involved in the regulation of their proliferation and differentiation *in vitro*. *Blood*, *101*(3), 856–862. https://doi.org/10.1182/blood-2002-05-1344.

Nilsson, S. K., Johnston, H. M., Whitty, G. A., Williams, B., Webb, R. J., Denhardt, D. T., et al. (2005). Osteopontin, a key component of the hematopoietic stem cell niche and regulator of primitive hematopoietic progenitor cells. *Blood*, *106*(4), 1232–1239. https://doi.org/10.1182/blood-2004-11-4422.

Norén-Nyström, U., Roos, G., Bergh, A., Botling, J., Lönnerholm, G., Porwit, A., et al. (2008). Bone marrow fibrosis in childhood acute lymphoblastic leukemia correlates to biological factors, treatment response and outcome. *Leukemia*, *22*(3), 504–510. https://doi.org/10.1038/sj.leu.2405072.

Nwabo Kamdje, A. H., Mosna, F., Bifari, F., Lisi, V., Bassi, G., Malpeli, G., et al. (2011). Notch-3 and Notch-4 signaling rescue from apoptosis human B-ALL cells in contact with human bone marrow-derived mesenchymal stromal cells. *Blood*, *118*(2), 380–389. https://doi.org/10.1182/blood-2010-12-326694.

Ogasawara, T., Yasuyama, M., & Kawauchi, K. (2003). Constitutive activation of extracellular signal-regulated kinase and p38 mitogen-activated protein kinase in B-cell lymphoproliferative disorders. *International Journal of Hematology*, 77(4), 364–370. https://doi.org/10.1007/BF02982645.

Parmar, K., Mauch, P., Vergilio, J. A., Sackstein, R., & Down, J. D. (2007). Distribution of hematopoietic stem cells in the bone marrow according to regional hypoxia. *Proceedings of the National Academy of Sciences of the United States of America*, *104*(13), 5431–5436. https://doi.org/10.1073/pnas.0701152104.

Pepinsky, R. B., Mumford, R. A., Chen, L. L., Leone, D., Amo, S. E., Van Riper, G., et al. (2002). Comparative assessment of the ligand and metal ion binding properties of integrins alpha 9 beta 1 and alpha 4 beta 1. *Biochemistry*, *41*(22), 7125–7141. https://doi.org/10.1021/Bi020024d.

Perez-Atayde, A. R., Sallan, S. E., Tedrow, U., Connors, S., Allred, E., & Folkman, J. (1997). Spectrum of tumor angiogenesis in the bone marrow of children with acute lymphoblastic leukemia. *American Journal of Pathology*, *150*(3), 815–821. Retrieved from https://www.ncbi.nlm.nih.gov/pmc/articles/PMC1857903/.

Petit, C., Gouel, F., Dubus, I., Heuclin, C., Roget, K., & Vannier, J. P. (2016). Hypoxia promotes chemoresistance in acute lymphoblastic leukemia cell lines by modulating death signaling pathways. *BMC Cancer*, *16*(1), 746. https://doi.org/10.1186/s12885-016-2776-1.

Pierini, A., Nishikii, H., Baker, J., Kimura, T., Kwon, H.-S., Pan, Y., et al. (2017). Foxp3+ regulatory T cells maintain the bone marrow microenvironment for B cell lymphopoiesis. *Nature Communications*, *8*(1), 15068. https://doi.org/10.1038/ncomms15068.

Potapnev, M. P., Petyovka, N. V., Belevtsev, M. V., Savitskiy, V. P., & Migal, N. V. (2003). Plasma level of tumor necrosis factor-alpha (TNF-alpha) correlates with leukocytosis and biological features of leukemic cells, but not treatment response of children with acute lymphoblastic leukemia. *Leukemia & Lymphoma*, *44*(6), 1077–1079. https://doi.org/10.1080/1042819031000068025.

Poyer, F., Coquerel, B., Pegahi, R., Cazin, L., Norris, V., Vannier, J. P., et al. (2009). Secretion of MMP-2 and MMP-9 induced by VEGF autocrine loop correlates with clinical features in childhood acute lymphoblastic leukemia. *Leukemia Research, 33*(3), 407–417. https://doi.org/10.1016/j.leukres.2008.08.019.

Pui, C. H., & Evans, W. E. (2006). Drug therapy—Treatment of acute lymphoblastic leukemia. *New England Journal of Medicine, 354*(2), 166–178. https://doi.org/10.1056/Nejmra052603.

Pulè, M. A., Gullmann, C., Dennis, D., McMahon, C., Jeffers, M., & Smith, O. P. (2002). Increased angiogenesis in bone marrow of children with acute lymphoblastic leukemia has no prognostic significance. *British Journal of Hematology, 118*(4), 991–998. https://doi.org/10.1046/j.1365-2141.2002.03761.x.

Quagliano, A., Gopalakrishnapillai, A., & Barwe, S. P. (2017). Epigenetic drug combination overcomes osteoblast-induced chemoprotection in pediatric acute lymphoid leukemia. *Leukemia Research, 56*, 36–43. https://doi.org/10.1016/j.leukres.2017.01.030.

Rajakumar, S. A., Papp, E., Grandal, I., Merico, D., Liu, C. C., Allo, B., et al. (2016). RANK-RANKL mediated bone destruction in B-cell acute lymphoblastic leukemia. *Blood, 128*(22), 908. https://doi.org/10.1182/blood.V128.22.908.908.

Rajakumar, S. A., Papp, E., Lee, K. K., Grandal, I., Merico, D., Liu, C. C., et al. (2020). B cell acute lymphoblastic leukemia cells mediate RANK-RANKL-dependent bone destruction. *Science Translational Medicine, 12*(561). https://doi.org/10.1126/scitranslmed.aba5942.

Ramasamy, R., Lam, E. W., Soeiro, I., Tisato, V., Bonnet, D., & Dazzi, F. (2007). Mesenchymal stem cells inhibit proliferation and apoptosis of tumor cells: impact on *in vivo* tumor growth. *Leukemia, 21*(2), 304–310. https://doi.org/10.1038/sj.leu.2404489.

Recchia, A. G., De Francesco, E. M., Vivacqua, A., Sisci, D., Panno, M. L., Andò, S., et al. (2011). The G protein-coupled receptor 30 is up-regulated by hypoxia-inducible factor-1alpha (HIF-1alpha) in breast cancer cells and cardiomyocytes. *Journal of Biological Chemistry, 286*(12), 10773–10782. https://doi.org/10.1074/jbc.M110.172247.

Rehe, K., Wilson, K., Bomken, S., Williamson, D., Irving, J., den Boer, M. L., et al. (2013). Acute B lymphoblastic leukemia-propagating cells are present at high frequency in diverse lymphoblast populations. *EMBO Molecular Medicine, 5*(1), 38–51. https://doi.org/10.1002/emmm.201201703.

Ryan, D. H., Nuccie, B. L., Abboud, C. N., & Winslow, J. M. (1991). Vascular cell adhesion molecule-1 and the integrin VLA-4 mediate adhesion of human B cell precursors to cultured bone marrow adherent cells. *The Journal of Clinical Investigation, 88*(3), 995–1004. https://doi.org/10.1172/JCI115403.

Rytelewski, M., Harutyunyan, K., Baran, N., Mallampati, S., Zal, M. A., Cavazos, A., et al. (2020). Inhibition of oxidative phosphorylation reverses bone marrow hypoxia visualized in imageable syngeneic B-ALL mouse model. *Frontiers in Oncology, 10*(991). https://doi.org/10.3389/fonc.2020.00991.

Sakaguchi, S., Yamaguchi, T., Nomura, T., & Ono, M. (2008). Regulatory T cells and immune tolerance. *Cell, 133*(5), 775–787. https://doi.org/10.1016/j.cell.2008.05.009.

Sala-Torra, O., Gundacker, H. M., Stirewalt, D. L., Ladne, P. A., Pogosova-Agadjanyan, E. L., Slovak, M. L., et al. (2007). Connective tissue growth factor (CTGF) expression and outcome in adult patients with acute lymphoblastic leukemia. *Blood, 109*(7), 3080–3083. https://doi.org/10.1182/blood-2006-06-031096.

Sargent, J. T. S., & Smith, O. P. (2010). Hematological emergencies managing hypercalcaemia in adults and children with hematological disorders. *British Journal of Hematology, 149*(4), 465–477. https://doi.org/10.1111/j.1365-2141.2010.08173.x.

Sarmadi, V. H., Tong, C. K., Vidyadaran, S., Abdullah, M., Seow, H. F., & Ramasamy, R. (2010). Mesenchymal stem cells inhibit proliferation of lymphoid origin hematopoietic tumor cells by inducing cell cycle arrest. *Medical Journal of Malaysia, 65*(3), 209–214. Retrieved from http://www.e-mjm.org/2010/v65n3/Mesenchymal_Stem_Cells.pdf.

Sasson, S. C., Smith, S., Seddiki, N., Zaunders, J. J., Bryant, A., Koelsch, K. K., et al. (2010). IL-7 receptor is expressed on adult pre-B-cell acute lymphoblastic leukemia and other B-cell derived neoplasms and correlates with expression of proliferation and survival markers. *Cytokine*, *50*(1), 58–68. https://doi.org/10.1016/j.cyto.2009.12.001.

Scharff, B. F. S. S., Modvig, S., Marquart, H. V., & Christensen, C. (2020). Integrin-mediated adhesion and chemoresistance of acute lymphoblastic leukemia cells residing in the bone marrow or the central nervous system. *Frontiers in Oncology*, *10*(775). https://doi.org/10.3389/fonc.2020.00775.

Schneider, P., Dubus, I., Gouel, F., Legrand, E., Vannier, J. P., & Vasse, M. (2011). What role for angiogenesis in childhood acute lymphoblastic leukemia? *Advances in Hematology*, *2011*, 274,628. https://doi.org/10.1155/2011/274628.

Schneider, P., Vasse, M., Al Bayati, A., Lenormand, B., & Vannier, J. P. (2002). Is high expression of the chemokine receptor CXCR-4 of predictive value for early relapse in childhood acute lymphoblastic leukemia? *British Journal of Hematology*, *119*(2), 579–580. https://doi.org/10.1046/j.1365-2141.2002.03835_6.x.

Schreiber, T. D., Steinl, C., Essl, M., Abele, H., Geiger, K., Muller, C. A., et al. (2009). The integrin α9β1 on hematopoietic stem and progenitor cells: involvement in cell adhesion, proliferation and differentiation. *Haematologica*, *94*(11), 1493–1501. https://doi.org/10.3324/haematol.2009.006072.

Shalapour, S., Hof, J., Kirschner-Schwabe, R., Bastian, L., Eckert, C., Prada, J., et al. (2011). High VLA-4 expression is associated with adverse outcome and distinct gene expression changes in childhood B-cell precursor acute lymphoblastic leukemia at first relapse. *Haematologica*, *96*(11), 1627–1635. https://doi.org/10.3324/haematol.2011.047993.

Sheng, X., Tucci, J., Parmentier, J.-H., Ji, L., Behan, J. W., Heisterkamp, N., et al. (2016). Adipocytes cause leukemia cell resistance to daunorubicin *via* oxidative stress response. *Oncotarget*, 7(45), 73147–73159. https://doi.org/10.18632/oncotarget.12246.

Shi, P. J., Xu, L. H., Lin, K. Y., Weng, W. J., & Fang, J. P. (2016). Synergism between the mTOR inhibitor rapamycin and FAK down-regulation in the treatment of acute lymphoblastic leukemia. *Journal of Hematology & Oncology*, *9*, 12. https://doi.org/10.1186/s13045-016-0241-x.

Si, M., Jiao, X., Li, Y., Chen, H., He, P., & Jiang, F. (2018). The role of cytokines and chemokines in the microenvironment of the blood-brain barrier in leukemia central nervous system metastasis. *Cancer Management and Research*, *10*, 305–313. https://doi.org/10.2147/CMAR.S152419.

Silver, I. A., Murrills, R. J., & Etherington, D. J. (1988). Microelectrode studies on the acid microenvironment beneath adherent macrophages and osteoclasts. *Experimental Cell Research*, *175*(2), 266–276. https://doi.org/10.1016/0014-4827(88)90191-7.

Simmons, P. J., Masinovsky, B., Longenecker, B. M., Berenson, R., Torok-Storb, B., & Gallatin, W. M. (1992). Vascular cell adhesion molecule-1 expressed by bone marrow stromal cells mediates the binding of hematopoietic progenitor cells. *Blood*, *80*(2), 388–395. https://doi.org/10.1182/blood.V80.2.388.388.

Sipkins, D. A., Wei, X., Wu, J. W., Runnels, J. M., Cote, D., Means, T. K., et al. (2005). *In vivo* imaging of specialized bone marrow endothelial microdomains for tumor engraftment. *Nature*, *435*(7044), 969–973. https://doi.org/10.1038/nature03703.

Sison, E. A. R., & Brown, P. (2011). The bone marrow microenvironment and leukemia: Biology and therapeutic targeting. *Expert Review of Hematology*, *4*(3), 271–283. https://doi.org/10.1586/Ehm.11.30.

Sison, E. A., Magoon, D., Li, L., Annesley, C. E., Rau, R. E., Small, D., et al. (2014). Plerixafor as a chemosensitizing agent in pediatric acute lymphoblastic leukemia: efficacy and potential mechanisms of resistance to CXCR4 inhibition. *Oncotarget*, *5*(19), 8947–8958. https://doi.org/10.18632/oncotarget.2407.

Sison, E. A. R., Rau, R. E., McIntyre, E., Li, L., Small, D., & Brown, P. (2013). MLL-rearranged acute lymphoblastic leukemia stem cell interactions with bone marrow stroma promote survival and therapeutic resistance that can be overcome with CXCR4 antagonism. *British Journal of Hematology, 160*(6), 785–797. https://doi.org/10.1111/bjh.12205.

Song, N., Gao, L., Qiu, H., Huang, C., Cheng, H., Zhou, H., et al. (2015). Mouse bone marrow-derived mesenchymal stem cells inhibit leukemia/lymphoma cell proliferation *in vitro* and in a mouse model of allogeneic bone marrow transplant. *International Journal of Molecular Medicine, 36*(1), 139–149. https://doi.org/10.3892/ijmm.2015.2191.

Spiegel, A., Kollet, O., Peled, A., Abel, L., Nagler, A., Bielorai, B., et al. (2004). Unique SDF-1-induced activation of human precursor-B ALL cells as a result of altered CXCR4 expression and signaling. *Blood, 103*(8), 2900–2907. https://doi.org/10.1182/blood-2003-06-1891.

Stary, J., Zimmermann, M., Campbell, M., Castillo, L., Dibar, E., Donska, S., et al. (2014). Intensive chemotherapy for childhood acute lymphoblastic leukemia: Results of the randomized intercontinental trial ALL IC-BFM 2002. *Journal of Clinical Oncology, 32*(3), 174–184. https://doi.org/10.1200/jco.2013.48.6522.

Suda, T., Takahashi, N., Udagawa, N., Jimi, E., Gillespie, M. T., & Martin, T. J. (1999). Modulation of osteoclast differentiation and function by the new members of the tumor necrosis factor receptor and ligand families. *Endocrine Reviews, 20*(3), 345–357. https://doi.org/10.1210/edrv.20.3.0367.

Suda, T., Takubo, K., & Semenza, G. L. (2011). Metabolic regulation of hematopoietic stem cells in the hypoxic niche. *Cell Stem Cell, 9*(4), 298–310. https://doi.org/10.1016/j.stem.2011.09.010.

Tabe, Y., & Konopleva, M. (2014). Advances in understanding the leukemia microenvironment. *British Journal of Hematology, 164*(6), 767–778. https://doi.org/10.1111/bjh.12725.

Taichman, R. S., & Emerson, S. G. (1998). The role of osteoblasts in the hematopoietic microenvironment. *Stem Cells, 16*(1), 7–15. Retrieved from http://www.ncbi.nlm.nih.gov/entrez/query.fcgi?cmd=Retrieve&db=PubMed&dopt=Citation&list_uids=9474743.

Takubo, K., Goda, N., Yamada, W., Iriuchishima, H., Ikeda, E., Kubota, Y., et al. (2010). Regulation of the HIF-1alpha level is essential for hematopoietic stem cells. *Cell Stem Cell,* 7(3), 391–402. https://doi.org/10.1016/j.stem.2010.06.020.

Tanaka, S., Takahashi, N., Tamura, T., Akatsu, T., Stanley, E. R., Kurokawa, T., et al. (1993). Macrophage colony-stimulating factor is indespensable for both proliferation and differentiation of osteoclast progenitors. *Journal of Clinical Investigation, 91*, 257–263. https://doi.org/10.1172/jci116179.

Udensi, U. K., & Tchounwou, P. B. (2014). Dual effect of oxidative stress on leukemia cancer induction and treatment. *Journal of Experimental & Clinical Cancer Research, 33*, 106. https://doi.org/10.1186/s13046-014-0106-5.

Vadillo, E., Dorantes-Acosta, E., Pelayo, R., & Schnoor, M. (2018). T cell acute lymphoblastic leukemia (T-ALL): New insights into the cellular origins and infiltration mechanisms common and unique among hematologic malignancies. *Blood Reviews, 32*(1), 36–51. https://doi.org/10.1016/j.blre.2017.08.006.

Valadbeigi, S., Javadian, S., Ebrahimi-Rad, M., Khatami, S., & Saghiri, R. (2019). Assessment of trace elements in serum of acute lymphoblastic and myeloid leukemia patients. *Experimental Oncology, 41*(1), 69–71. https://doi.org/10.32471/exp-oncology.2312-8852.vol-41-no-1.12730.

van den Berk, L. C., van der Veer, A., Willemse, M. E., Theeuwes, M. J., Luijendijk, M. W., Tong, W. H., et al. (2014). Disturbed CXCR4/CXCL12 axis in pediatric precursor B-cell acute lymphoblastic leukemia. *British Journal of Hematology, 166*(2), 240–249. https://doi.org/10.1111/bjh.12883.

van der Loo, J. C., Xiao, X., McMillin, D., Hashino, K., Kato, I., & Williams, D. A. (1998). VLA-5 is expressed by mouse and human long-term repopulating hematopoietic cells and mediates adhesion to extracellular matrix protein fibronectin. *Journal of Clinical Investigation, 102*(5), 1051–1061. https://doi.org/10.1172/jci3687.

Veiga, J. P., Costa, L. F., Sallan, S. E., Nadler, L. M., & Cardoso, A. A. (2006). Leukemia-stimulated bone marrow endothelium promotes leukemia cell survival. *Experimental Hematology, 34*(5), 610–621. https://doi.org/10.1016/j.exphem.2006.01.013.

Verma, D., Zanetti, C., Godavarthy, P. S., Kumar, R., Minciacchi, V. R., Pfeiffer, J., et al. (2020). Bone marrow niche-derived extracellular matrix-degrading enzymes influence the progression of B-cell acute lymphoblastic leukemia. *Leukemia, 34*(6), 1540–1552. https://doi.org/10.1038/s41375-019-0674-7.

Vorwerk, P., Wex, H., Hohmann, B., Oh, Y., Rosenfeld, R. G., & Mittler, U. (2000). CTGF (IGFBP-rP2) is specifically expressed in malignant lymphoblasts of patients with acute lymphoblastic leukemia (ALL). *British Journal of Cancer, 83*(6), 756–760. https://doi.org/10.1054/bjoc.2000.1364.

Vukovic, M., Sepulveda, C., Subramani, C., Guitart, A. V., Mohr, J., Allen, L., et al. (2016). Adult hematopoietic stem cells lacking Hif-1alpha self-renew normally. *Blood, 127*(23), 2841–2846. https://doi.org/10.1182/blood-2015-10-677138.

Wang, W., Zimmerman, G., Huang, X., Yu, S., Myers, J., Wang, Y., et al. (2016). Aberrant notch signaling in the bone marrow microenvironment of acute lymphoid leukemia suppresses osteoblast-mediated support of hematopoietic niche function. *Cancer Research, 76*(6), 1641–1652. https://doi.org/10.1158/0008-5472.Can-15-2092.

Wellmann, S., Guschmann, M., Griethe, W., Eckert, C., von Stackelberg, A., Lottaz, C., et al. (2004). Activation of the HIF pathway in childhood ALL, prognostic implications of VEGF. *Leukemia, 18*(5), 926–933. https://doi.org/10.1038/sj.leu.2403332.

Wells, J. E., Howlett, M., Halse, H. M., Heng, J., Ford, J., Cheung, L. C., et al. (2016). High expression of connective tissue growth factor accelerates dissemination of leukemia. *Oncogene, 35*(35), 4591–4600. https://doi.org/10.1038/onc.2015.525.

Williams, M. T. S., Yousafzai, Y., Cox, C., Blair, A., Carmody, R., Sai, S., et al. (2014). Interleukin-15 enhances cellular proliferation and upregulates CNS homing molecules in pre-B acute lymphoblastic leukemia. *Blood, 123*(20), 3116–3127. https://doi.org/10.1182/blood-2013-05-499970.

Winkler, I. G., Barbier, V., Wadley, R., Zannettino, A. C., Williams, S., & Lévesque, J. P. (2010). Positioning of bone marrow hematopoietic and stromal cells relative to blood flow *in vivo*: Serially reconstituting hematopoietic stem cells reside in distinct non-perfused niches. *Blood, 116*(3), 375–385. https://doi.org/10.1182/blood-2009-07-233437.

Witkowski, M. T., Kousteni, S., & Aifantis, I. (2020). Mapping and targeting of the leukemic microenvironment. *The Journal of Experimental Medicine, 217*(2), e20190589. https://doi.org/10.1084/jem.20190589.

Włodek, J., & Pituch-Noworolska, A. (2018). The influence of fibronectin on proliferation and apoptosis of acute lymphoblastic leukemia cells *in vitro*. *Polish Journal of Pathology, 69*(1), 62–66. https://doi.org/10.5114/pjp.2018.75338.

Wu, C. P., Qing, X., Wu, C. Y., Zhu, H., & Zhou, H. Y. (2012). Immunophenotype and increased presence of CD4(+)CD25(+) regulatory T cells in patients with acute lymphoblastic leukemia. *Oncology Letters, 3*(2), 421–424. https://doi.org/10.3892/ol.2011.499.

Xu, X., Sun, Q., Liang, X., Chen, Z., Zhang, X., Zhou, X., et al. (2019). Mechanisms of relapse after CD19 CAR T-cell therapy for acute lymphoblastic leukemia and its prevention and treatment strategies. *Frontiers in Immunology, 10*, 2664. https://doi.org/10.3389/fimmu.2019.02664.

Yang, G.-C., Xu, Y.-H., Chen, H.-X., & Wang, X.-J. (2015). Acute lymphoblastic leukemia cells inhibit the differentiation of bone mesenchymal stem cells into osteoblasts in vitro by activating notch signaling. *Stem Cells International, 2015*, 162410. https://doi.org/10.1155/2015/162410.

Yao, H., Price, T. T., Cantelli, G., Ngo, B., Warner, M. J., Olivere, L., et al. (2018). Leukemia hijacks a neural mechanism to invade the central nervous system. *Nature, 560*(7716), 55–60. https://doi.org/10.1038/s41586-018-0342-5.

Zhou, X., Zhou, S., Li, B., Li, Q., Gao, L., Li, D., et al. (2015). Transmembrane TNF-α preferentially expressed by leukemia stem cells and blasts is a potent target for antibody therapy. *Blood, 126*(12), 1433–1442. https://doi.org/10.1182/blood-2015-01-624833.

Zuo, X. L., Chen, J. M., Zhou, X., Li, X. Z., & Mei, G. Y. (2006). Levels of selenium, zinc, copper, and antioxidant enzyme activity in patients with leukemia. *Biological Trace Element Research, 114*(1–3), 41–53. https://doi.org/10.1385/bter:114:1:41.

CHAPTER TWO

Stem cell niches in bone and their roles in cancer metastasis

Ingunn Holen and Lewis Quayle*
University of Sheffield, Medical School, Sheffield, United Kingdom
*Corresponding author: e-mail address: l.quayle@sheffield.ac.uk

Contents

1. Introduction

Metastatic disease remains a leading cause of cancer-related death. The two most common cancer types, breast and prostate cancer, frequently metastasize to the skeleton. Metastatic foci are initiated by a latent sub-population of disseminated tumor cells (DTCs) that escape the primary tumor and colonize distal sites, frequently entering into a protracted state of mitotic arrest before eventual emergence to reinitiate tumor growth (Aguirre-Ghiso & Sosa, 2018). The inability to remove this population has proved to be the major obstacle in developing truly curative treatments for metastatic disease, with 15-year recurrence rates reaching as high as 40% (Pan et al., 2017). A substantial body of evidence supports that the reciprocal

Advances in Stem Cells and their Niches, Volume 5
ISSN 2468-5097
https://doi.org/10.1016/bs.asn.2020.09.002

cross talk between disseminated cancer cells and those of the bone marrow microenvironment plays a crucial role in regulating both the terminal stages of the metastatic dissemination cascade and the eventual progression from growth arrested micrometastatic foci to overt bone lesions (Ghajar et al., 2013; Sleeman, 2012; Sosa, Bragado, & Aguirre-Ghiso, 2014).

Bone marrow is comprised of cells of the hematopoietic and reticuloendothelial systems, as well as several cell types of mesenchymal origin. The specific niche microenvironment in which hematopoietic stem cells (HSCs) reside is composed of osteoblasts, adipocytes, endothelial cells and nerves, and the interactions amongst some of these cells and other cell types are critical for regulating osteogenesis, bone remodeling, and hematopoiesis (Arai et al., 2004; Calvi et al., 2003; Nombela-Arrieta et al., 2013; Zhang et al., 2003; Zhou et al., 2017). Stephen Paget famously hypothesized in his "seed and soil" theory that metastatic cancer cells (the seeds) specifically seek a post-metastatic site (the soil) in order to survive outside of the primary tumor (Paget, 1989). Bone-tropic DTCs have since been shown in multiple studies to commandeer niche microenvironments such as the HSC niche in order to persist within bone. However, while some of the molecular mechanisms utilized by DTCs under these circumstances have been extensively studied, many of the tumor-supportive roles of the bone marrow microenvironment and the major molecular mechanisms governing the pathogenesis of bone metastasis are still poorly understood. Increased understanding of these will likely provide the key to unlocking vastly improved prognosis for patients with skeletal metastases. In this chapter we will discuss the different mechanisms that are currently proposed to govern the pathogenesis of bone metastasis in the context of the niche microenvironment in which DTCs reside.

2. Niches within the bone marrow

Bone marrow contains several partially overlapping physiological niche microenvironments, including the endosteal, perivascular, HSC, adipocyte and immune cell niches (Mendez-Ferrer et al., 2010) (Fig. 1). The endosteal niche is that which forms the periphery of both the trabecular and endocortical bone surface, consisting of bone-resident fibroblasts, osteocytes, bone-forming osteoblasts and bone-resorptive osteoclasts; the coordinated actions of these latter two cell types is tightly coupled under physiological conditions to allow the continual balanced process of bone remodeling. In addition to their role in bone remodeling, the cellular

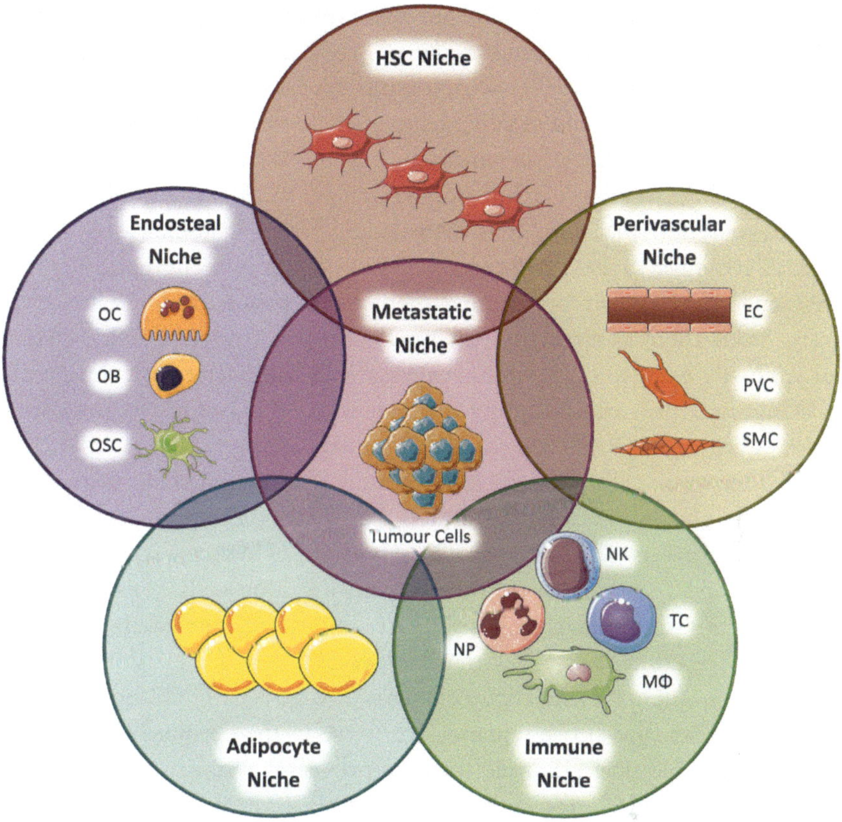

Fig. 1 Niche microenvironments within bone. The bone metastatic niche in which disseminated tumor cells reside is comprised of a number of overlapping components which form physiological niche microenvironments: the hematopoietic stem cell (HSC) niche; the endosteal niche (composed of osteoclasts (OC), osteoblasts (OB), osteocytes (OSC), and fibroblasts); the perivascular niche (composed of endothelial cells (EC), pericytes (PVC) and vascular smooth muscle cells (SMC) that make up the microvasculature); the immune niche which is comprised of both transient and bone-resident populations of immune cells (e.g., macrophages (MΦ) natural killer cells (NK), neutrophils (NP) and T-cells (TC)) and their precursors; and adipocytes, which constitute their own niche microenvironment according to emerging evidence. The exact degree of overlap and interaction between these niches, as well as the degree to which cancer cells are influenced by each remains to be fully elucidated. However, this complex environment, known as the metastatic niche, is thought to regulate homing, survival, dormancy and outgrowth of disseminated tumor cells in bone.

constituents of the endosteal niche are also proposed to aid maintenance of HSCs in a mitotically quiescent state (Sims & Martin, 2015). By contrast, the cells of the perivascular niche, which consists of vascular endothelial cells, pericytes and smooth muscle cells, have been shown to play a key role in promoting HSC mobilization, proliferation and differentiation, as well as secretion of angiocrine growth factors which act to recruit endothelial progenitors, mesenchymal stem cells (MSCs) and HSCs (Butler, Kobayashi, & Rafii, 2010; Park, Sykes, & Scadden, 2012). Although the endosteal and perivascular niches can be compartmentalized and considered independently, both their constituent cell types and physiological functions are non-mutually exclusive, playing an intricately linked series of roles pivotal for HSC regulation and coupling the processes of angiogenesis and osteogenesis (Park et al., 2012; Stegen & Carmeliet, 2018).

The metastatic niche in bone is comprised of several overlapping constituents of the physiological hematopoietic, endosteal, and perivascular niche microenvironments (Fig. 1). A growing body of evidence also increasingly indicates that bone marrow adipocytes and immune cells constitute their own microenvironmental niches and also comprise part of the niche in which DTCs come to reside (Malladi et al., 2016; Templeton et al., 2015). This complex niche has been shown to govern DTC homing; disseminated breast and prostate cancer cells are both recruited to the HSC niche via physiological HSC recruitment processes and subsequently compete directly with HSCs for niche occupancy (Price et al., 2016). Upon successfully engrafting within the bone microenvironment, it is the reciprocal interplay between DTCs and the cells of the niche in which they come to reside that governs the molecular mechanisms that dictate whether DTCs actively proliferate, become or remain mitotically quiescent, or die. Therefore, it is ultimately the cellular constituents of the bone metastatic niche that determine the temporal dynamics and eventual course of disease pathogenesis. This means that therapeutic targeting of components of the bone microenvironment may complement anti-cancer agents to improve patient outcome.

3. Niche constituents originating from mesenchymal stem cells

3.1 Osteoblasts

Osteoblasts are derived from MSCs under the tight regulation of numerous canonical signaling pathways and transcription factors, key amongst which are RUNX2 and Osterix, which promote the initial commitment of MSCs to osteoprogenitor cells that further differentiate to form mature osteoblasts (Zaidi, 2007). Mature osteoblasts are the cells primarily responsible for new bone formation through secretion of type-I collagen, alkaline phosphatase and osteocalcin, and also play a key regulatory role in homing and maintenance of HSCs (Shiozawa & Taichman, 2012). A large body of evidence from in vivo model systems support a key role for osteoblasts in the bone metastasis sequelae. Osteoblasts are suggested as potential mediators of DTC homing to bone, primarily due to the frequent observation that DTCs locate within osteoblast-rich regions (Haider, Holen, Dear, Hunter, & Brown, 2014). For example, following intracardiac inoculation of the bone-tropic PC-3 sub-clonal lineage PC-3NW1 into athymic nude mice, significantly greater numbers of DTCs were identified on osteoblast-rich lateral endocortical surfaces of the tibia compared with osteoblast-deficient medial endocortical surfaces (Wang et al., 2014). In another study in which PC-3 and HSCs were simultaneously injected into irradiated severe combined immunodeficient (SCID) mice (via intracardiac and intravenous routes, respectively) both PC-3 cells and HSCs homed specifically to osteoblast-rich regions. Moreover, both PC-3 and C4—2B human prostate cancer cells were able to prevent binding of HSCs to murine primary calvarial osteoblasts, demonstrating that disseminated cancer cells directly compete for the HSC niche formed partially by osteoblast (Shiozawa et al., 2011). Similar patterns of osteoblast-specific dissemination have been reported in breast cancer. In two distinct in vivo models systems, both MDA-MB-231 and MDA-MB-435 breast cancer cells were shown to preferentially locate to the metaphyseal region of the murine distal femur, especially in close proximity to osteoblast-rich trabecular bone surfaces (Haider et al., 2014; Phadke et al., 2006). It is suggested that this phenomenon is mediated by osteoblast secreted CXCL12 and RANKL, factors shown to favor cancer cell dissemination through binding their cognate receptors (CXCR4 and RANK, respectively) expressed on the DTC surface (Price et al., 2016; Sun et al., 2014). When PC-3NW1-inoculated animals were treated with the CXCR4 antagonist AMD3100, the differential pattern of PC-3NW1 cell dissemination between lateral and medial endocortical surfaces was lost, supporting that ligation of CXCR4 by osteoblast secreted CXCL12 drives DTC homing to specific bone microenvironments (Wang et al., 2014). This hypothesis was further supported by reports from breast cancer models in which both newly formed and established metastatic foci were found to be anchored to the bone marrow via CXCL12-CXCR4 interactions, and early metastatic foci in the MMTV-PyMT model were consistently shown to emerge adjacent to the trabecular region below the growth plate cartilage which is rich in CXCL12-high osteoprogenitor cells (Devignes et al., 2018; Price et al., 2016). Observations made in other reports have extended our understanding of potential mechanisms by which osteoblasts facilitate DTC homing and residency beyond that of the CXCL12-CXCR4 axis. For example, when cells of the MCF-7 human breast cancer lineage were inoculated into the iliac artery of athymic nude mice, DTCs were found localized to osteoblasts through interactions between DTC E-cadherin and osteoblastic N-cadherin. This interaction was shown to activate the downstream mammalian target of rapamycin (mTOR) pathway, with subsequent pharmacological inhibition of this pathway delaying early osteoblast-specific colonization (Wang et al., 2015). Similarly, both the human prostate cancer cell line C4—2B4 and human breast cancer cell line MDA-MB-231 have been shown to bind murine pre-osteoblasts through homophilic cadherin-11 to cadherin-11 interactions (Huang et al., 2010; Tamura, Hiraga, Myoui, Yoshikawa, & Yoneda, 2008).

While only a limited number of the diverse array of findings relating to the role of osteoblasts in early cancer cell dissemination are described here, all of the studies discussed demonstrate that osteoblasts play a critical in DTC homing to bone, and that osteoblasts form an integral part of the metastatic niche that is required for DTCs to gain a foothold at this distal site. DTCs can remain in an indolent state of mitotic quiescence for decades (a phenomenon referred to clinically as dormancy) before emerging to re-enter the cell cycle and establish overt metastatic disease. An ever-growing body of experimental evidence has implicated osteoblast-driven mechanisms as playing a significantly contributory role at this stage of disease pathogenesis. One of the best-defined mechanistic regulators of osteoblast-mediated dormancy in bone metastasis is the annexin-II-GAS6-AXL-TGF-β axis. The initial indication for the importance of this came from a series of studies that showed significantly increased binding of PC-3 cells to murine calvarial osteoblasts and greater overall metastatic burden in annexin-II or growth arrest-specific-6 (GAS6) wild-type mice versus knockout animals (Shiozawa et al., 2016, 2008). In subsequent studies, the presence of osteoblast-secreted annexin-II in the cellular microenvironment was shown to induce increased expression of AXL (the GAS6 receptor) by DTCs, while osteoblastic secretion of GAS6 was enhanced in the presence of the DTCs themselves (Shiozawa et al., 2011). The ligation of AXL express on

DTCs by osteoblastic GAS6 induced expression of transforming growth factor-β1 (TGFβ1), transforming growth factor-β2 (TGFβ2) and the TGF-β receptors TGFBR2 and TGFBR3 in DTCs, leading to an autocrine feedback loop resulting in dormancy (Yumoto et al., 2016). A series of similar studies in breast cancer have elucidated several well-defined mechanisms that relate to TGFβ2-regulated cellular dormancy, especially that mediated by a TGFβ2-induced high p38:ERK signaling ratio, subsequent induction of the metastatic suppressor proteins BHLHE41 (also known as DEC2 or SHARP1) and p27, eventual repression of CDK4 activity causing entry into dormancy (Bragado et al., 2013; Marlow et al., 2013). Osteoblastic GAS6 was also later shown to interact with the MER receptor on the DTC cell surface, resulting in activation of the mTOR pathway and enhancing acquisition of the $CD133^+$ $CD44^+$ "stem-like" phenotype by disseminated PC-3 cells in the murine bone marrow which coincided with their entry into dormancy (Shiozawa et al., 2016). All of these studies illustrate that further to their role in early tumor cell dissemination and colonization of bone, osteoblasts play a significant contributory role in regulation of metastatic dormancy.

It is important to note that, while the vast majority of mechanisms by which cancer cells reawaken and progress from dormant micrometastases to actively proliferating secondary tumors remain to be elucidated, the synergistic tumor cell and osteoblast-driven secretion of pro-inflammatory molecules, recruitment of bone-resorptive osteoclasts, and the resultant activation of the "vicious cycle of bone metastasis" is well established; this eventually leads to the disruption of the fine-tuned balance of bone turnover and distinctive macroscopic changes that are characteristic of bone metastasis (Bussard, Venzon, & Mastro, 2010). For example, a number of studies in breast cancer have demonstrated enhanced osteoblastic secretion of pro-inflammatory cytokines such as IL-6, IL-8, monocyte chemo-attractant protein-1 (MCP-1), macrophage-inflammatory protein 2 (MIP-2), and vascular endothelial growth factor (VEGF) in response to the presence of metastatic breast cancer cells (Bussard et al., 2010; Kinder, Chislock, Bussard, Shuman, & Mastro, 2008). In one such study, induction of osteoblastic IL-6 secretion in the presence of JAG1-overexpressing bone-tropic MDA-MB-231 breast cancer cells resulted in the rapid expansion of the tumoral mass in response to an IL-6-driven activation of Notch signaling (Sethi, Dai, Winter, & Kang, 2011). Similarly, the rapid expansion of disseminated NT2.5 breast cancer cells in the fibroblasts accelerate stromal-supported tumorigenesis (FASST) mouse model was indirectly induced by osteoblast-derived IL-6 stimulating osteoclastogenesis and osteolytic release of bone stromal growth factors (Luo et al., 2016). Though far from an exhaustive account, these studies exemplify the important role that osteoblasts within the metastatic niche play in facilitating the later stages of disease pathogenesis.

3.2 Osteocytes

Osteocytes are the terminally differentiated cells of the osteoblastic lineage and arise when mature osteoblasts become embedded within the bone matrix. Approximately 90–95% of adult skeletal tissue contains osteocytes, which function predominantly in mechanosensory and stromal maintenance roles, particularly regulation of bone remodeling (Bonewald, 2007, 2011; Chen, Wang, Zhao, & Wang, 2018). Although osteocytes were originally not included in conceptual models of bone metastasis, thought simply to form part of the microenvironment in which DTCs reside, more recently several experimental reports have initiated new interest in their role in disease progression.

Bone remodeling itself is controlled by tightly balancing osteoblastogenesis, osteoclastogenesis and the respective activities of osteoblasts and osteoclasts. Osteocytes negatively regulate osteoblast activity by releasing DKK1 and sclerostin (inhibitors of the Wnt pathway that are important for osteoblastic differentiation) and enhance osteoclastic resorptive activity by releasing RANKL when they undergo apoptosis (Bonewald, 2007, 2011; Chen, Wang, Zhao, & Wang, 2018; Sims & Martin, 2015). In one study, the conditioned medium obtained from a co-culture of the JJN3 human multiple myeloma cell line and the human pre-osteocyte cell line HOB-01 stimulated formation of osteoclasts from human primary peripheral blood mononuclear cells by enhancing the secretion of IL-11 and CCL3. The same study found that osteocyte death, levels of IL-11 and osteoclast activity were significantly elevated in the osteolytic bone lesions of patients with multiple myeloma (Giuliani et al., 2012). These data suggest that osteocytes create a favorable environment for bone metastatic cancer cell outgrowth during the latter stages of disease progression by promoting both osteoclastogenesis and resultant osteolytic bone damage that then acts to drive cancer progression through release of matrix-bound growth factors. While these findings are reported from models of multiple myeloma, there may be comparative biological mechanism in the advanced stages of other bone-tropic cancers, such as breast and prostate.

Some evidence for the direct pro-metastatic and pro-tumorigenic role of osteocyte-derived factors has been reported in breast and prostate cancer models, suggesting that osteocytes could also directly affect both the early colonization of bone and the eventual outgrowth of disseminated cancer cells. In one such study, the conditioned medium from murine MLO-Y4 osteocytes had significantly stimulatory effects on the proliferation of human breast (MDA-MB-231 and MCF-7) and prostate cancer (DU145 and PC-3) cells, as well as increasing the migration of MDA-MB-231 cells (Cui, Evans, & Jiang, 2016). Separate studies have shown that MLO-Y4 conditioned medium enhanced the anchorage-independent growth, migration, and invasion of MDA-MB-231 cells and murine breast cancer cell line Py8119, and that MLO-Y4 conditioned medium induced the growth, migration, and invasion of PC-3 and C4—2B human prostate cancer cells by up-regulating EGR1 expression (Wang et al., 2019; Zhou et al., 2016). When MLO-Y4 cells were mechanically stimulated by application of hydrostatic pressure, the MLO-Y4 conditioned medium potentiated growth, migration, and invasion of PC-3, DU145, and LNCaP, which was demonstrated to be partially attributable to up-regulation of the chemotactic cytokine CCL5 and the pro-metastatic enzymes MMP2 and MMP9 (Sottnik, Dai, Zhang, Campbell, & Keller, 2015).

Despite the array of findings that support the prospective pro-metastatic role of osteocytes in bone metastasis, some studies have reported contrasting findings. For example, the conditioned medium from the murine MLO-A5 osteocyte lineage was found to reduce the invasive capability of murine 4T1.2 breast cancer cells, through type I collagen by down-regulating the epithelial-to-mesenchymal transition-regulatory transcription factor SNAIL and increasing AKT activation (Chen et al., 2018; Liu et al., 2018). Similarly, another study that used the conditioned medium from mechanically stimulated (fluid flow-induced shear stress) MLO-Y4 cells showed that adhesion of metastatic MDA-MB-231 breast cancer cells to human umbilical vein endothelial cells (HUVECs) via ICAM-1, endothelial cell permeability, tumor cell invasive capability and secretion of MMP-9 were all significantly reduced in the presence of osteocyte conditioned medium (Ma, Xu, Mei, Middleton, & You, 2018). These findings seemingly strengthen the purported importance of mechanical loading (as would be the case during exercise) in the prevention of bone metastases and highlight a potential key role of osteocytes in negatively regulating the establishment of metastatic foci in bone-tropic cancer.

Although the role of osteocytes the pathogenesis of bone metastasis remain to be fully elucidated, the critical importance of osteocytes to regulation of bone remodeling supports the plausibility that osteocytes do affect tumor cell dormancy and metastatic outgrowth in bone, either via presently hypothetical direct interactions with DTCs or via their influence over more active bone niche constituents (e.g., endothelial cells, osteoblasts, and osteoclasts) (Bellido, 2014). Further studies are warranted to allow us to more definitively conclude the precise contribution of osteocytes in bone metastasis.

3.3 Bone marrow adipocytes

Adipocytes, also known as lipocytes or fat cells, primarily specialize in storing energy as fat. Bone marrow contains a significant proportion of adipocytes, comprising approximately 15% of bone marrow volume in children and increasing with age to account for approximately 70% of marrow space in adolescents and adults (Fazeli et al., 2013). As a result of their prolific numbers, bone marrow adipocytes serve as one of the major components of the niche microenvironment for hematopoietic stem cells but have historically been largely considered outside of conceptual models of bone metastasis, in much the same way as osteocytes (Naveiras et al., 2009; Zhou et al., 2017). However, a high fat diet and obesity have long been reported as independent prognostic indicators associated with increased tumor growth and bone destruction in metastatic disease and a multitude of recent studies have demonstrated that bone marrow adipose tissue acts to both store and secrete adipokines and cytokines which strongly modulate the behavior of cells in the bone marrow and influence metastatic disease progression through paracrine and endocrine signaling (Diedrich et al., 2016; Herroon et al., 2013; Herroon, Rajagurubandara, Diedrich, Heath, & Podgorski, 2018; Morris & Edwards, 2016).

Experimental and clinical evidence points to adipocytes contributing to metastatic dissemination and initial recruitment of cells to the bone metastatic niche. When human bone marrow stromal cells were transformed into adipocytes by treatment with omega-6 polyunsaturated fatty acid (ω-6FAs), significantly more PC-3 prostate cancer cells migrated toward these than non-treated bone marrow stromal cells (Brown, Hart, Gazi, Bagley, & Clarke, 2006; Brown et al., 2010). Likewise, co-culture of MDA-MB-231 breast cancer cells with bone fragments obtained from patients who underwent hip replacement surgery showed extensive tumor cell migration toward and colonization of the bone marrow adipose tissue compartment by MDA-MB-231 cells (Templeton et al., 2015).

As a consequence of their primary role as fat storage cells, adipocytes contain lipid droplets that are a source of fatty acids. When located adjacent to tumor cells, adipocytes have been shown to transfer lipids directly to the cancer cells. Close proximity of PC-3 cells to ω-6Fas-transformed stromal cells resulted in significantly increased lipid intake (Brown et al., 2006, 2010). This lipid transfer may not only act as a ready source of energy, but has also been shown to serve as a stimulus for metabolic reprogramming (Brown et al., 2006). In one study, murine bone marrow adipocytes were shown to induce a shift toward glycolytic metabolism in human prostate cancer cell lines PC-3 and ARCaP cells by up-regulating glycolytic enzymes (ENO2, LDH, HK2, PDK1 and activated PDH), increasing lactate release, reducing the rate of oxygen consumption and decreasing net ATP levels. These effects were mediated by increased expression of HIF-1α target genes CA9, VEGF and GLUT1 (Diedrich et al., 2016). In two similar studies, adipocyte-induced shifting of breast and prostate cancer cell metabolism toward glycolysis increased the expression of FABP4, HMOX1 and IL-1, resulting in an increase in cellular proliferation and invasiveness (Dirat et al., 2011; Herroon et al., 2013). Likewise, co-culture of PC-3 and ARCaP cells with murine bone marrow adipocytes significantly increased expression of HOMX1 and GRP78 (both involved in protecting against cellular oxidative stress) and promoted increased cancer cell invasiveness. Subsequent over-expression of HMOX1 induced increased levels of GRP78 and pro-survival factors Bcl-xL and survivin, as well as promoting enhanced growth of PC-3 and ARCaP cells within the bone following intra-tibial injection, compared to control cells. Moreover, bone marrow aspirates obtained from prostate cancer patients with bone metastases showed higher levels of HMOX1 compared to primary prostate tumors or samples collected from other metastatic sites (Herroon et al., 2018). Taken together, these studies suggest that adipocyte-induced metabolic reprogramming and up-regulation of the oxidative stress response in particular play a prominent role in the way in which adipocytes influence formation of bone metastatic foci.

The release of pro-inflammatory cytokines and growth factors such as IL-6, TNF-α and CXCL12 by adipocytes has been shown to support tumor development and inhibit apoptosis in multiple myeloma (Caers et al., 2007; Morris & Edwards, 2018; Wang et al., 2017). This tumor promoting function has also been observed in prostate cancer, in which adipocyte secretion of CXCL1 and CXCL2 was shown to enhance osteoclastogenesis and result in increased cancer cell survival (Hardaway, Herroon, Rajagurubandara, & Podgorski, 2015). However, the findings relating to the tumor effects of adiponectin and leptin, the most important adipocyte-secreted adipokines, on are contradictory. Leptin has a documented pro-tumor effect and expression levels of leptin have been associated with increased probability of metastasis (Snoussi et al., 2006). In one study, adipocyte-derived leptin and IL-1 were shown to increase breast cancer cell colonization and migration to adipocyte-rich regions of the skeleton (Chu et al., 2019; Templeton et al., 2015). Leptin has been shown to increase expression of VEGF by breast cancer cells via HIF-1α and NF-κB signaling, thereby promoting cancer progression (Gonzalez-Perez et al., 2010). Adiponectin is primarily involved in glucose uptake and catabolic fatty acid metabolism, but has also been shown to possess anti-tumoral activity within the bone microenvironment by both decreasing DTC proliferation and stimulating apoptosis via mTOR and NF-κB signaling. Paradoxically, this endogenous anti-tumoral effect is negated in obesity due to reduced expression of adiponectin receptors, leading to chemoresistance in the tumor cell population (Chu et al., 2019).

Adipocytes have emerged as mediators of metastatic bone disease and all of the studies described above indicate that bone marrow adipocytes play an important role in the regulation of metastatic progression of solid tumors within the bone. However, the molecular and cellular interactions between adipocytes and metastatic cancer cells within the bone microenvironment remain quite poorly understood and therefore offer an interesting and potentially bountiful area for future research.

4. Niche constituents originating from hematopoietic stem cells

4.1 Osteoclasts

Osteoclasts are large multinucleate cells derived from the monocyte/macrophage lineage of hematopoietic precursors and are primarily responsible for bone resorption. Excessive osteoclastic bone resorption is the hallmark of late-stage metastatic bone disease and the molecular mechanisms by which osteoclastic activity potentiates the growth of overt metastatic foci, widely referred to as the "vicious cycle of bone metastasis", is extensively documented (Cook, Shay, Araujo, & Lynch, 2014; Guise, 2002); during this stage of metastatic progression, tumor cells stimulate osteoclasts to resorb the surrounding bone matrix, thereby creating sufficient space into which newly formed tumor cells can grow and releasing a myriad of matrix-bound growth factors which further fuel tumor growth in bone (Fig. 2).

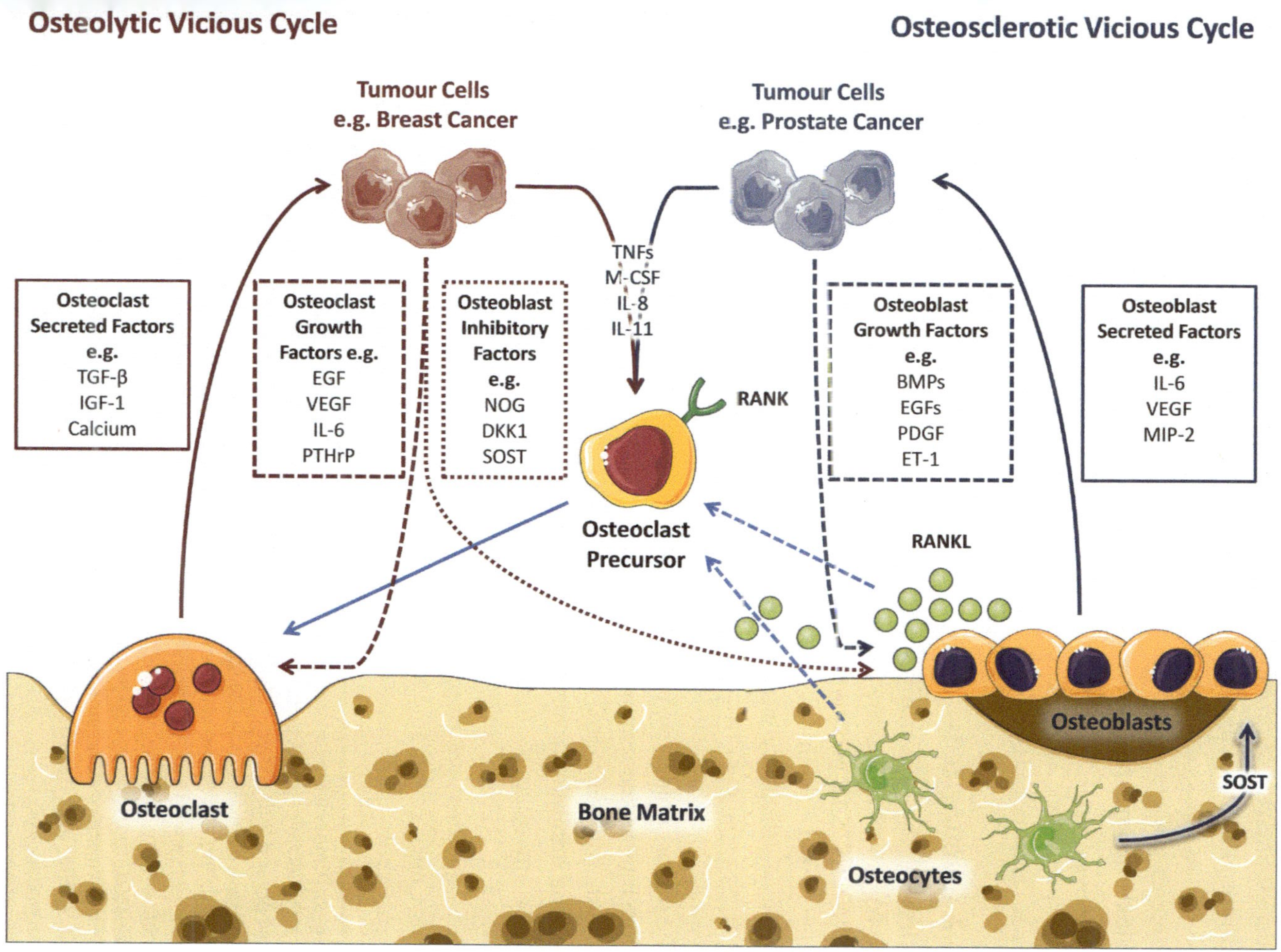

Fig. 2 See figure legend on next page.

As a result of their key role in the later stages of disease pathogenesis, the current standard of care treatments for metastatic bone disease, such as the bisphosphonates and the RANKL inhibitor denosumab, were developed to inhibit osteoclast activity. By reducing osteoclastic bone resorption, these agents are capable of slowing down the tumor-induced osteolysis which in-turn provides a pro-proliferative stimulus for cancer cells by releasing matrix-bound growth factors (Hanley, Adachi, Bell, & Brown, 2012; Kimmel, 2007). Unfortunately, once osteolytic lesions have developed, treatment is restricted to palliative care.

4.2 Immune cells

Although the bone marrow is the site of origin of all immune cell types and contains various mature immune cell populations, the reciprocal cross talk between DTCs and bone marrow resident immune cells remains largely unexplored. The limited experimental findings to date appear to suggest that immune cells play contrasting roles in the early and late stages of metastasis; playing a role in instructing metastatic dormancy early on while later fostering rapid progression of metastatic bone disease. Prospective evidence for the tumor suppressive role of the immune system in early metastasis comes from reports that recipients of organs from disease-free donors who had previously had melanoma developed metastases in the transplanted organ following immunosuppression, suggesting that dormant cells were prevented from emerging to form overt metastatic foci by the immune system (MacKie, Reid, & Junor, 2003). It remains unclear whether DTC dormancy was dependent on immune suppression or whether it resulted from other tissue microenvironmental alterations arising from transplantation. At present, direct evidence for the tumor-suppressive role of bone marrow

Fig. 2 Tumor cell-bone cell interactions that drive cancer-induced bone disease. Under the influence of physical factors within the bone microenvironment (including hypoxia, low pH and high extracellular calcium levels), cancer cells that disseminate to the bone secrete a number of soluble factors which disrupt the tightly balanced process of bone remodeling. Osteolytic lesions are generated when factors released by cancer cells (e.g., in breast cancer) inhibit osteoblast activity, stimulate osteoclast formation and enhance their activity, thereby stimulating bone resorption. In osteosclerotic disease, factors released from cancer cells (e.g., prostate cancer cells) act on osteoblasts, stimulating bone formation, along with the stimulation of osteoblast-mediated osteoclastogenesis via RANKL expression, inducing osteolytic lesions. Factors released from stimulated bone cells and those released from the bone matrix by osteolytic activity enhance the growth and activity of the disseminated tumor cells, hence creating a "vicious cycle" that results in the formation of an overt secondary tumor and destruction of the bone.

immune cells in the very early stages of bone metastasis pathogenesis is lacking (Rhodes & Hillen, 2019). However, it is well established that the HSC niche is an immune privileged site, offering protection to HSCs from environmental insults and the resultant immune response. One study found that $CD8^+$ (cytotoxic) T-cell-suppressing regulatory T-cells (T-regs) co-localize with HSCs on endosteal surfaces of trabecular regions of the bone marrow (Fujisaki et al., 2011). These findings suggest that T-regs create a stem cell-supportive niche and mediate protection from immune attack; such an environment would clearly also benefit bone-tropic DTCs by preventing their elimination and promoting their survival in bone.

In contrast to our relatively limited knowledge of the role of the immune niche in the very early stages of bone metastasis, some evidence does exist for the role of the immune system in suppressing metastatic outgrowth later in disease. In one study, overt bone metastases from immunocompetent mice inoculated with murine 4T1.2 breast cancer cells were shown to express suppressed levels of IRF7 (a prominent regulator of interferon (IFN) induction and NK-cell-mediated anti-tumor immunity), compared to expression levels in cells from matched primary tumors. Over-expression of IRF7 and treatment with exogenous IFN-γ restored the anti-metastatic immune response and reduced the overall metastatic burden in tumor-bearing animals. Conversely, the inoculation of IRF7 over-expressing 4T1.2 cells into mice with depleted NK (and $CD8^+$ T-cell) populations led to accelerated development of bone metastasis when compared to immunocompetent control animals (Bidwell et al., 2012). Similarly, a defective anti-tumor $CD8^+$ T-cell response in immunocompromised tumor-bearing PLCγ2 −/− mice was shown to result in increased tumor growth in bone, despite the impaired osteoclast function that occurs in this model (Zhang et al., 2011). These data suggest that the IFN-γ-mediated NK-cell and CD8+ T-cell immune responses contribute to the immune suppression of overt bone metastasis. Indeed, high expression of IRF7-regulated genes in primary tumors was associated with prolonged bone metastasis-free survival in a cohort of over 800 patients (Bidwell et al., 2012).

A growing body of evidence also exists for the pro-tumorigenic role of immune cells in the later stages of disease progression. In one study, significantly higher numbers of $CD4^+FOXP3^+$ regulatory T-cells were found in the bone marrow of prostate cancer patients with bone metastases compared to those without. These regulatory T-cells were shown to inhibit osteoclast activity, tipping bone remodeling toward excessive formation of new bone tissue without preceding osteoclastic resorption, resulting in osteoblastic bone lesions that are common in bone metastatic prostate cancer (Zhao et al., 2012). In an experimental model of breast cancer, RANKL levels in $CD3^+$ T-cells within the bone marrow of BALB/c nude mice orthotopically inoculated with murine 4T1 breast cancer cells were significantly enhanced compared to levels in mice inoculated with the non-metastatic 67NR 4T1 sub-line. This elevated level of RANKL led to increased octeoclastic activity and an overall greater number of metastatic foci within bone (Monteiro et al., 2013). Intracardiac injection of 4T1 cells into BALB/c nude mice was shown to increase the numbers of plasmacytoid dendritic cells and $CD4^+$ helper T-cells within the bone marrow, which in-turn elicited osteolytic bone lesions and increased overall tumor burden (Sawant et al., 2012).

The apparently pro-tumorigenic role of some bone-resident immune cells is not limited to those of lymphoid origin, but has also been observed in cell types of the myeloid lineage. Significantly increased numbers of $CD11b^+Gr\text{-}1^+$ myeloid-derived suppressor cells (MDSCs) were observed in the bone marrow of C57BL6/KaLwRij mice intravenously injected with 5TGM1 murine multiple myeloma cells compared to non-tumor-bearing mice. These MDSCs showed greater potential to differentiate into mature and functional osteoclasts than those from control mice and were associated with increased incidence of osteolytic bone lesions (Zhuang et al., 2012). Almost identical results were reported in a separate study in which BALB/c nude mice were inoculated with 4T1 murine breast cancer cells (Sawant et al., 2013). In bone samples obtained from prostate cancer patients with metastatic bone lesions, a significantly greater number of $CD68^+$ tumor-associated macrophages (TAMs) were observed in proximity to the tumor mass than in other regions of the bone (Wu et al., 2016). When TAMs were depleted in a study using the MAFIA mouse model, both growth of murine RM-1 prostate cancer cells within the bone and resultant osteolytic lesions were significantly reduced. Similar results were observed for both RM-1 murine prostate cancer and PC-3 human prostate cancer cell lines inoculated into C57BL6 or athymic nude mice, respectively, following depletion of TAMs using liposomal clodronate (Soki et al., 2015). While these studies demonstrate the very limited direct evidence for a role in bone metastases, macrophages are heterogeneous and highly plastic cells with a phenotypic spectrum determined by their location and activation status, rather than existing as discrete sub-populations with specific functions (Geissmann et al., 2008; Michalski & McCauley, 2017). This variability, as well as a lack of consensus over marker signatures with which to accurately identify their multiple sub-types, makes macrophages inherently difficult to identify and their role in disease complex to characterize. While some emerging evidence does suggests that various context-specific macrophage sub-populations play important roles at different stages of metastatic disease pathogenesis, their role(s) in the early stages of bone homing, niche engraftment, and metastatic dormancy still remain to be fully elucidated (Lin, Xu, & Lan, 2019; Qian et al., 2009).

5. Other niche constituent cell types

5.1 Endothelial cells

Bone is a highly vascularized tissue, where an intricate network of vessels not only provide oxygen, nutrients and endocrine signaling molecules (e.g., hormones, growth factors and neurotransmitters) but also act to remove waste metabolites and cytokines produced in the marrow, serve as a transport conduit for various cell types, including hematopoietic and immune cells formed within bone, and DTCs in metastatic cancer (Kusumbe, Ramasamy, & Adams, 2014). Bone-tropic cancer cells preferentially localize to the metaphyseal region of the long bones, often taking up residency in a perivascular location adjacent to endothelial cells (Allocca et al., 2019). The bone metaphysis has extensive arterial vascularity, providing a high local oxygen tension and growth-permissive microenvironment through the ready provision of nutrients and growth factors. However, while the vasculature is known to be crucial for overt tumor growth in later stages of metastatic bone disease, relatively little is known about the role of bone marrow endothelial cells as a tumor microenvironment. In contrast to the obvious role of the vasculature in facilitating tumor growth and expansion, emerging reports support that the bone vasculature plays a growth-inhibitory role in early stages of metastatic disease pathogenesis by strongly influencing DTC dormancy.

Orthotopically inoculated human MDA-MB-231 and intracardiac injected HMT-3522-T4-2 breast cancer cell lines both disseminated to bones of NOD/SCID mice, becoming dormant in close proximity to the bone microvascular endothelium. HMT-3522-T4-2 cells also remained dormant in an organotypic bone microvascular mimetic co-culture system of HUVECs and murine primary bone marrow stromal cells and more dormant cells were observed when in proximity to thrombospondin-1-expressing mature endothelium, while TGF-β1 and periostin-expressing immature neovascular tips accelerated HMT-3522-T4-2 cell proliferation (Ghajar et al., 2013). Similarly, murine 4T07 breast cancer cells disseminated to bone following orthotopic implantation and located in closer to the perisinusoidal vasculature than to megakaryocytes or osteoblasts within the marrow of BALB/c nude mice. In addition, perivascular-resident 4T07 cells were shown to be markedly more resistant to chemotherapy (doxorubicin and cyclophosphamide in combination) than cells located elsewhere in the bone marrow. Similar findings were reported when HMT-3522-T4-2 were grown in an organotypic bone microvascular mimetic co-culture system. However, in contrast to the dogmatic purported hypothesis that chemoresistance in dormancy is a result of mitotic quiescence, vascular-induced resistance to doxorubicin was not due to cell cycle arrest but instead a result of the integrin-β1-von Willebrand factor and the integrin αvβ3-VCAM-1 axes that facilitate vascular binding. Indeed, the resistance of 4T07 cells in the marrow of BALB/c nude mice to the doxorubicin and cyclophosphamide regimen was reversed by both shRNA and antibody-induced blockade of integrin-mediated vascular binding (Carlson et al., 2019). These findings strongly suggest that bone marrow endothelial cells are involved in the regulation of DTC dormancy and chemoresistance in early stages of bone metastasis (Fig. 3); how these molecular signaling axes become altered or superseded to allow dormant DTCs to begin overt outgrowth and formation of secondary metastatic foci is, however, still unknown.

5.2 Nerves

The bone marrow is highly innervated and the potential biological importance of neuronal cells (nerves) within the bone marrow microenvironmental niche for cancer progression has recently come to light (comprehensively reviewed by Elefteriou, 2018). A number of reports suggest that the sympathetic nervous system in particular plays a strong regulatory role in progression of cancer bone metastasis and that suppression of nervous system function can reduce metastasis (Hayakawa et al., 2017; Magnon et al., 2013; Saloman et al., 2016; Zhao et al., 2014). Further studies have demonstrated that bone-resident neurones play a role in the initial establishment of metastatic foci. Both chronic immobilization stress and non-selective agonistic stimulation of β-adrenergic receptors with isoproterenol are shown to induce local expression of RANKL within the bone marrow, resulting in enhanced colonization of the bones of athymic nude mice by disseminated MDA-MB-231 human breast cancer cells (Campbell et al., 2012). This pharmacologically elevated sympathetic neuronal activity was later

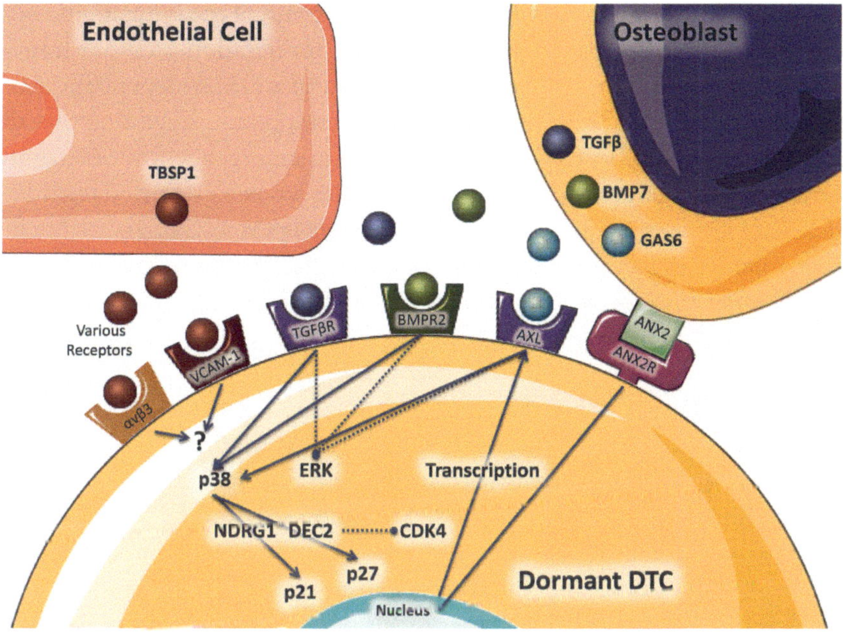

Fig. 3 Niche cell regulation of tumor cell dormancy in bone. Ligation of ANX2R on DTCs by ANX2 on the osteoblast cell membrane causes up-regulation of the AXL receptor on the cancer cell and enhances binding of GAS6 secreted by the osteoblast. Binding of osteoblast-derived TGFβ, GAS6 and BMPR2 to their respective receptors on the cancer cell surface all result in inhibition of ERK1/2 and enhanced signaling by p38, creating the high p38:Erk signaling ratio that activates metastatic suppressor proteins DEC2 (also known as BHLHE41 or SHARP1), p21 and p27, causing repression of CDK4 activity and ultimately leading to entry into mitotic quiescence (dormancy). Niche perivascular endothelial cells secrete TBSP1 that interacts with various receptors, including integrin-β 1, integrin αvβ3, von Willebrand factor and VCAM-1, to facilitate vascular binding and cause entry into dormancy by currently undefined mechanisms.

reported to facilitate bone colonization by increasing neovascularization within tumor-bearing bones through stimulating osteoblastic and stromal secretion of VEGF (Mulcrone et al., 2017). In another study, isoproterenol treatment promoted adhesion of MDA-MB-231 cells to murine primary bone marrow endothelial cells and murine endothelial cell line C166 in vitro by inducing IL-1β production and downstream increased expression of endothelial adhesion molecules E- and P-selectin (Clement-Demange, Mulcrone, Tabarestani, Sterling, & Elefteriou, 2018). These data suggest that bone-resident nerve cells play a role in initial metastatic colonization by DTCs in bone-tropic cancers. Interestingly, noradrenaline (the predominant neurotransmitter secreted by neurones of the sympathetic nervous system) has also been shown to induce proliferation of PC-3 human prostate cancer cells that were G_0/G_1-arrested in a GAS6-dependent manner when co-cultured with murine calvarial osteoblasts, suggesting that the sympathetic nervous system could play a role in emergence from metastatic dormancy (Decker et al., 2017). While little else has been reported regarding the role of nerve cells on DTC dormancy, bone is a richly innervated organ and further investigation of the biological role of bone-resident neurones on metastatic progression is therefore clearly warranted.

5.3 Megakaryocytes and platelets

Megakaryocytes are a rare, platelet-producing cell type that reside within the bone marrow in close proximity to bone marrow sinusoids. In addition to platelet production, megakaryocytes have a broad role in bone marrow homeostasis through interactions with HSCs, endothelial cells, adipocytes, osteoblasts and osteoclasts, and as a result are able to indirectly influence bone metastatic disease pathogenesis (Malara et al., 2015). For example, one study demonstrated that megakaryocytes were able to stimulate osteoblast proliferation in vitro through production of BMPs and TGFβ and inhibit osteoclastogenesis via production of osteoprotegerin, IL-10 and GM-CSF (Beeton, Bord, Ireland, & Compston, 2006). In addition, megakaryocytes contain three types of secretory granules, α-granules, dense granules and lysosomes, which are released in response to injury. The α-granules may be of particular significance to bone metastasis, as they contain over 800 proteins, including; pro- and anti-angiogenic factors, growth factors (including TGFβ), matrix metalloproteinase and various pro-necrotic factors, many of which are shown to affect tumor cell biology and modulate the extracellular matrix (Zufferey, Fontana, Reny, Nolli, & Sanchez, 2012). The platelets that megakaryocytes produce also enable these cells to exert an indirect influence on metastatic disease progression via production of lysophosphatidic acid, which has been shown to stimulate bone metastasis formation by eliciting a number of pro-tumorigenic effects and enhancing osteoclast activity (Boucharaba et al., 2004, 2006; Leblanc & Peyruchaud, 2016a, 2016b).

Limited evidence also indicates that megakaryocytes are able to directly affect the development of bone metastasis through their influence over the HSC niche. When megakaryocytes were depleted in vivo, a specific loss of megakaryocyte derived CXCL4 coincided with emergence of HSCs from quiescence. This was shown to be due to the loss of megakaryocyte-derived CXCL4 (Bruns et al., 2014). Given that DTCs are proposed to reside within the HSC niche, there is a distinct possibility that CXCL4 could be at least partially responsible for maintaining DTCs in a state of quiescence within this niche. Whether or not CXCL4 does regulate the mitotic behavior of DTCs resident within the HSC niche and whether megakaryocytes have a broader influence on the early stages of bone metastasis remains to be established.

6. Future perspectives

The development of bone metastases in advanced cancers such as breast and prostate foreshadows a very poor prognosis. While established treatments such as the anti-RANKL human monoclonal antibody denosumab and osteoclast activity-suppressing bisphosphonates are able to reduce cancer-induced bone disease, they ultimately fail to improve overall survival of patients with established skeletal metastases (Fizazi et al., 2011; Stopeck et al., 2010). Therefore, therapeutic approaches that act by means other than modulating bone remodeling are urgently required, in order to eliminate DTCs or halt metastatic disease at an earlier stage, thereby lowering the mortality of cancer patients at risk of disease progression. As described in this chapter, the roles that the various bone niche cell types that constitute the tumor cell microenvironment, as well as the reciprocal interactions between these cells and DTCs that may influence disease pathogenesis, have increasingly been elucidated. A particular challenge has been the realization that the bone microenvironment comprises multiple transient, context-specific and highly plastic cellular sub-populations (in particular of immune cells), for which there are no clearly defined markers. Despite significant research effort to date, further studies are still warranted to expand our relatively limited understanding of the complex temporospatial interactions that underpin tumor colonization and progression in bone, leading to identification of novel therapeutic targets. In particular, the intense research focused on harnessing the anti-tumor potential of the immune system holds great promise and may ultimately allow elimination of DTCs before they become a clinical problem. We predict that over the next few years, the development of agents that can modify specific cell types within bone marrow niches to disrupt their interactions with DTCs will provide a multitude of new potential strategies for prevention and treatment of bone metastasis, ultimately improving the survival of cancer patients for which there is currently no truly curative treatment.

References

Aguirre-Ghiso, J. A., & Sosa, M. S. (2018). Emerging topics on disseminated Cancer cell dormancy and the paradigm of metastasis. *Annual Review of Cancer Biology*, *2*(1), 377–393.

Allocca, G., Hughes, R., Wang, N., Brown, H. K., Ottewell, P. D., Brown, N. J., et al. (2019). The bone metastasis niche in breast cancer-potential overlap with the haematopoietic stem cell niche in vivo. *Journal of bone oncology*, *17*, 100244.

Arai, F., Hirao, A., Ohmura, M., Sato, H., Matsuoka, S., Takubo, K., et al. (2004). Tie2/angiopoietin-1 signaling regulates hematopoietic stem cell quiescence in the bone marrow niche. *Cell*, *118*(2), 149–161.

Beeton, C. A., Bord, S., Ireland, D., & Compston, J. E. (2006). Osteoclast formation and bone resorption are inhibited by megakaryocytes. *Bone*, *39*(5), 985–990.

Bellido, T. (2014). Osteocyte-driven bone remodeling. *Calcified Tissue International*, *94*(1), 25–34.

Bidwell, B. N., Slaney, C. Y., Withana, N. P., Forster, S., Cao, Y., Loi, S., et al. (2012). Silencing of Irf7 pathways in breast cancer cells promotes bone metastasis through immune escape. *Nature Medicine*, *18*(8), 1224–1231.

Bonewald, L. F. (2007). Osteocytes as dynamic multifunctional cells. *Annals of the New York Academy of Sciences*, *1116*, 281–290.

Bonewald, L. F. (2011). The amazing osteocyte. *Journal of Bone and Mineral Research*, *26*(2), 229–238.

Boucharaba, A., Serre, C. M., Grès, S., Saulnier-Blache, J. S., Bordet, J. C., Guglielmi, J., et al. (2004). Platelet-derived lysophosphatidic acid supports the progression of osteolytic bone metastases in breast cancer. *The Journal of Clinical Investigation*, *114*(12), 1714–1725.

Boucharaba, A., Serre, C. M., Guglielmi, J., Bordet, J. C., Clézardin, P., & Peyruchaud, O. (2006). The type 1 lysophosphatidic acid receptor is a target for therapy in bone metastases. *Proceedings of the National Academy of Sciences of the United States of America*, *103*(25), 9643–9648.

Bragado, P., Estrada, Y., Parikh, F., Krause, S., Capobianco, C., Farina, H. G., et al. (2013). TGF-beta2 dictates disseminated tumour cell fate in target organs through TGF-beta-RIII and p38alpha/beta signalling. *Nature Cell Biology*, *15*(11), 1351–1361.

Brown, M. D., Hart, C. A., Gazi, E., Bagley, S., & Clarke, N. W. (2006). Promotion of prostatic metastatic migration towards human bone marrow stoma by omega 6 and its inhibition by omega 3 PUFAs. *British Journal of Cancer*, *94*(6), 842–853.

Brown, M. D., Hart, C., Gazi, E., Gardner, P., Lockyer, N., & Clarke, N. (2010). Influence of omega-6 PUFA arachidonic acid and bone marrow adipocytes on metastatic spread from prostate cancer. *British Journal of Cancer*, *102*(2), 403–413.

Bruns, I., Lucas, D., Pinho, S., Ahmed, J., Lambert, M. P., Kunisaki, Y., et al. (2014). Megakaryocytes regulate hematopoietic stem cell quiescence through CXCL4 secretion. *Nature Medicine*, *20*(11), 1315–1320.

Bussard, K. M., Venzon, D. J., & Mastro, A. M. (2010). Osteoblasts are a major source of inflammatory cytokines in the tumor microenvironment of bone metastatic breast cancer. *Journal of Cellular Biochemistry*, *111*(5), 1138–1148.

Butler, J. M., Kobayashi, H., & Rafii, S. (2010). Instructive role of the vascular niche in promoting tumour growth and tissue repair by angiocrine factors. *Nature Reviews. Cancer*, *10*(2), 138–146.

Caers, J., Deleu, S., Belaid, Z., De Raeve, H., Van Valckenborgh, E., De Bruyne, E., et al. (2007). Neighboring adipocytes participate in the bone marrow microenvironment of multiple myeloma cells. *Leukemia, 21*(7), 1580–1584.

Calvi, L. M., Adams, G. B., Weibrecht, K. W., Weber, J. M., Olson, D. P., Knight, M. C., et al. (2003). Osteoblastic cells regulate the haematopoietic stem cell niche. *Nature, 425*(6960), 841–846.

Campbell, J. P., Karolak, M. R., Ma, Y., Perrien, D. S., Masood-Campbell, S. K., Penner, N. L., et al. (2012). Stimulation of host bone marrow stromal cells by sympathetic nerves promotes breast cancer bone metastasis in mice. *PLoS Biology, 10*(7), e1001363.

Carlson, P., Dasgupta, A., Grzelak, C. A., Kim, J., Barrett, A., Coleman, I. M., et al. (2019). Targeting the perivascular niche sensitizes disseminated tumour cells to chemotherapy. *Nature Cell Biology, 21*(2), 238–250.

Chen, A., Wang, L., Liu, S., Wang, Y., Liu, Y., Wang, M., et al. (2018). Attraction and compaction of migratory breast Cancer cells by bone matrix proteins through tumor-osteocyte interactions. *Scientific Reports, 8*(1), 5420.

Chen, X., Wang, L., Zhao, K., & Wang, H. (2018). Osteocytogenesis: Roles of physicochemical factors, collagen cleavage, and exogenous molecules. *Tissue Engineering. Part B, Reviews, 24*(3), 215–225.

Chu, D. T., Phuong, T. N. T., Tien, N. L. B., Tran, D. K., Nguyen, T. T., Thanh, V. V., et al. (2019). The effects of adipocytes on the regulation of breast Cancer in the tumor microenvironment: An update. *Cell, 8*(8).

Clement-Demange, L., Mulcrone, P. L., Tabarestani, T. Q., Sterling, J. A., & Elefteriou, F. (2018). beta2ARs stimulation in osteoblasts promotes breast cancer cell adhesion to bone marrow endothelial cells in an IL-1beta and selectin-dependent manner. *J Bone Oncol, 13*, 1–10.

Cook, L. M., Shay, G., Araujo, A., & Lynch, C. C. (2014). Integrating new discoveries into the "vicious cycle" paradigm of prostate to bone metastases. *Cancer Metastasis Reviews, 33*(2–3), 511–525.

Cui, Y. X., Evans, B. A., & Jiang, W. G. (2016). New roles of osteocytes in proliferation, migration and invasion of breast and prostate Cancer cells. *Anticancer Research, 36*(3), 1193–1201.

Decker, A. M., Jung, Y., Cackowski, F. C., Yumoto, K., Wang, J., & Taichman, R. S. (2017). Sympathetic signaling reactivates quiescent disseminated prostate Cancer cells in the bone marrow. *Molecular Cancer Research, 15*(12), 1644–1655.

Devignes, C. S., Aslan, Y., Brenot, A., Devillers, A., Schepers, K., Fabre, S., et al. (2018). HIF signaling in osteoblast-lineage cells promotes systemic breast cancer growth and metastasis in mice. *Proceedings of the National Academy of Sciences of the United States of America, 115*(5), E992–E1001.

Diedrich, J. D., Rajagurubandara, E., Herroon, M. K., Mahapatra, G., Huttemann, M., & Podgorski, I. (2016). Bone marrow adipocytes promote the Warburg phenotype in metastatic prostate tumors via HIF-1alpha activation. *Oncotarget*, 7(40), 64854–64877.

Dirat, B., Bochet, L., Dabek, M., Daviaud, D., Dauvillier, S., Majed, B., et al. (2011). Cancer-associated adipocytes exhibit an activated phenotype and contribute to breast cancer invasion. *Cancer Research, 71*(7), 2455–2465.

Elefteriou, F. (2018). Impact of the autonomic nervous system on the skeleton. *Physiological Reviews, 98*(3), 1083–1112.

Fazeli, P. K., Horowitz, M. C., MacDougald, O. A., Scheller, E. L., Rodeheffer, M. S., Rosen, C. J., et al. (2013). Marrow fat and bone- -new perspectives. *The Journal of Clinical Endocrinology and Metabolism, 98*(3), 935–945.

Fizazi, K., Carducci, M., Smith, M., Damiao, R., Brown, J., Karsh, L., et al. (2011). Denosumab versus zoledronic acid for treatment of bone metastases in men with castration-resistant prostate cancer: A randomised, double-blind study. *Lancet, 377*(9768), 813–822.

Fujisaki, J., Wu, J., Carlson, A. L., Silberstein, L., Putheti, P., Larocca, R., et al. (2011). In vivo imaging of Treg cells providing immune privilege to the haematopoietic stem-cell niche. *Nature*, *474*(7350), 216–219.

Geissmann, F., Auffray, C., Palframan, R., Wirrig, C., Ciocca, A., Campisi, L., et al. (2008). Blood monocytes: Distinct subsets, how they relate to dendritic cells, and their possible roles in the regulation of T-cell responses. *Immunology and Cell Biology*, *86*(5), 398–408.

Ghajar, C. M., Peinado, H., Mori, H., Matei, I. R., Evason, K. J., Brazier, H., et al. (2013). The perivascular niche regulates breast tumour dormancy. *Nature Cell Biology*, *15*(7), 807–817.

Giuliani, N., Ferretti, M., Bolzoni, M., Storti, P., Lazzaretti, M., Dalla Palma, B., et al. (2012). Increased osteocyte death in multiple myeloma patients: Role in myeloma-induced osteoclast formation. *Leukemia*, *26*(6), 1391–1401.

Gonzalez-Perez, R. R., Xu, Y., Guo, S., Watters, A., Zhou, W., & Leibovich, S. J. (2010). Leptin upregulates VEGF in breast cancer via canonic and non-canonical signalling pathways and NFkappaB/HIF-1alpha activation. *Cellular Signalling*, *22*(9), 1350–1362.

Guise, T. A. (2002). The vicious cycle of bone metastases. *Journal of Musculoskeletal & Neuronal Interactions*, *2*(6), 570–572.

Haider, M. T., Holen, I., Dear, T. N., Hunter, K., & Brown, H. K. (2014). Modifying the osteoblastic niche with zoledronic acid in vivo-potential implications for breast cancer bone metastasis. *Bone*, *66*, 240–250.

Hanley, D. A., Adachi, J. D., Bell, A., & Brown, V. (2012). Denosumab: Mechanism of action and clinical outcomes. *International Journal of Clinical Practice*, *66*(12), 1139–1146.

Hardaway, A. L., Herroon, M. K., Rajagurubandara, E., & Podgorski, I. (2015). Marrow adipocyte-derived CXCL1 and CXCL2 contribute to osteolysis in metastatic prostate cancer. *Clinical & Experimental Metastasis*, *32*(4), 353–368.

Hayakawa, Y., Sakitani, K., Konishi, M., Asfaha, S., Niikura, R., Tomita, H., et al. (2017). Nerve growth factor promotes gastric tumorigenesis through aberrant cholinergic signaling. *Cancer Cell*, *31*(1), 21–34.

Herroon, M. K., Rajagurubandara, E., Diedrich, J. D., Heath, E. I., & Podgorski, I. (2018). Adipocyte-activated oxidative and ER stress pathways promote tumor survival in bone via upregulation of Heme oxygenase 1 and Survivin. *Scientific Reports*, *8*(1), 40.

Herroon, M. K., Rajagurubandara, E., Hardaway, A. L., Powell, K., Turchick, A., Feldmann, D., et al. (2013). Bone marrow adipocytes promote tumor growth in bone via FABP4-dependent mechanisms. *Oncotarget*, *4*(11), 2108–2123.

Huang, C. F., Lira, C., Chu, K., Bilen, M. A., Lee, Y. C., Ye, X., et al. (2010). Cadherin-11 increases migration and invasion of prostate cancer cells and enhances their interaction with osteoblasts. *Cancer Research*, *70*(11), 4580–4589.

Kimmel, D. B. (2007). Mechanism of action, pharmacokinetic and pharmacodynamic profile, and clinical applications of nitrogen-containing bisphosphonates. *Journal of Dental Research*, *86*(11), 1022–1033.

Kinder, M., Chislock, E., Bussard, K. M., Shuman, L., & Mastro, A. M. (2008). Metastatic breast cancer induces an osteoblast inflammatory response. *Experimental Cell Research*, *314*(1), 173–183.

Kusumbe, A. P., Ramasamy, S. K., & Adams, R. H. (2014). Coupling of angiogenesis and osteogenesis by a specific vessel subtype in bone. *Nature*, *507*(7492), 323–328.

Leblanc, R., & Peyruchaud, O. (2016a). Metastasis: New functional implications of platelets and megakaryocytes. *Blood*, *128*(1), 24–31.

Leblanc, R., & Peyruchaud, O. (2016b). The role of platelets and megakaryocytes in bone metastasis. *J Bone Oncol*, *5*(3), 109–111.

Lin, Y., Xu, J., & Lan, H. (2019). Tumor-associated macrophages in tumor metastasis: Biological roles and clinical therapeutic applications. *Journal of Hematology & Oncology*, *12*(1), 76.

Liu, S., Fan, Y., Chen, A., Jalali, A., Minami, K., Ogawa, K., et al. (2018). Osteocyte-driven downregulation of Snail restrains effects of Drd 2 inhibitors on mammary tumor cells. *Cancer Research*, *78*(14), 3865–3876.

Luo, X., Fu, Y., Loza, A. J., Murali, B., Leahy, K. M., Ruhland, M. K., et al. (2016). Stromal-initiated changes in the bone promote metastatic niche development. *Cell Reports*, *14*(1), 82–92.

Ma, Y. V., Xu, L., Mei, X., Middleton, K., & You, L. (2018). Mechanically stimulated osteocytes reduce the bone-metastatic potential of breast cancer cells in vitro by signaling through endothelial cells. *Journal of Cellular Biochemistry*, *120*(5), 7590–7601.

MacKie, R. M., Reid, R., & Junor, B. (2003). Fatal melanoma transferred in a donated kidney 16 years after melanoma surgery. *The New England Journal of Medicine*, *348*(6), 567–568.

Magnon, C., Hall, S. J., Lin, J., Xue, X., Gerber, L., Freedland, S. J., et al. (2013). Autonomic nerve development contributes to prostate cancer progression. *Science*, *341*(6142), 1236361.

Malara, A., Abbonante, V., Di Buduo, C. A., Tozzi, L., Currao, M., & Balduini, A. (2015). The secret life of a megakaryocyte: Emerging roles in bone marrow homeostasis control. *Cellular and Molecular Life Sciences*, *72*(8), 1517–1536.

Malladi, S., Macalinao, D. G., Jin, X., He, L., Basnet, H., Zou, Y., et al. (2016). Metastatic latency and immune evasion through autocrine inhibition of WNT. *Cell*, *165*(1), 45–60.

Marlow, R., Honeth, G., Lombardi, S., Cariati, M., Hessey, S., Pipili, A., et al. (2013). A novel model of dormancy for bone metastatic breast cancer cells. *Cancer Research*, *73*(23), 6886–6899.

Mendez-Ferrer, S., Michurina, T. V., Ferraro, F., Mazloom, A. R., Macarthur, B. D., Lira, S. A., et al. (2010). Mesenchymal and haematopoietic stem cells form a unique bone marrow niche. *Nature*, *466*(7308), 829–834.

Michalski, M. N., & McCauley, L. K. (2017). Macrophages and skeletal health. *Pharmacology & Therapeutics*, *174*, 43–54.

Monteiro, A. C., Leal, A. C., Goncalves-Silva, T., Mercadante, A. C., Kestelman, F., Chaves, S. B., et al. (2013). T cells induce pre-metastatic osteolytic disease and help bone metastases establishment in a mouse model of metastatic breast cancer. *PLoS ONE*, *8*(7), e68171.

Morris, E. V., & Edwards, C. M. (2016). The role of bone marrow adipocytes in bone metastasis. *J Bone Oncol*, *5*(3), 121–123.

Morris, E. V., & Edwards, C. M. (2018). Adipokines, adiposity, and bone marrow adipocytes: Dangerous accomplices in multiple myeloma. *Journal of Cellular Physiology*, *233*(12), 9159–9166.

Mulcrone, P. L., Campbell, J. P., Clement-Demange, L., Anbinder, A. L., Merkel, A. R., Brekken, R. A., et al. (2017). Skeletal colonization by breast Cancer cells is stimulated by an osteoblast and beta2AR-dependent neo-Angiogenic switch. *Journal of Bone and Mineral Research*, *32*(7), 1442–1454.

Naveiras, O., Nardi, V., Wenzel, P. L., Hauschka, P. V., Fahey, F., & Daley, G. Q. (2009). Bone-marrow adipocytes as negative regulators of the haematopoietic microenvironment. *Nature*, *460*(7252), 259–263.

Nombela-Arrieta, C., Pivarnik, G., Winkel, B., Canty, K. J., Harley, B., Mahoney, J. E., et al. (2013). Quantitative imaging of haematopoietic stem and progenitor cell localization and hypoxic status in the bone marrow microenvironment. *Nature Cell Biology*, *15*(5), 533–543.

Paget, S. (1989). The distribution of secondary growths in cancer of the breast. 1889. *Cancer Metastasis Reviews*, *8*(2), 98–101.

Pan, H., Gray, R., Braybrooke, J., Davies, C., Taylor, C., McGale, P., et al. (2017). 20-year risks of breast-Cancer recurrence after stopping endocrine therapy at 5 years. *The New England Journal of Medicine, 377*(19), 1836–1846.

Park, D., Sykes, D. B., & Scadden, D. T. (2012). The hematopoietic stem cell niche. *Frontiers in Bioscience (Landmark Ed), 17*, 30–39.

Phadke, P. A., Mercer, R. R., Harms, J. F., Jia, Y., Frost, A. R., Jewell, J. L., et al. (2006). Kinetics of metastatic breast cancer cell trafficking in bone. *Clinical Cancer Research, 12*(5), 1431–1440.

Price, T. T., Burness, M. L., Sivan, A., Warner, M. J., Cheng, R., Lee, C. H., et al. (2016). Dormant breast cancer micrometastases reside in specific bone marrow niches that regulate their transit to and from bone. *Science Translational Medicine, 8*(340), 340ra373.

Qian, B., Deng, Y., Im, J. H., Muschel, R. J., Zou, Y., Li, J., et al. (2009). A distinct macrophage population mediates metastatic breast cancer cell extravasation, establishment and growth. *PLoS ONE, 4*(8), e6562.

Rhodes, A., & Hillen, T. (2019). A mathematical model for the immune-mediated theory of metastasis. *Journal of Theoretical Biology, 482*, 109999.

Saloman, J. L., Albers, K. M., Li, D., Hartman, D. J., Crawford, H. C., Muha, E. A., et al. (2016). Ablation of sensory neurons in a genetic model of pancreatic ductal adenocarcinoma slows initiation and progression of cancer. *Proceedings of the National Academy of Sciences of the United States of America, 113*(11), 3078–3083.

Sawant, A., Deshane, J., Jules, J., Lee, C. M., Harris, B. A., Feng, X., et al. (2013). Myeloid-derived suppressor cells function as novel osteoclast progenitors enhancing bone loss in breast cancer. *Cancer Research, 73*(2), 672–682.

Sawant, A., Hensel, J. A., Chanda, D., Harris, B. A., Siegal, G. P., Maheshwari, A., et al. (2012). Depletion of plasmacytoid dendritic cells inhibits tumor growth and prevents bone metastasis of breast cancer cells. *Journal of Immunology, 189*(9), 4258–4265.

Sethi, N., Dai, X., Winter, C. G., & Kang, Y. (2011). Tumor-derived JAGGED1 promotes osteolytic bone metastasis of breast cancer by engaging notch signaling in bone cells. *Cancer Cell, 19*(2), 192–205.

Shiozawa, Y., Berry, J. E., Eber, M. R., Jung, Y., Yumoto, K., Cackowski, F. C., et al. (2016). The marrow niche controls the cancer stem cell phenotype of disseminated prostate cancer. *Oncotarget, 7*(27), 41217–41232.

Shiozawa, Y., Havens, A. M., Jung, Y., Ziegler, A. M., Pedersen, E. A., Wang, J., et al. (2008). Annexin II/annexin II receptor axis regulates adhesion, migration, homing, and growth of prostate cancer. *Journal of Cellular Biochemistry, 105*(2), 370–380.

Shiozawa, Y., Pedersen, E. A., Havens, A. M., Jung, Y., Mishra, A., Joseph, J., et al. (2011). Human prostate cancer metastases target the hematopoietic stem cell niche to establish footholds in mouse bone marrow. *The Journal of Clinical Investigation, 121*(4), 1298–1312.

Shiozawa, Y., & Taichman, R. S. (2012). Getting blood from bone: An emerging understanding of the role that osteoblasts play in regulating hematopoietic stem cells within their niche. *Experimental Hematology, 40*(9), 685–694.

Sims, N. A., & Martin, T. J. (2015). Coupling signals between the osteoclast and osteoblast: How are messages transmitted between these temporary visitors to the bone surface? *Frontiers in Endocrinology, 6*, 1–5. article 41.

Sleeman, J. P. (2012). The metastatic niche and stromal progression. *Cancer Metastasis Reviews, 31*(3–4), 429–440.

Snoussi, K., Strosberg, A. D., Bouaouina, N., Ben Ahmed, S., Helal, A. N., & Chouchane, L. (2006). Leptin and leptin receptor polymorphisms are associated with increased risk and poor prognosis of breast carcinoma. *BMC Cancer, 6*, 38.

Soki, F. N., Cho, S. W., Kim, Y. W., Jones, J. D., Park, S. I., Koh, A. J., et al. (2015). Bone marrow macrophages support prostate cancer growth in bone. *Oncotarget, 6*(34), 35782–35796.

Sosa, M. S., Bragado, P., & Aguirre-Ghiso, J. A. (2014). Mechanisms of disseminated cancer cell dormancy: An awakening field. *Nature Reviews. Cancer, 14*(9), 611–622.

Sottnik, J. L., Dai, J., Zhang, H., Campbell, B., & Keller, E. T. (2015). Tumor-induced pressure in the bone microenvironment causes osteocytes to promote the growth of prostate cancer bone metastases. *Cancer Research, 75*(11), 2151–2158.

Stegen, S., & Carmeliet, G. (2018). The skeletal vascular system - breathing life into bone tissue. *Bone, 115*, 50–58.

Stopeck, A. T., Lipton, A., Body, J. J., Steger, G. G., Tonkin, K., de Boer, R. H., et al. (2010). Denosumab compared with zoledronic acid for the treatment of bone metastases in patients with advanced breast cancer: A randomized, double-blind study. *Journal of Clinical Oncology, 28*(35), 5132–5139.

Sun, Y., Mao, X., Fan, C., Liu, C., Guo, A., Guan, S., et al. (2014). CXCL12-CXCR4 axis promotes the natural selection of breast cancer cell metastasis. *Tumour Biology, 35*(8), 7765–7773.

Tamura, D., Hiraga, T., Myoui, A., Yoshikawa, H., & Yoneda, T. (2008). Cadherin-11-mediated interactions with bone marrow stromal/osteoblastic cells support selective colonization of breast cancer cells in bone. *International Journal of Oncology, 33*(1), 17–24.

Templeton, Z. S., Lie, W. R., Wang, W., Rosenberg-Hasson, Y., Alluri, R. V., Tamaresis, J. S., et al. (2015). Breast Cancer cell colonization of the human bone marrow adipose tissue niche. *Neoplasia, 17*(12), 849–861.

Wang, J., Chen, G. L., Cao, S., Zhao, M. C., Liu, Y. Q., Chen, X. X., et al. (2017). Adipogenic niches for melanoma cell colonization and growth in bone marrow. *Laboratory Investigation, 97*(6), 737–745.

Wang, N., Docherty, F. E., Brown, H. K., Reeves, K. J., Fowles, A. C., Ottewell, P. D., et al. (2014). Prostate cancer cells preferentially home to osteoblast-rich areas in the early stages of bone metastasis: Evidence from in vivo models. *Journal of Bone and Mineral Research, 29*(12), 2688–2696.

Wang, W., Yang, X., Dai, J., Lu, Y., Zhang, J., & Keller, E. T. (2019). Prostate cancer promotes a vicious cycle of bone metastasis progression through inducing osteocytes to secrete GDF15 that stimulates prostate cancer growth and invasion. *Oncogene, 38*(23), 4540–4559.

Wang, H., Yu, C., Gao, X., Welte, T., Muscarella, A. M., Tian, L., et al. (2015). The osteogenic niche promotes early-stage bone colonization of disseminated breast cancer cells. *Cancer Cell, 27*(2), 193–210.

Wu, A. C., He, Y., Broomfield, A., Paatan, N. J., Harrington, B. S., Tseng, H. W., et al. (2016). CD169(+) macrophages mediate pathological formation of woven bone in skeletal lesions of prostate cancer. *The Journal of Pathology, 239*(2), 218–230.

Yumoto, K., Eber, M. R., Wang, J., Cackowski, F. C., Decker, A. M., Lee, E., et al. (2016). Axl is required for TGF-beta2-induced dormancy of prostate cancer cells in the bone marrow. *Scientific Reports, 6*, 36520.

Zaidi, M. (2007). Skeletal remodeling in health and disease. *Nature Medicine, 13*(7), 791–801.

Zhang, K., Kim, S., Cremasco, V., Hirbe, A. C., Collins, L., Piwnica-Worms, D., et al. (2011). CD8 + T cells regulate bone tumor burden independent of osteoclast resorption. *Cancer Research, 71*(14), 4799–4808.

Zhang, J., Niu, C., Ye, L., Huang, H., He, X., Tong, W. G., et al. (2003). Identification of the haematopoietic stem cell niche and control of the niche size. *Nature, 425*(6960), 836–841.

Zhao, C. M., Hayakawa, Y., Kodama, Y., Muthupalani, S., Westphalen, C. B., Andersen, G. T., et al. (2014). Denervation suppresses gastric tumorigenesis. *Science Translational Medicine, 6*(250), 250ra115.

Zhao, E., Wang, L., Dai, J., Kryczek, I., Wei, S., Vatan, L., et al. (2012). Regulatory T cells in the bone marrow microenvironment in patients with prostate cancer. *Oncoimmunology, 1*(2), 152–161.

Zhou, J. Z., Riquelme, M. A., Gu, S., Kar, R., Gao, X., Sun, L., et al. (2016). Osteocytic connexin hemichannels suppress breast cancer growth and bone metastasis. *Oncogene*, *35*(43), 5597–5607.

Zhou, B. O., Yu, H., Yue, R., Zhao, Z., Rios, J. J., Naveiras, O., et al. (2017). Bone marrow adipocytes promote the regeneration of stem cells and haematopoiesis by secreting SCF. *Nature Cell Biology*, *19*(8), 891–903.

Zhuang, J., Zhang, J., Lwin, S. T., Edwards, J. R., Edwards, C. M., Mundy, G. R., et al. (2012). Osteoclasts in multiple myeloma are derived from gr-1+CD11b+myeloid-derived suppressor cells. *PLoS ONE*, 7(11), e48871.

Zufferey, A., Fontana, P., Reny, J. L., Nolli, S., & Sanchez, J. C. (2012). Platelet proteomics. *Mass Spectrometry Reviews*, *31*(2), 331–351.

CHAPTER THREE

The role of vasculature in cancer stem cell niches

Unnikrishnan Sivan[a,†], Jessica De Angelis[a,†], Saravana K. Ramasamy[b,c], and Anjali P. Kusumbe[a,*]

[a]Tissue and Tumor Microenvironments Group, The Kennedy Institute of Rheumatology, University of Oxford, Oxford, United Kingdom
[b]Institute of Clinical Sciences, Imperial College London, London, United Kingdom
[c]MRC London Institute of Medical Sciences, Imperial College London, London, United Kingdom
*Corresponding author: e-mail address: anjali.kusumbe@kennedy.ox.ac.uk

Contents

1. Introduction

Cancerous growth involves a heterogeneous population of cells that originate after an unusual event in the life cycle. Within the cancerous growth, a rare fraction of cells possessing stemness and can initiate tumorigenesis or carcinogenesis and contribute to the remission after therapy are termed as cancer stem cells (CSCs) (Al-Hajj et al., 2003; Ayob & Ramasamy, 2018; Lapidot et al., 1994; Yu et al., 2012). Analogous to normal stem cells, CSCs are dependent on their niche to maintain their unique properties and stemness (Batlle & Clevers, 2017; Plaks et al., 2015; Vermeulen et al., 2010). The paradigm of the tumor

† These authors contributed equally.

Advances in Stem Cells and their Niches, Volume 5
ISSN 2468-5097
https://doi.org/10.1016/bs.asn.2020.09.003

microenvironment originated from the studies showing that tumorigenesis preferentially originates from specialized niches within healthy tissue and reports about the premetastatic niches (Hart & Fidler, 1980; Kaplan et al., 2005; Magnon et al., 2013; Peterson et al., 2015; Psaila & Lyden, 2009). A cancer microenvironment is created by the influence of both the secreted factors from primary cancer and from the host cells, which aids the dissemination and survival of CSCs (Chin & Wang, 2016; Kaplan et al., 2005; Singh et al., 2019; Wong & Reiter, 2011; Zheng et al., 2018). The tumor microenvironment plays a crucial role in therapeutic resistance by providing a quiescence promoting niche and by enhancing tissue invasion (T. Huang et al., 2019; Qu et al., 2019; Ye et al., 2012; Zhou et al., 2015). Targeting the tumor microenvironment opens a window to study the pathological dynamics of carcinogenesis and metastasis and opens up a horizon of new therapeutic strategies to eradicate the primary tumors and obstruct metastasis (Ordóñez-Morán & Huelsken, 2014). This chapter aims to review the literature on the role of the vasculature, perivascular cells, and angiocrine factors within the tumor microenvironment in tumor growth, invasion and metastasis.

2. Niche functions of vasculature

The vasculature is composed of a heterogeneous population of Endothelial Cells (ECs) and associated pericytes that interacts with the surrounding microenvironment to maintain normal homeostasis. In healthy tissues, ECs secrete paracrine factors, termed as angiocrine factors, to maintain tissue homeostasis. ECs secrete or express membrane-bound factors to modulate the surrounding extracellular matrix or directly communicate with neighboring cells. These angiocrine factors can reciprocally communicate with neighboring cells to create a vascular niche that supports the maintenance of stem and progenitor cells during homeostasis and regeneration. For instance, liver sinusoidal endothelial cells (LSECs) promotes liver organogenesis, polarisation and regeneration (Ding et al., 2014). The bone marrow endothelial cells and pericytes provide a vascular niche for the self-renewal, differentiation and localization of hemopoietic stem cells (HSCs) (Avecilla et al., 2004; de Bruijn et al., 2000; Kopp et al., 2005).

3. Links between vasculature and cancer cell stemness

Vasculature plays a major role in the establishment of a CSC niche. In the primary tumor, this is achieved due to an anatomical location, which is higher in nutrient availability and signaling gradient (Kumar et al., 2019).

This is usually achieved by residing close to a highly vascularized bed (Beck et al., 2011; Cogle et al., 2016; Ping et al., 2011; Yan et al., 2014). However, CSCs can also mediate angiogenesis (Bao et al., 2006; Folkins et al., 2009a; Gilbertson & Rich, 2007). In some highly angiogenetic cancers, it is reported that CSCs can even cross lineage-differentiation boundaries to form different vascular cells (Liu et al., 2017; Ricci-Vitiani et al., 2010; Shangguan et al., 2017). CSCs can differentiate into functional endothelial cells to form blood vessels through a process termed as vascular mimicry (Maniotis et al., 1999; Patan, 2004). Specifically, up-regulation of the Jagged1, a Notch ligand via overexpression of Twist1 mediates the trans-differentiation of cancer cells to endothelial cells (Chen et al., 2014a). Therapeutic stress could also lead to the transdifferentiation of CSCs into endothelial cells; for example, ionization radiation can induce the endothelial differentiation of CSC through Tie2 signaling (Deshors et al., 2019). In some other cases, ECs produce CXCL12 (SDF-1/IL-8) and TGFβ to induce the differentiation of CSCs to pericytes to facilitate angiogenesis and cancer growth (Cheng et al., 2013).

4. Angiocrine crosstalk in the tumor microenvironment

Cancer cells hijack the normal signaling pathways and homeostasis mechanisms for their survival and replenishment by producing signaling molecules and growth factors (Table 1). Under physiological conditions, angiogenesis is stimulated through vascular endothelial growth factor signaling (Hoeben et al., 2004; Nör et al., 1999). Vascular endothelial growth factor (VEGF) plays a crucial role in cancer angiogenesis as well. Inhibiting the VEGF signaling using an inhibitor substantially reduced the micro-vessel density in high CSC tumors (Beck et al., 2011a; Folkins et al., 2009b; Jia et al., 2013; Mahecha & Wang, 2017). In glioma, cancer stem cells express VEGF more than other cell populations (Bao et al., 2006). One of the key members in the VEGF growth factor family is VEGFA, produced by CSC, is a key player in the cancer angiogenesis (Fig. 1) (Sun et al., 2020b; Treps et al., 2017; Yang et al., 2017; Zhao et al., 2017). VEGFA modifies the niche in several ways which include, enhancing the migration of cells through chemotaxis, and allowing macromolecules to cross the blood vessel by enhancing vascular permeability (Heinolainen et al., 2017; Lakshmikanthan et al., 2018; Nagy et al., 2008; Ourradi et al., 2017). Non-physiological tissue levels of acidosis, hypoxia, and reactive oxygen species (ROS) help CSCs to enhance

Table 1 Factors driving angiocrine crosstalk in different malignancies.

Sl. No	Factors in angiocrine crosstalk	Function	Cell type	References
1	CXCL12	Differentiation into vascular pericytes	Endothelial cell	Cheng et al., (2013), Würth et al., (2014)
2	Shh	CSC stemness	Endothelial cell	Yan et al., (2014)
3	DLL4	CSC self-renewal and maintenance	Endothelial cell	Zhu et al. (2011)
4	JAG1	CSC phenotype	Endothelial cell	Lu et al. (2013), Zhu et al. (2011)
5	Osteopontin	CSC stemness, radiation, chemotherapy resistance, EMT, and angiogenesis	Perivascular cells	Cao et al. (2019), Chakraborty et al. (2008), Leali et al. (2003), Liu et al. (2016), Pietras et al. (2014)
6	TGFβ	Differentiation to vascular pericytes	Endothelial cell	Cheng et al. (2013)
7	CXCL12	Angiogenesis, lymphangiogenesis, tumor proliferation, chemo-resistance, and immunomodulation	Endothelial cells, Perivascular cells, and macrophages	Shi and Wei (2016) Meng et al. (2018)
8	VEGF-A	Resistance to chemotherapy	Endothelial cell	Poulos et al. (2014b)
9	FGF2	Self-renewal	Endothelial cell	Otte et al. (2019)
10	Epiregulin	CSC like properties and invasion	Endothelial cell	Riese and Cullum (2014), Sun et al. (2016)
11	PDGF-B	Susceptibility to radiation and chemotherapy	Pericytes	Singh et al. (2019)
12	VEGF-C	Immune escape	Lymphatic vessel endothelial cells	Tacconi et al. (2019)

13	CCL8	Stemness and tumor invasion	Tumor-associated macrophages	Zhang et al. (2020)
14	CCL5	CSC self-renewal, migration, EMT, and invasion	Tumor-associated macrophages	Huang et al. (2020)
15	CX3CL1	Metastasis	Endothelial cell	Sun et al. (2020a, 2020b)
16	Thrombo-spondin-1	Tumor dormancy	Endothelial cell	Ghajar et al. (2013)
17	Angiostatin	Tumor dormancy	Cancer mediated conversion of plasminogen	O'Reilly et al. (1996)
18	VEGFA	Stemness, angiogenesis, lymphangiogenesis, vascular hyper permeability, Matrix modulation	CSC	Beck et al. (2011a), Belotti et al. (2003), Hirakawa et al. (2005), Ping et al. (2011), Tawada et al. (2014), Treps et al. (2017)
19	VEGFC	Stemness, Angiogenesis, lymphangiogenesis, vascular hyper permeability,	CSC	Chen et al. (2012), Pan et al. (2018)
20	MMPs	Matrix modulation, VEGF production	CSC	Belotti et al. (2003)
21	Neuropilin-1	Stemness and renewal	CSC	Beck et al. (2011b), Goel et al. (2013)
22	MiRNA-21	Angiogenesis, VEGF production	CSC	Sun et al. (2017)
23	PlGF	Angiogenesis and Cancer invasion *PlGF-1 inhibit angiogenesis in Fibrosarcoma cells	CSC	Chen et al. (2014c) Eriksson et al. (2002)

Continued

Table 1 Factors driving angiocrine crosstalk in different malignancies.—cont'd

Sl. No	Factors in angiocrine crosstalk	Function	Cell type	References
24	CXCR4	VEGF production and angiogenesis Differentiation to vascular pericytes	CSC	Ping et al. (2011), Cheng et al. (2013)
25	CXCR1	CSC properties and invasion and metastasis	CSC	Chen et al. (2014b)
26	NOTCH1&2	CSC self-renewal with the help of EC derived Notch ligands and cancer invasion	CSC	Wang et al. (2019), Zhu et al. (2011)
27	VEGFR-2	Endothelial differentiation of poorly differentiated cancer cells	CSC	Liu et al. (2017)
28	CXCL12	CSC properties, angiogenesis, and vasculogenesis	CSC	Corrò et al. (2019), Folkins et al. (2009a)
29	Periostin	Malignant growth	CSC	Zhou et al. (2015)

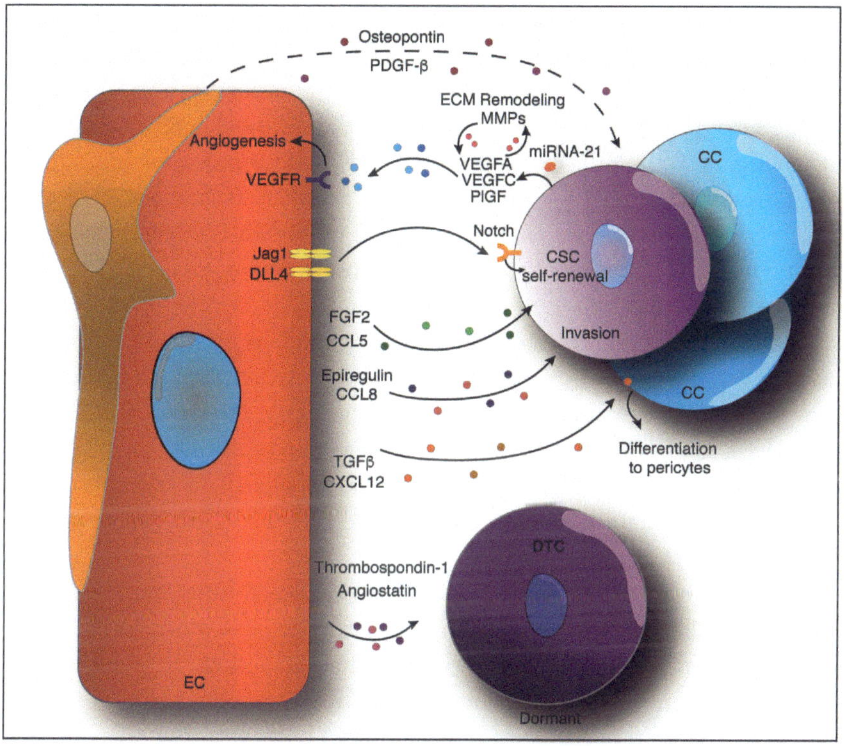

Fig. 1 Cancer stem cell and endothelial cell crosstalk. Endothelial cells (ECs) secrete angiocrine factors to promote tumor growth by providing nurturing niches for cancer stem cell (CSC) self-renewal and invasion. In response to angiocrine signals such as TGF-b and CXCL12 a subset of CSCs undergo epithelial-to-pericyte transition (EPT), inducing the expression of pericyte markers and increasing CSC motility and invasiveness. Concurrently, CSC localised in the perivascular niche secrete angiogenic factors to promote vessel growth which supports the growth of tumour cells. The perivascular niche can also actively induce and support dormancy by modulating the surrounding environment. Extracellular factors produced by ECs such as Thrombospondin and Angiostatin induces quiescence of the disseminated tumor cells (DTCs). VEGF: Vascular Endothelial Growth Factor; PIGF; Placental Growth Factor; VEGFR: Vascular Endothelial Growth Factor Receptor; Jag1:Jagged-1; DLL4: Delta-like Ligand 4; PDGF-b: Platelet-derived growth factor Beta; ECM; Extracellular Matrix; MMP: Matrix Metalloprotease; CC: Cancer Cell; FGF: Fibroblast Growth Factor; CCL; Chemokine Ligands; TGF-b; Transforming Growth Factor Beta; CXCL12: C-X-C motif chemokine 12; miRNA-21: MicroRNA 21.

VEGF production (Boedtkjer & Pedersen, 2020; Plaks et al., 2015). Several miRNAs and long non-coding RNAs (LncRNAs) produced by CSCs are essential for inducing angiogenesis by VEGF dependent and independent pathways (Chen et al., 2020; Sun et al., 2020b; Sun et al., 2017). VEGF induces the production of collagenases and matrix metalloproteinases

(MMPs), thereby modifying the tumor microenvironment (Pufe et al., 2004; Shan et al., 2013; Unemori et al., 1992). The production of MMPs further enhances the production of VEGF by cancer cells and thereby, angiogenesis (Figure 1) (Belotti et al., 2003; Mahecha & Wang, 2017; Rundhaug, 2003). VEGFA from CSCs also induces lymphangiogenesis (the formation of lymphatic vessels) in cancer (Chen et al., 2012; Hirakawa et al., 2005; Pan et al., 2018). The possible role of VEGFA to recruit lymphatic endothelial cells from the bone marrow to the tumor microenvironment is also reported recently (Tawada et al., 2014). VEGFC, a potent lymphangiogenesis inducer in tumors and its production by CSCs together with VEGFA contributes toward maintenance of CSC stemness, invasion, and therapeutic resistance by remodeling the microenvironment (B. S. Beck et al., 2011a; Li & Li, 2014; Maeda et al., 2008, p. 133; C.-A. Wang & Tsai, 2015). Placenta growth factor (PlGF), a member of the VEGF family, enhances cancer invasion in esophageal cancers by activating MMP9 (Chen et al., 2014c).

In glioblastomas, CD44 and osteopontin interactions mark the CSC population. Osteopontin secreted by perivascular cells binds to the CD44 receptor to promote the stemness in Glioblastoma (Fig. 1) (Pietras et al., 2014). Osteopontin also promotes breast tumor angiogenesis in a VEGF dependent manner (Chakraborty et al., 2008). In the case of pancreatic cancers, pancreatic stellate cells produce high levels of Osteopontin under hypoxic conditions activated by ROS. This hypoxia driven Osteopontin production by pancreatic stellate cells promotes epithelial to mesenchymal transition (EMT) and CSC properties of the pancreatic cancer cells (J. Cao et al., 2019). Osteopontin expression in hepatocellular carcinoma promotes chemoresistance and CSC phenotype (Liu et al., 2016). Osteopontin induces the production of angiogenic factors such as fibroblast growth factor FGF-2 production from ECs (Leali et al., 2003). VEGF-A protects the leukemic cells from chemotherapy (Poulos et al., 2014b). Another member of the family VEGFC, from lymphatic ECs, promotes colorectal cancer cells to escape from the immune system (Tacconi et al., 2019). Self-renewal of colon cancer cells is influenced by FGF2 signaling (Otte et al., 2019). The ECs within tumors provide Notch ligands DLL4 and JAG1, which binds to the NOTCH1 and NOTCH2 receptors on the CSCs to maintain the CSC pool and self-renewal properties in different cancer types (Lu et al., 2013; Zhu et al., 2011). Hedgehog signaling is an essential pathway to maintain stemness of CSCs and has been reviewed (Cochrane et al., 2015). ECs produce Sonic hedgehog (Shh) to enhance the stemness of CSCs (Yan et al., 2014).

The acidic microenvironment within the tumors induces angiogenesis. Acidosis enhances the CXCL12 production from ECs and induces angiogenesis and lymphangiogenesis ((Nakanishi et al., 2016; Shi & Wei, 2016). Moreover, perivascular cells and macrophages also secrete CXCL12 (Garza Treviño et al., 2019). CXCL12 within the tumor niche plays a multitude of roles, including angiogenesis, tumor proliferation, chemoresistance, and immunomodulation (Meng et al., 2018). Secreted factors from CSCs recruit various niche cells to aid tumor survival and growth. Periostin, secreted by CSCs in glioblastomas, is known to promote malignancy by attracting macrophages (Zhou et al., 2015). Macrophages within the tumor niche promote stemness of cancer cells, invasion, and metastasis through CCL8 and CCL5 secretion (Huang et al., 2020; Zhang et al., 2020). Rather than depending on niche cells for CXCL12, renal carcinoma CSCs upregulates CXCL12, which functions in an autocrine manner (Conò et al., 2019).

Epidermal growth factor (EGF) family of proteins contributes to the tumor microenvironment by modulating angiogenesis and vascular remodeling (Riese & Cullum, 2014). Epiregulin is an important member of the EGF family with great implications in cancer, metastasis, and survival (Kohsaka et al., 2014; Riese & Cullum, 2014). Esophageal cancer-associated ECs produce Epiregulin, which influences metastasis and promote CSC like phenotype (Sun et al., 2016). EGF is well known for its role in EMT and promoting the stemness properties of cancer cells (Xu et al., 2016). The expression of EGFR by tumor endothelial cells equips them to respond to the EGF produced by tumor cells and other niche associated cells (Amin et al., 2006). Targeting EGFR in ECs is found to be enough to inhibit tumor growth (Amin et al., 2008).

5. Maladaptation of vasculature in the tumor microenvironment

The vasculature of tumors is leaky and malformed. Tumor microenvironment exhibit perturbations such as acidosis, hypoxia, nutritional deficiency, non-physiological levels of ROS, chemokines and growth factors. The vasculature and vascular cells inside a tumor adapt to a different program of gene expression and development (Boedtkjer & Pedersen, 2020; Hardee & Zagzag, 2012; Pasquier et al., 2020). As expected, ECs trans-differentiated from cancer stem cells have entirely different gene expression and properties compared to normal endothelium (Ricci-Vitiani et al., 2010). In addition to CSC induced trans-differentiation via secreted factors, chemotherapeutic stress (Baisiwala et al., 2019) and ionizing radiation (Deshors et al., 2019) induce cross lineage differentiation to promote vascular mimicry.

6. Vascular microenvironments supporting the cancer stem cell quiescence

Tumor dormancy is a net result of the CSC's ability to remain quiescent and/or of the simultaneous regression of tumor mass due to cell death, where the CSC niche plays a crucial role (Rak et al., 2008). The vascular niche is also a contributor to the tumor dormancy (Almog et al., 2006). Cancer cells in close physical contact with the abluminal surface of the brain vasculature can become dormant in a VEGF-A dependent manner (Kienast et al., 2010). The bone and the bone marrow offer suitable sites for homing and dormancy of the disseminated tumor cells (DTCs). Bone marrow-derived hemopoietic cells expressing VEGFR1 forms clusters that act as a pre-metastatic niche that attracts tumor cells (Kaplan et al., 2005). The heterogeneity in the blood vessels in terms of size, shape, and architecture together with the molecular profile may offer a selective advantage to cancer cells in bone (Kusumbe, 2016a). The bone marrow offers a complex vascular niche to regulate the quiescence of long-term hematopoietic stem cells while allowing the proliferation of progenitors according to the demands such as infection, inflammation and growth. The absence of E-selectin on ECs from the niche has shown to enhance quiescence and chemoresistance (Winkler et al., 2012a). The same selectivity by the endothelium maintains the dormancy of tumor cells according to their location in the bone marrow (Ghajar et al., 2013). Thrombospondin-1, expressed by the mature microvascular endothelial cells, can induce tumor dormancy (Ghajar et al., 2013). Another angiocrine factor, angiostatin, is a long-known molecule for inducing and sustaining dormancy (O'Reilly et al., 1996). The angiocrine factors capable of stimulating dormancy are comprehensively reviewed recently elsewhere (Jahanban-Esfahlan et al., 2019).

7. Vascular niches for disseminated tumor cells and CSCs in the bone marrow

The bone is one of the most common sites for cancer metastasis. In pathological conditions, ECs can stimulate the growth of leukemic cells (Dias et al., 2001). Furthermore, spatial analysis of transplanted human leukemic cells shows that these cells initially localize to osteoblasts and gradually migrate to the vasculature and diaphyseal region. Following cancer immunotherapy, remaining cancer cells localize in the perivascular niches and trabecular endosteum, demonstrating that the bone marrow act as a protective vascular niche for cancer cells (Ninomiya et al., 2007). VEGFA increases the proliferation and adhesion of leukemic cells with specific inhibition of this signaling pathway improving the efficacy of chemotherapy (Poulos et al., 2014a). B lymphoma cells upregulate FGF4, which increases Jagged-1 in ECs, initiating a reciprocal Notch signaling cascade with lymphoma cells that increases the invasiveness and enhances chemoresistance (Fig. 2) (Cao et al., 2014). The bone marrow vasculature is a heterogeneous population of ECs, whereby a specialized subset of ECs close to the growth plate (termed Type H vessels) are associated with osteoprogenitors and hemopoietic stem cells (Kusumbe et al., 2014, 2016), which provide a cytokine rich environment that may promote tumor cell survival. Intrinsic properties of the bone marrow endothelium, such as the high density and low blood flow of sinusoids (Type L vessels) in the bone marrow increase the crosstalk between ECs and CSCs creating an optimal environment for tumor metastasis (Kusumbe et al., 2014; Virk & Lieberman, 2007). C-X3-C motif chemokine ligand 1 (CX3CL1) secretion mediated by EC secretion of ADAM17 increases spinal metastasis in hepatocellular carcinoma (Fig. 2) (Sun et al., 2020a)

Disseminated tumor cells (DTCs) located close to the vasculature persist in a long-term state of quiescence in the bone marrow (Ghajar et al., 2013). The large diameter and reduced blood flow in bone marrow sinusoidal endothelium allow for DTCs to be dormant near the vasculature (Kopp et al., 2005; Kusumbe, 2016b). ECs secrete thrombospondin-1 that creates a favorable vascular niche for DTC dormancy. These dormant DTCs can be reactivated in response to the remodeling of the vasculature (Ghajar et al., 2013). Secreted angiocrine factors could act as a chemoattractant for circulating CSCs, thereby facilitating their migration into the bone marrow cavity (Kusumbe, 2016b; Winkler et al., 2012b). Additionally, ECs protect DTCs from chemotherapy via inhibition of VWF. While VCAM-1, can sensitize DTCs to chemotherapy (Carlson et al., 2019). Age-related microenvironmental changes, in particular, a loss of pericytes and associated PDGF-B signaling, promotes DTC proliferation (Singh et al., 2019). Thus, the alterations in the vascular microenvironment can trigger the switch between DTC dormancy and reactivation.

Acute Myeloid Leukemia (AML) is the most frequent acute leukemia with a high incidence of relapse (Bertoli et al., 2017). AML cells localize and integrate into the bone marrow vasculature, and these cells can produce

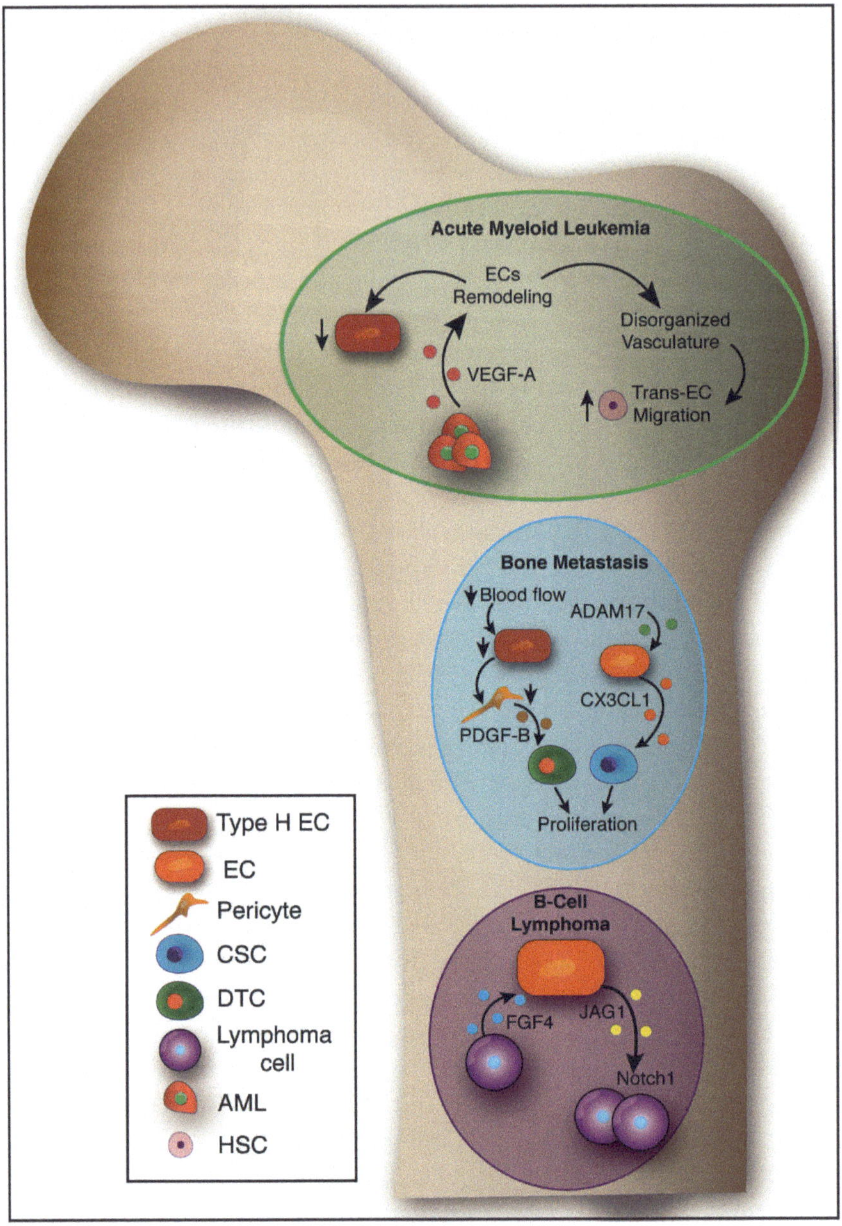

Fig. 2 Angiocrine signalling in bone malignancies. Bone is one of the most frequent sites of metastasis. In Acute Myeloid Leukaemia (AML) cancer cells secrete VEGF that initiates Endothelial cell (EC) remodeling, resulting in increased trans-endothelial migration of Haemopoietic Stem Cells (HSC) and a reduction in the type H ECs. In Bone metastasis, a reduction in Type H ECs leads to reduced pericytes, which increases CSC proliferation. Additionally, modulation of the extracellular matrix by ADAM17 leads to the release of CX3CL1 by sinusoidal ECs that increases metastasis. In lymphomagenesis lymphoma cells secrete factors that trigger the NOTCH pathway activation, signalling to lymphoma cells and increasing proliferation. VEGF: Vascular Endothelial Growth Factor; PDGF-b: Platelet-derived growth factor Beta; FGF4: Fibroblast Growth Factor 4; Jag1: Jagged-1; ADAM17: A Disintegrin and Metalloproteinase 17; CX3CL1; chemokine (C-X3-C motif) ligand 1; DTC: Disseminated Tumor Cells

endothelial-like gene expression signatures, allowing them to acquire dormancy (Cogle et al., 2014). AML cells bind to ECs via E-selectin to integrate into the chemoprotective vascular niche (Winkler et al., 2014). The vasculature secretes high levels of VEGF in AML patients, promoting bone marrow niche remodeling and increasing the microvessel density in AML patients (Fiedler et al., 1997; Padró et al., 2000). In mice, a reduction in type H vessels and vascular leakiness are hallmarks of AML (Duarte et al., 2018; Passaro et al., 2017). AML mediated vessel remodeling shows increased vascular permeability, which in turn increases trans-endothelial migration (Fig. 2) (Duarte et al., 2018). The increase in vessel density correlates with an increase in the severity of the disease and a decrease in survival rates (Kuzu et al., 2004). Broad inhibition of VEGFRs, PDGFRs and c-Kit removes the protective effects from ECs and sensitizes AML cells to chemotherapeutic agents (Drusbosky et al., 2015). Thus, vascular remodeling and angiogenesis provide a preferential vasculature microenvironment for AML cells.

8. Conclusions

Traditional treatments targeting cancer have focused on the cell-intrinsic properties of CSCs. However, this view has been challenged as growing evidence demonstrates that their microenvironment influences CSCs. Vasculature and blood vessel derived angiocrine factors are key components of the tumor microenvironment. CSCs express angiogenic factors to promote vascular growth and create a supportive vascular niche for tumor growth. Simultaneously, through angiocrine signaling ECs create vascular niches that regulate CSC behavior and can provide a chemoprotective microenvironment for CSCs and DTCs. Further, alterations in the vascular microenvironments can reactivate the dormant DTCs and lead to relapse. In conjunction with traditional chemotherapy, treatments that disrupt angiocrine crosstalk holds the potential to chemosensitize otherwise chemoresistant CSCs to improve the efficacy of treatment.

References

Al-Hajj, M., Wicha, M. S., Benito-Hernandez, A., Morrison, S. J., & Clarke, M. F. (2003). Prospective identification of tumorigenic breast cancer cells. *Proceedings of the National Academy of Sciences of the United States of America*, *100*(7), 3983–3988. https://doi.org/10.1073/pnas.0530291100.

Almog, N., Henke, V., Flores, L., Hlatky, L., Kung, A. L., Wright, R. D., et al. (2006). Prolonged dormancy of human liposarcoma is associated with impaired tumor angiogenesis. *The FASEB Journal*, *20*(7), 947–949. https://doi.org/10.1096/fj.05-3946fje.

Amin, D. N., Bielenberg, D. R., Lifshits, E., Heymach, J. V., & Klagsbrun, M. (2008). Targeting EGFR activity in blood vessels is sufficient to inhibit tumor growth and is accompanied by an increase in VEGFR-2 dependence in tumor endothelial cells. *Microvascular Research*, *76*(1), 15–22. https://doi.org/10.1016/j.mvr.2008.01.002.

Amin, D. N., Hida, K., Bielenberg, D. R., & Klagsbrun, M. (2006). Tumor endothelial cells express epidermal growth factor receptor (EGFR) but not ErbB3 and are responsive to EGF and to EGFR kinase inhibitors. *Cancer Research*, *66*(4), 2173–2180. https://doi.org/10.1158/0008-5472.CAN-05-3387.

Avecilla, S. T., Hattori, K., Heissig, B., Tejada, R., Liao, F., Shido, K., et al. (2004). Chemokine-mediated interaction of hematopoietic progenitors with the bone marrow vascular niche is required for thrombopoiesis. *Nature Medicine*, *10*(1), 64–71. https://doi.org/10.1038/nm973.

Ayob, A. Z., & Ramasamy, T. S. (2018). Cancer stem cells as key drivers of tumour progression. *Journal of Biomedical Science*, *25*(1), 20. https://doi.org/10.1186/s12929-018-0426-4.

Baisiwala, S., Auffinger, B., Caragher, S. P., Shireman, J. M., Ahsan, R., Lee, G., et al. (2019). Chemotherapeutic stress induces transdifferentiation of glioblastoma cells to endothelial cells and promotes vascular mimicry. *Stem Cells International*, *2019*. https://doi.org/10.1155/2019/6107456.

Bao, S., Wu, Q., Sathornsumetee, S., Hao, Y., Li, Z., Hjelmeland, A. B., et al. (2006). Stem cell-like glioma cells promote tumor angiogenesis through vascular endothelial growth factor. *Cancer Research*, *66*(16), 7843–7848. https://doi.org/10.1158/0008-5472.CAN-06-1010.

Batlle, E., & Clevers, H. (2017). Cancer stem cells revisited. *Nature Medicine*, *23*(10), 1124–1134. https://doi.org/10.1038/nm.4409.

Beck, B., Driessens, G., Goossens, S., Youssef, K. K., Kuchnio, A., Caauwe, A., et al. (2011). A vascular niche and a VEGF–Nrp1 loop regulate the initiation and stemness of skin tumours. *Nature*, *478*(7369), 399–403. https://doi.org/10.1038/nature10525.

Beck, B. S., Driessens, G., Goossens, S., Youssef, K. K., Kuchnio, A., Caauwe, A., et al. (2011a). A vascular niche and a VEGF–Nrp1 loop regulate the initiation and stemness of skin tumours. *Nature*. https://doi.org/10.1038/nature10525.

Beck, B. S., Driessens, G., Goossens, S., Youssef, K. K., Kuchnio, A., Caauwe, A., et al. (2011b). A vascular niche and a VEGF–Nrp1 loop regulate the initiation and stemness of skin tumours. *Nature*. https://doi.org/10.1038/nature10525.

Belotti, D., Paganoni, P., Manenti, L., Garofalo, A., Marchini, S., Taraboletti, G., et al. (2003). Matrix metalloproteinases (MMP9 and MMP2) induce the release of vascular endothelial growth factor (VEGF) by ovarian carcinoma cells: Implications for ascites formation. *Cancer Research*, *63*(17), 5224–5229.

Bertoli, S., Tavitian, S., Huynh, A., Borel, C., Guenounou, S., Luquet, I., et al. (2017). Improved outcome for AML patients over the years 2000-2014. *Blood Cancer Journal*, 7(12), 635. https://doi.org/10.1038/s41408-017-0011-1.

Boedtkjer, E., & Pedersen, S. F. (2020). The acidic tumor microenvironment as a driver of cancer. *Annual Review of Physiology*, *82*(1), 103–126. https://doi.org/10.1146/annurev-physiol-021119-034627.

Cao, J., Li, J., Sun, L., Qin, T., Xiao, Y., Chen, K., et al. (2019). Hypoxia-driven paracrine osteopontin/integrin αvβ3 signaling promotes pancreatic cancer cell epithelial–mesenchymal transition and cancer stem cell-like properties by modulating forkhead box protein M1. *Molecular Oncology*, *13*(2), 228–245. https://doi.org/10.1002/1878-0261.12399.

Cao, Z., Ding, B.-S., Guo, P., Lee, S. B., Butler, J. M., Casey, S. C., et al. (2014). Angiocrine factors deployed by tumor vascular niche induce B cell lymphoma invasiveness and chemoresistance. *Cancer Cell*, *25*(3), 350–365. https://doi.org/10.1016/j.ccr.2014.02.005.

Carlson, P., Dasgupta, A., Grzelak, C. A., Kim, J., Barrett, A., Coleman, I. M., et al. (2019). Targeting the perivascular niche sensitizes disseminated tumour cells to chemotherapy. *Nature Cell Biology*, *21*(2), 238–250. https://doi.org/10.1038/s41556-018-0267-0.

Chakraborty, G., Jain, S., & Kundu, G. C. (2008). Osteopontin promotes vascular endothelial growth factor–dependent breast tumor growth and angiogenesis via autocrine and paracrine mechanisms. *Cancer Research*, *68*(1), 152–161. https://doi.org/10.1158/0008-5472.CAN-07-2126.

Chen, H.-F., Huang, C.-H., Liu, C.-J., Hung, J.-J., Hsu, C.-C., Teng, S.-C., et al. (2014a). Twist1 induces endothelial differentiation of tumour cells through the Jagged1-KLF4 axis. *Nature Communications*, *5*(1), 1–13. https://doi.org/10.1038/ncomms5697.

Chen, J., Chen, S., Zhuo, L., Zhu, Y., & Zheng, H. (2020). Regulation of cancer stem cell properties, angiogenesis, and vasculogenic mimicry by miR-450a-5p/SOX2 axis in colorectal cancer. *Cell Death & Disease*, *11*(3), 1–13. https://doi.org/10.1038/s41419-020-2361-z.

Chen, J.-C., Chang, Y.-W., Hong, C.-C., Yu, Y.-H., & Su, J.-L. (2012). The role of the VEGF-C/VEGFRs axis in tumor progression and therapy. *International Journal of Molecular Sciences*, *14*(1), 88–107. https://doi.org/10.3390/ijms14010088.

Chen, L., Fan, J., Chen, H., Meng, Z., Chen, Z., Wang, P., et al. (2014b). The IL-8/CXCR1 axis is associated with cancer stem cell-like properties and correlates with clinical prognosis in human pancreatic cancer cases. *Scientific Reports*, *4*(1), 1–7. https://doi.org/10.1038/srep05911.

Chen, Y., Jiang, T., Mao, A., & Xu, J. (2014c). Esophageal cancer stem cells express PLGF to increase cancer invasion through MMP9 activation. *Tumor Biology*, *35*(12), 12749–12755. https://doi.org/10.1007/s13277-014-2601-x.

Cheng, L., Huang, Z., Zhou, W., Wu, Q., Donnola, S., Liu, J. K., et al. (2013). Glioblastoma stem cells generate vascular pericytes to support vessel function and tumor growth. *Cell*, *153*(1), 139–152. https://doi.org/10.1016/j.cell.2013.02.021.

Chin, A. R., & Wang, S. E. (2016). Cancer tills the premetastatic field: Mechanistic basis and clinical implications. *Clinical Cancer Research: An Official Journal of the American Association for Cancer Research*, *22*(15), 3725–3733. https://doi.org/10.1158/1078-0432.CCR-16-0028.

Cochrane, C. R., Szczepny, A., Watkins, D. N., & Cain, J. E. (2015). Hedgehog signaling in the maintenance of cancer stem cells. *Cancers*, 7(3), 1554–1585. https://doi.org/10.3390/cancers7030851.

Cogle, C. R., Bosse, R. C., Brewer, T., Migdady, Y., Shirzad, R., Kampen, K. R., et al. (2016). Acute myeloid leukemia in the vascular niche. *Cancer Letters*, *380*(2), 552–560. https://doi.org/10.1016/j.canlet.2015.05.007.

Cogle, C. R., Goldman, D. C., Madlambayan, G. J., Leon, R. P., Masri, A. A., Clark, H. A., et al. (2014). Functional integration of acute myeloid leukemia into the vascular niche. *Leukemia*, *28*(10), 1978–1987. https://doi.org/10.1038/leu.2014.109.

Corrò, C., Healy, M. E., Engler, S., Bodenmiller, B., Li, Z., Schraml, P., et al. (2019). IL-8 and CXCR1 expression is associated with cancer stem cell-like properties of clear cell renal cancer. *The Journal of Pathology*, *248*(3), 377–389. https://doi.org/10.1002/path.5267.

de Bruijn, M. F. T. R., Speck, N. A., Peeters, M. C. E., & Dzierzak, E. (2000). Definitive hematopoietic stem cells first develop within the major arterial regions of the mouse embryo. *The EMBO Journal*, *19*(11), 2465–2474. https://doi.org/10.1093/emboj/19.11.2465.

Deshors, P., Toulas, C., Arnauduc, F., Malric, L., Siegfried, A., Nicaise, Y., et al. (2019). Ionizing radiation induces endothelial transdifferentiation of glioblastoma stem-like cells through the Tie2 signaling pathway. *Cell Death & Disease*, *10*(11), 1–15. https://doi.org/10.1038/s41419-019-2055-6.

Dias, S., Hattori, K., Heissig, B., Zhu, Z., Wu, Y., Witte, L., et al. (2001). Inhibition of both paracrine and autocrine VEGF/ VEGFR-2 signaling pathways is essential to induce long-term remission of xenotransplanted human leukemias. *Proceedings of the National Academy of Sciences*, *98*(19), 10857–10862. https://doi.org/10.1073/pnas.191117498.

Ding, B.-S., Cao, Z., Lis, R., Nolan, D. J., Guo, P., Simons, M., et al. (2014). Divergent angiocrine signals from vascular niche balance liver regeneration and fibrosis. *Nature*, *505*(7481), 97–102. https://doi.org/10.1038/nature12681.

Drusbosky, L., Gars, E., Trujillo, A., McGee, C., Meacham, A., Wise, E., et al. (2015). Endothelial cell derived angiocrine support of acute myeloid leukemia targeted by receptor tyrosine kinase inhibition. *Leukemia Research*, *39*(9), 984–989. https://doi.org/10.1016/j.leukres.2015.05.015.

Duarte, D., Hawkins, E. D., Akinduro, O., Ang, H., De Filippo, K., Kong, I. Y., et al. (2018). Inhibition of endosteal vascular niche remodeling rescues hematopoietic stem cell loss in AML. *Cell Stem Cell*, *22*(1), 64–77.e6. https://doi.org/10.1016/j.stem.2017.11.006.

Eriksson, A., Cao, R., Pawliuk, R., Berg, S. M., Tsang, M., Zhou, D., et al. (2002). Placenta growth factor-1 antagonizes VEGF-induced angiogenesis and tumor growth by the formation of functionally inactive PlGF-1/VEGF heterodimers. *Cancer Cell*, *1*(1), 99–108. https://doi.org/10.1016/s1535-6108(02)00028-4.

Fiedler, W., Graeven, U., Ergün, S., Verago, S., Kilic, N., Stockschläder, M., et al. (1997). Vascular endothelial growth factor, a possible paracrine growth factor in human acute myeloid leukemia. *Blood*, *89*(6), 1870–1875.

Folkins, C., Shaked, Y., Man, S., Tang, T., Lee, C. R., Zhu, Z., et al. (2009a). Glioma tumor stem-like cells promote tumor angiogenesis and vasculogenesis via vascular endothelial growth factor and stromal-derived factor 1. *Cancer Research*, *69*(18), 7243–7251. https://doi.org/10.1158/0008-5472.CAN-09-0167.

Folkins, C., Shaked, Y., Man, S., Tang, T., Lee, C. R., Zhu, Z., et al. (2009b). Glioma tumor stem-like cells promote tumor angiogenesis and vasculogenesis via vascular endothelial growth factor and stromal-derived factor 1. *Cancer Research*, *69*(18), 7243–7251. https://doi.org/10.1158/0008-5472.CAN-09-0167.

Garza Treviño, E. N., González, P. D., Valencia Salgado, C. I., & Martinez Garza, A. (2019). Effects of pericytes and colon cancer stem cells in the tumor microenvironment. *Cancer Cell International*, *19*(1), 173. https://doi.org/10.1186/s12935-019-0888-9.

Ghajar, C. M., Peinado, H., Mori, H., Matei, I. R., Evason, K. J., Brazier, H., et al. (2013). The perivascular niche regulates breast tumour dormancy. *Nature Cell Biology*, *15*(7), 807–817. https://doi.org/10.1038/ncb2767.

Gilbertson, R. J., & Rich, J. N. (2007). Making a tumour's bed: Glioblastoma stem cells and the vascular niche. *Nature Reviews. Cancer*, 7(10), 733–736. https://doi.org/10.1038/nrc2246.

Goel, H. L., Pursell, B., Chang, C., Shaw, L. M., Mao, J., Simin, K., et al. (2013). GLI1 regulates a novel neuropilin-2/α6β1 integrin based autocrine pathway that contributes to breast cancer initiation. *EMBO Molecular Medicine*, *5*(4), 488–508. https://doi.org/10.1002/emmm.201202078.

Hardee, M. E., & Zagzag, D. (2012). Mechanisms of glioma-associated neovascularization. *The American Journal of Pathology*, *181*(4), 1126–1141. https://doi.org/10.1016/j.ajpath.2012.06.030.

Hart, I. R., & Fidler, I. J. (1980). Role of organ selectivity in the determination of metastatic patterns of B16 melanoma. *Cancer Research*, *40*(7), 2281–2287.

Heinolainen, K., Karaman, S., D'Amico, G., Tammela, T., Sormunen, R., Eklund, L., et al. (2017). VEGFR3 modulates vascular permeability by controlling VEGF/ VEGFR2 signaling. *Circulation Research*, *120*(9), 1414–1425. https://doi.org/10.1161/CIRCRESAHA.116.310477.

Hirakawa, S., Kodama, S., Kunstfeld, R., Kajiya, K., Brown, L. F., & Detmar, M. (2005). VEGF-A induces tumor and sentinel lymph node lymphangiogenesis and promotes lymphatic metastasis. *The Journal of Experimental Medicine*, *201*(7), 1089–1099. https://doi.org/10.1084/jem.20041896.

Hoeben, A., Landuyt, B., Highley, M. S., Wildiers, H., Van Oosterom, A. T., & De Bruijn, E. A. (2004). Vascular endothelial growth factor and angiogenesis. *Pharmacological Reviews*, *56*(4), 549–580. https://doi.org/10.1124/pr.56.4.3.

Huang, R., Wang, S., Wang, N., Zheng, Y., Zhou, J., Yang, B., et al. (2020). CCL5 derived from tumor-associated macrophages promotes prostate cancer stem cells and metastasis via activating β-catenin/STAT3 signaling. *Cell Death & Disease*, *11*(4), 1–20. https://doi.org/10.1038/s41419-020-2435-y.

Huang, T., Song, C., Zheng, L., Xia, L., Li, Y., & Zhou, Y. (2019). The roles of extracellular vesicles in gastric cancer development, microenvironment, anti-cancer drug resistance, and therapy. *Molecular Cancer*, *18*(1), 62. https://doi.org/10.1186/s12943-019-0967-5.

Jahanban-Esfahlan, R., Seidi, K., Manjili, M. H., Jahanban-Esfahlan, A., Javaheri, T., & Zare, P. (2019). Tumor cell dormancy: Threat or opportunity in the fight against cancer. *Cancers*, *11*(8), 1207. https://doi.org/10.3390/cancers11081207.

Jia, W., Gao, X.-J., Zhang, Z.-D., Yang, Z.-X., & Zhang, G. (2013). S100A4 silencing suppresses proliferation, angiogenesis and invasion of thyroid cancer cells through down-regulation of MMP-9 and VEGF. *European Review for Medical and Pharmacological Sciences*, *17*(11), 1495–1508.

Kaplan, R. N., Riba, R. D., Zacharoulis, S., Bramley, A. H., Vincent, L., Costa, C., et al. (2005). VEGFR1-positive haematopoietic bone marrow progenitors initiate the pre-metastatic niche. *Nature*, *438*(7069), 820–827. https://doi.org/10.1038/nature04186.

Kienast, Y., von Baumgarten, L., Fuhrmann, M., Klinkert, W. E. F., Goldbrunner, R., Herms, J., et al. (2010). Real-time imaging reveals the single steps of brain metastasis formation. *Nature Medicine*, *16*(1), 116–122. https://doi.org/10.1038/nm.2072.

Kohsaka, S., Hinohara, K., Wang, L., Nishimura, T., Urushido, M., Yachi, K., et al. (2014). Epiregulin enhances tumorigenicity by activating the ERK/MAPK pathway in glioblastoma. *Neuro-Oncology*, *16*(7), 960–970. https://doi.org/10.1093/neuonc/not315.

Kopp, H.-G., Avecilla, S. T., Hooper, A. T., & Rafii, S. (2005). The bone marrow vascular niche: Home of HSC differentiation and mobilization. *Physiology (Bethesda, Md.)*, *20*, 349–356. https://doi.org/10.1152/physiol.00025.2005.

Kumar, S., Sharife, H., Kreisel, T., Mogilevsky, M., Bar-Lev, L., Grunewald, M., et al. (2019). Intra-tumoral metabolic zonation and resultant phenotypic diversification are dictated by blood vessel proximity. *Cell Metabolism*, *30*(1), 201–211.e6. https://doi.org/10.1016/j.cmet.2019.04.003.

Kusumbe, A. P. (2016a). Vascular niches for disseminated tumour cells in bone. *Journal of Bone Oncology*, *5*(3), 112–116. https://doi.org/10.1016/j.jbo.2016.04.003.

Kusumbe, A. P. (2016b). Vascular niches for disseminated tumour cells in bone. *Journal of Bone Oncology*, *5*(3), 112–116. https://doi.org/10.1016/j.jbo.2016.04.003.

Kusumbe, A. P., Ramasamy, S. K., & Adams, R. H. (2014). Coupling of angiogenesis and osteogenesis by a specific vessel subtype in bone. *Nature*, *507*(7492), 323–328. https://doi.org/10.1038/nature13145.

Kusumbe, A. P., Ramasamy, S. K., Itkin, T., Mäe, M. A., Langen, U. H., Betsholtz, C., et al. (2016). Age-dependent modulation of vascular niches for haematopoietic stem cells. *Nature*, *532*(7599), 380–384. https://doi.org/10.1038/nature17638.

Kuzu, I., Beksac, M., Arat, M., Celebi, H., Elhan, A. H., & Erekul, S. (2004). Bone marrow microvessel density (MVD) in adult acute myeloid leukemia (AML): Therapy induced changes and effects on survival. *Leukemia & Lymphoma*, *45*(6), 1185–1190. https://doi.org/10.1080/1042819032000159915.

Lakshmikanthan, S., Sobczak, M., Calzi, S. L., Shaw, L., Grant, M. B., & Chrzanowska-Wodnicka, M. (2018). Rap1B promotes VEGF-induced endothelial permeability and is required for dynamic regulation of the endothelial barrier. *Journal of Cell Science, 131*(1). https://doi.org/10.1242/jcs.207605.

Lapidot, T., Sirard, C., Vormoor, J., Murdoch, B., Hoang, T., Caceres-Cortes, J., et al. (1994). A cell initiating human acute myeloid leukaemia after transplantation into SCID mice. *Nature, 367*(6464), 645–648. https://doi.org/10.1038/367645a0.

Leali, D., Dell'Era, P., Stabile, H., Sennino, B., Chambers, A. F., Naldini, A., et al. (2003). Osteopontin (Eta-1) and fibroblast growth factor-2 cross-talk in angiogenesis. *The Journal of Immunology, 171*(2), 1085–1093. https://doi.org/10.4049/jimmunol.171.2.1085.

Li, S., & Li, Q. (2014). Cancer stem cells and tumor metastasis (Review). *International Journal of Oncology, 44*(6), 1806–1812. https://doi.org/10.3892/ijo.2014.2362.

Liu, G., Fan, X., Tang, M., Chen, R., Wang, H., Jia, R., et al. (2016). Osteopontin induces autophagy to promote chemo-resistance in human hepatocellular carcinoma cells. *Cancer Letters, 383*(2), 171–182. https://doi.org/10.1016/j.canlet.2016.09.033.

Liu, Z., Qi, L., Li, Y., Zhao, X., & Sun, B. (2017). VEGFR2 regulates endothelial differentiation of colon cancer cells. *BMC Cancer, 17*(1), 593. https://doi.org/10.1186/s12885-017-3578-9.

Lu, J., Ye, X., Fan, F., Xia, L., Bhattacharya, R., Bellister, S., et al. (2013). Endothelial cells promote the colorectal cancer stem cell phenotype through a soluble form of jagged-1. *Cancer Cell, 23*(2), 171–185. https://doi.org/10.1016/j.ccr.2012.12.021.

Maeda, S., Shinchi, H., Kurahara, H., Mataki, Y., Maemura, K., Sato, M., et al. (2008). CD133 expression is correlated with lymph node metastasis and vascular endothelial growth factor-C expression in pancreatic cancer. *British Journal of Cancer, 98*(8), 1389–1397. https://doi.org/10.1038/sj.bjc.6604307.

Magnon, C., Hall, S. J., Lin, J., Xue, X., Gerber, L., Freedland, S. J., et al. (2013). Autonomic nerve development contributes to prostate cancer progression. *Science (New York, N.Y.), 341*(6142), 1236361. https://doi.org/10.1126/science.1236361.

Mahecha, A. M., & Wang, H. (2017). The influence of vascular endothelial growth factor-A and matrix metalloproteinase-2 and -9 in angiogenesis, metastasis, and prognosis of endometrial cancer. *OncoTargets and Therapy, 10*, 4617–4624. https://doi.org/10.2147/OTT.S132558.

Maniotis, A. J., Folberg, R., Hess, A., Seftor, E. A., Gardner, L. M., Pe'er, J., et al. (1999). Vascular channel formation by human melanoma cells in vivo and in vitro: Vasculogenic mimicry. *The American Journal of Pathology, 155*(3), 739–752. https://doi.org/10.1016/S0002-9440(10)65173-5.

Meng, W., Xue, S., & Chen, Y. (2018). The role of CXCL12 in tumor microenvironment. *Gene, 641*, 105–110. https://doi.org/10.1016/j.gene.2017.10.015.

Nagy, J. A., Benjamin, L., Zeng, H., Dvorak, A. M., & Dvorak, H. F. (2008). Vascular permeability, vascular hyperpermeability and angiogenesis. *Angiogenesis, 11*(2), 109–119. https://doi.org/10.1007/s10456-008-9099-z.

Nakanishi, M., Morita, Y., Hata, K., & Muragaki, Y. (2016). Acidic microenvironments induce lymphangiogenesis and IL-8 production via TRPV1 activation in human lymphatic endothelial cells. *Experimental Cell Research, 345*(2), 180–189. https://doi.org/10.1016/j.yexcr.2016.06.006.

Ninomiya, M., Abe, A., Katsumi, A., Xu, J., Ito, M., Arai, F., et al. (2007). Homing, proliferation and survival sites of human leukemia cells in vivo in immunodeficient mice. *Leukemia, 21*(1), 136–142. https://doi.org/10.1038/sj.leu.2404432.

Nör, J. E., Christensen, J., Mooney, D. J., & Polverini, P. J. (1999). Vascular endothelial growth factor (VEGF)-mediated angiogenesis is associated with enhanced endothelial cell survival and induction of Bcl-2 expression. *The American Journal of Pathology, 154*(2), 375–384. https://doi.org/10.1016/S0002-9440(10)65284-4.

Ordóñez-Morán, P., & Huelsken, J. (2014). Complex metastatic niches: Already a target for therapy? *Current Opinion in Cell Biology*, *31*, 29–38. https://doi.org/10.1016/j.ceb.2014.06.012.

O'Reilly, M. S., Holmgren, L., Chen, C., & Folkman, J. (1996). Angiostatin induces and sustains dormancy of human primary tumors in mice. *Nature Medicine*, *2*(6), 689–692. https://doi.org/10.1038/nm0696-689.

Otte, J., Dizdar, L., Behrens, B., Goering, W., Knoefel, W. T., Wruck, W., et al. (2019). fgf signalling in the self-renewal of colon cancer organoids. *Scientific Reports*, *9*(1), 1–10. https://doi.org/10.1038/s41598-019-53907-7.

Ourradi, K., Blythe, T., Jarrett, C., Barratt, S. L., Welsh, G. I., & Millar, A. B. (2017). VEGF isoforms have differential effects on permeability of human pulmonary microvascular endothelial cells. *Respiratory Research*, *18*(1), 116. https://doi.org/10.1186/s12931-017-0602-1.

Padró, T., Ruiz, S., Bieker, R., Bürger, H., Steins, M., Kienast, J., et al. (2000). Increased angiogenesis in the bone marrow of patients with acute myeloid leukemia. *Blood*, *95*(8), 2637–2644.

Pan, Z., Lu, X., Zhao, J., Gao, Q., & Wang, J. (2018). VEGF-C is positively associated with lymphangiogenesis and lymphatic metastasis in rectal cancer. *International Journal of Clinical and Experimental Pathology*, *11*(3), 1777–1783.

Pasquier, J., Ghiabi, P., Chouchane, L., Razzouk, K., Rafii, S., & Rafii, A. (2020). Angiocrine endothelium: From physiology to cancer. *Journal of Translational Medicine*, *18*(1), 52. https://doi.org/10.1186/s12967-020-02244-9.

Passaro, D., Di Tullio, A., Abarrategi, A., Rouault-Pierre, K., Foster, K., Ariza-McNaughton, L., et al. (2017). Increased vascular permeability in the bone marrow microenvironment contributes to disease progression and drug response in acute myeloid leukemia. *Cancer Cell*, *32*(3), 324–341.e6. https://doi.org/10.1016/j.ccell.2017.08.001.

Patan, S. (2004). Vasculogenesis and angiogenesis. *Cancer Treatment and Research*, *117*, 3–32. https://doi.org/10.1007/978-1-4419-8871-3_1.

Peterson, S. C., Eberl, M., Vagnozzi, A. N., Belkadi, A., Veniaminova, N. A., Verhaegen, M. E., et al. (2015). Basal cell carcinoma preferentially arises from stem cells within hair follicle and mechanosensory niches. *Cell Stem Cell*, *16*(4), 400–412. https://doi.org/10.1016/j.stem.2015.02.006.

Pietras, A., Katz, A. M., Ekström, E. J., Wee, B., Halliday, J. J., Pitter, K. L., et al. (2014). Osteopontin-CD44 signaling in the glioma perivascular niche enhances cancer stem cell phenotypes and promotes aggressive tumor growth. *Cell Stem Cell*, *14*(3), 357–369. https://doi.org/10.1016/j.stem.2014.01.005.

Ping, Y., Yao, X., Jiang, J., Zhao, L., Yu, S., Jiang, T., et al. (2011). The chemokine CXCL12 and its receptor CXCR4 promote glioma stem cell-mediated VEGF production and tumour angiogenesis via PI3K/AKT signalling. *The Journal of Pathology*, *224*(3), 344–354. https://doi.org/10.1002/path.2908.

Plaks, V., Kong, N., & Werb, Z. (2015). The cancer stem cell niche: How essential is the niche in regulating stemness of tumor cells? *Cell Stem Cell*, *16*(3), 225–238. https://doi.org/10.1016/j.stem.2015.02.015.

Poulos, M. G., Gars, E. J., Gutkin, M. C., Kloss, C. C., Ginsberg, M., Scandura, J. M., et al. (2014a). Activation of the vascular niche supports leukemic progression and resistance to chemotherapy. *Experimental Hematology*, *42*(11), 976–986.e3. https://doi.org/10.1016/j.exphem.2014.08.003.

Poulos, M. G., Gars, E. J., Gutkin, M. C., Kloss, C. C., Ginsberg, M., Scandura, J. M., et al. (2014b). Activation of the vascular niche supports leukemic progression and resistance to chemotherapy. *Experimental Hematology*, *42*(11), 976–986.e3. https://doi.org/10.1016/j.exphem.2014.08.003.

Psaila, B., & Lyden, D. (2009). The metastatic niche: Adapting the foreign soil. *Nature Reviews. Cancer*, *9*(4), 285–293. https://doi.org/10.1038/nrc2621.

Pufe, T., Harde, V., Petersen, W., Goldring, M. B., Tillmann, B., & Mentlein, R. (2004). Vascular endothelial growth factor (VEGF) induces matrix metalloproteinase expression in immortalized chondrocytes. *The Journal of Pathology*, *202*(3), 367–374. https://doi.org/10.1002/path.1527.

Qu, Y., Dou, B., Tan, H., Feng, Y., Wang, N., & Wang, D. (2019). Tumor microenvironment-driven non-cell-autonomous resistance to antineoplastic treatment. *Molecular Cancer*, *18*(1), 69. https://doi.org/10.1186/s12943-019-0992-4.

Rak, J., Milsom, C., & Yu, J. (2008). Vascular determinants of cancer stem cell dormancy—Do age and coagulation system play a role? *APMIS*, *116*(7-8), 660–676. https://doi.org/10.1111/j.1600-0463.2008.01058.x.

Ricci-Vitiani, L., Pallini, R., Biffoni, M., Todaro, M., Invernici, G., Cenci, T., et al. (2010). Tumour vascularization via endothelial differentiation of glioblastoma stem-like cells. *Nature*, *468*(7325), 824–828. https://doi.org/10.1038/nature09557.

Riese, D. J., & Cullum, R. L. (2014). Epiregulin: Roles in normal physiology and cancer. *Seminars in Cell & Developmental Biology*, *0*, 49–56. https://doi.org/10.1016/j.semcdb.2014.03.005.

Rundhaug, J. E. (2003). Matrix metalloproteinases, angiogenesis, and cancer: commentary re: A. C. Lockhart et al., reduction of wound angiogenesis in patients treated with BMS-275291, a broad spectrum matrix metalloproteinase inhibitor. *Clinical Cancer Research: An Official Journal of The American Association for Cancer Research*, *9*, 551–554.

Shan, B., Li, W., Yang, S.-Y., & Li, Z.-R. (2013). Estrogen up-regulates MMP2/9 expression in endometrial epithelial cell via VEGF-ERK1/2 pathway. *Asian Pacific Journal of Tropical Medicine*, *6*(10), 826–830. https://doi.org/10.1016/S1995-7645(13)60146-7.

Shangguan, W., Fan, C., Chen, X., Lu, R., Liu, Y., Li, Y., et al. (2017). Endothelium originated from colorectal cancer stem cells constitute cancer blood vessels. *Cancer Science*, *108*(7), 1357–1367. https://doi.org/10.1111/cas.13262.

Shi, J., & Wei, P.-K. (2016). Interleukin-8: A potent promoter of angiogenesis in gastric cancer. *Oncology Letters*, *11*(2), 1043–1050. https://doi.org/10.3892/ol.2015.4035.

Singh, A., Veeriah, V., Xi, P., Labella, R., Chen, J., Romeo, S. G., et al. (2019). Angiocrine signals regulate quiescence and therapy resistance in bone metastasis. *JCI Insight*, *4*(13). https://doi.org/10.1172/jci.insight.125679.

Sun, C., Hu, A., Wang, S., Tian, B., Jiang, L., Liang, Y., et al. (2020a). ADAM17-regulated CX3CL1 expression produced by bone marrow endothelial cells promotes spinal metastasis from hepatocellular carcinoma. *International Journal of Oncology*, *57*(1), 249–263. https://doi.org/10.3892/ijo.2020.5045.

Sun, L., Pan, J., Yu, L., Liu, H., Shu, X., Sun, L., et al. (2016). Tumor endothelial cells promote metastasis and cancer stem cell-like phenotype through elevated Epiregulin in esophageal cancer. *American Journal of Cancer Research*, *6*(10), 2277–2288.

Sun, S.-L., Shu, Y.-G., & Tao, M.-Y. (2020b). LncRNA CCAT2 promotes angiogenesis in glioma through activation of VEGFA signalling by sponging miR-424. *Molecular and Cellular Biochemistry*, *468*(1), 69–82. https://doi.org/10.1007/s11010-020-03712-y.

Sun, X., Ma, X., Wang, J., Zhao, Y., Wang, Y., Bihl, J. C., et al. (2017). Glioma stem cells-derived exosomes promote the angiogenic ability of endothelial cells through miR-21/VEGF signal. *Oncotarget*, *8*(22), 36137–36148. https://doi.org/10.18632/oncotarget.16661.

Tacconi, C., Ungaro, F., Correale, C., Arena, V., Massimino, L., Detmar, M., et al. (2019). Activation of the VEGFC/VEGFR3 pathway induces tumor immune escape in colorectal cancer. *Cancer Research*, *79*(16), 4196–4210. https://doi.org/10.1158/0008-5472.CAN-18-3657.

Tawada, M., Hayashi, S.-I., Ikegame, Y., Nakashima, S., & Yoshida, K. (2014). Possible involvement of tumor-producing VEGF-A in the recruitment of lymphatic endothelial progenitor cells from bone marrow. *Oncology Reports*, *32*(6), 2359–2364. https://doi.org/10.3892/or.2014.3499.

Treps, L., Perret, R., Edmond, S., Ricard, D., & Gavard, J. (2017). Glioblastoma stem-like cells secrete the pro-angiogenic VEGF-A factor in extracellular vesicles. *Journal of Extracellular Vesicles*, *6*(1), 1359479. https://doi.org/10.1080/20013078.2017.1359479.

Unemori, E. N., Ferrara, N., Bauer, E. A., & Amento, E. P. (1992). Vascular endothelial growth factor induces interstitial collagenase expression in human endothelial cells. *Journal of Cellular Physiology*, *153*(3), 557–562. https://doi.org/10.1002/jcp.1041530317.

Vermeulen, L., De Sousa, E., Melo, F., van der Heijden, M., Cameron, K., de Jong, J. H., et al. (2010). Wnt activity defines colon cancer stem cells and is regulated by the microenvironment. *Nature Cell Biology*, *12*(5), 468–476. https://doi.org/10.1038/ncb2048.

Virk, M. S., & Lieberman, J. R. (2007). Tumor metastasis to bone. *Arthritis Research & Therapy*, *9*(Suppl 1), S5. https://doi.org/10.1186/ar2169.

Wang, C.-A., & Tsai, S.-J. (2015). The non-canonical role of vascular endothelial growth factor-C axis in cancer progression. *Experimental Biology and Medicine*, *240*(6), 718–724. https://doi.org/10.1177/1535370215583802.

Wang, J., Xu, S.-L., Duan, J.-J., Yi, L., Guo, Y.-F., Shi, Y., et al. (2019). Invasion of white matter tracts by glioma stem cells is regulated by a NOTCH1 – SOX2 positive-feedback loop. *Nature Neuroscience*, *22*(1), 91–105. https://doi.org/10.1038/s41593-018-0285-z.

Winkler, I. G., Barbier, V., Nowlan, B., Jacobsen, R. N., Forristal, C. E., Patton, J. T., et al. (2012a). Vascular niche E-selectin regulates hematopoietic stem cell dormancy, self renewal and chemoresistance. *Nature Medicine*, *18*(11), 1651–1657. https://doi.org/10.1038/nm.2969.

Winkler, I. G., Barbier, V., Nowlan, B., Jacobsen, R. N., Forristal, C. E., Patton, J. T., et al. (2012b). Vascular niche E-selectin regulates hematopoietic stem cell dormancy, self renewal and chemoresistance. *Nature Medicine*, *18*(11), 1651–1657. https://doi.org/10.1038/nm.2969.

Winkler, I. G., Barbier, V., Pattabiraman, D. R., Gonda, T. J., Magnani, J. L., & Levesque, J.-P. (2014). Vascular niche E-selectin protects acute myeloid leukaemia stem cells from chemotherapy. *Blood*, *124*(21), 620. https://doi.org/10.1182/blood.V124.21.620.620.

Wong, S. Y., & Reiter, J. F. (2011). Wounding mobilizes hair follicle stem cells to form tumors. *Proceedings of the National Academy of Sciences of the United States of America*, *108*(10), 4093–4098. https://doi.org/10.1073/pnas.1013098108.

Würth, R., Bajetto, A., Harrison, J. K., Barbieri, F., & Florio, T. (2014). CXCL12 modulation of CXCR4 and CXCR7 activity in human glioblastoma stem-like cells and regulation of the tumor microenvironment. *Frontiers in Cellular Neuroscience*, *8*. https://doi.org/10.3389/fncel.2014.00144.

Xu, Q., Zhang, Q., Ishida, Y., Hajjar, S., Tang, X., Shi, H., et al. (2016). EGF induces epithelial-mesenchymal transition and cancer stem-like cell properties in human oral cancer cells via promoting Warburg effect. *Oncotarget*, *8*(6), 9557–9571. https://doi.org/10.18632/oncotarget.13771.

Yan, G.-N., Yang, L., Lv, Y.-F., Shi, Y., Shen, L.-L., Yao, X.-H., et al. (2014). Endothelial cells promote stem-like phenotype of glioma cells through activating the Hedgehog pathway. *The Journal of Pathology*, *234*(1), 11–22. https://doi.org/10.1002/path.4349.

Yang, H.-K., Chen, H., Mao, F., Xiao, Q.-G., Xie, R.-F., & Lei, T. (2017). Downregulation of LRIG2 expression inhibits angiogenesis of glioma via EGFR/VEGF-A pathway. *Oncology Letters*, *14*(4), 4021–4028. https://doi.org/10.3892/ol.2017.6671.

Ye, X., Xu, S., Xin, Y., Yu, S., Ping, Y., Chen, L., et al. (2012). Tumor-associated microglia/macrophages enhance the invasion of glioma stem-like cells via TGF-β1 signaling pathway. *The Journal of Immunology, 189*(1), 444–453. https://doi.org/10.4049/jimmunol.1103248.

Yu, Z., Pestell, T. G., Lisanti, M. P., & Pestell, R. G. (2012). Cancer stem cells. *The International Journal of Biochemistry & Cell Biology, 44*(12), 2144–2151. https://doi.org/10.1016/j.biocel.2012.08.022.

Zhang, X., Chen, L., Dang, W., Cao, M., Xiao, J., Lv, S., et al. (2020). CCL8 secreted by tumor-associated macrophages promotes invasion and stemness of glioblastoma cells via ERK1/2 signaling. *Laboratory Investigation, 100*(4), 619–629. https://doi.org/10.1038/s41374-019-0345-3.

Zhao, P., Chen, A., Qi, Q., Zhou, W., Feng, Z., Wang, J., et al. (2017). Impact of VEGFA polymorphisms on glioma risk in Chinese. *Oncotarget, 8*(48), 83712–83722. https://doi.org/10.18632/oncotarget.19380.

Zheng, J., Jia, L., Mori, S., & Kodama, T. (2018). Evaluation of metastatic niches in distant organs after surgical removal of tumor-bearing lymph nodes. *BMC Cancer, 18*. https://doi.org/10.1186/s12885-018-4538-8.

Zhou, W., Ke, S. Q., Huang, Z., Flavahan, W., Fang, X., Paul, J., et al. (2015). Periostin secreted by glioblastoma stem cells recruits M2 tumour-associated macrophages and promotes malignant growth. *Nature Cell Biology, 17*(2), 170–182. https://doi.org/10.1038/ncb3090.

Zhu, T. S., Costello, M. A., Talsma, C. E., Flack, C. G., Crowley, J. G., Hamm, L. L., et al. (2011). Endothelial cells create a stem cell niche in glioblastoma by providing NOTCH ligands that nurture self-renewal of cancer stem-like cells. *Cancer Research, 71*(18), 6061–6072. https://doi.org/10.1158/0008-5472.CAN-10-4269.

CHAPTER FOUR

The lung cancer stem cell niche

Ivan Bertoncello*
Lung Health Research Centre, Department of Pharmacology and Therapeutics, University of Melbourne, Melbourne, VIC, Australia
*Corresponding author: e-mail address: ivanb@unimelb.edu.au

Contents

1. Introduction

Lung cancer comprises a heterogeneous group of complex diseases classified on the basis of their histopathology, immunocytochemistry, and genetic profiling (Table 1) (Rodriguez-Canales, Parra-Cuentas, & Wistuba, 2016;

ISSN 2468-5097
https://doi.org/10.1016/bs.asn.2020.10.002

Table 1 2015 WHO classification of histologic types and subtypes of lung epithelial tumors.[a]

Adenocarcinoma	Small cell carcinoma
Lepidic adenocarcinoma	• Combined small cell carcinoma
Acinar adenocarcinoma	Large cell neuroendocrine carcinoma
Papillary adenocarcinoma	• Combined large cell neuroendocrine carcinoma
Micropapillary adenocarcinoma	Carcinoid tumors
Solid adenocarcinoma	• Typical carcinoid tumors
Invasive mucinous adenocarcinoma	• Atypical carcinoid tumors
• Mixed invasive mucinous	Large cell carcinoma
• Nonmucinous adenocarcinoma	Adenosquamous carcinoma
Colloid adenocarcinoma	Sarcomatid carcinomas
Fetal adenocarcinoma	• Pleiomorphic carcinomas
Minimally invasive adenocarcinoma	• Spindle cell carcinoma
• Nonmucinous	• Giant cell carcinoma
• Mucinous	• Carcinosarcoma
Preinvasive Lesions	• Pulmonary blastoma
• Atypical adenomatous hyperplasia	Other and unclassified carcinomas
• Adenocarcinoma in situ	• Lymphoepithelioma-like carcinoma
• Nonmucinous and mucinous	• NUT carcinoma
Squamous cell carcinoma	Salivary gland type tumors
Keratinizing squamous cell carcinoma	• Mucoepidermoidcarcinoma
Non-keratinizing squamous cell carcinoma	• Adenoid cystic carcinoma
Basaloid squamous cell carcinoma	• Epithelial-myoepithelial carcinoma
Preinvasive Lesion	• Pleiomorphic adenoma
• Squamous cell carcinoma in situ	
Neuroendocrine tumors	

[a]Abridged classification abstracted from Travis et al. (2015) omitting some rare epithelial tumor types. The International Classification of Diseases for Oncology (ICDO) lists more than 50 codes describing epithelial tumors of the lung.

Travis et al., 2015). Lung cancer is the most commonly diagnosed cancer, and the leading cause of cancer death worldwide, accounting for approximately 1 in 5 cancer deaths (Barta, Powell, & Wisnivesky, 2019; Bray et al., 2018). While the majority of lung cancers are caused by exposure to tobacco smoke, the significant proportion of lung cancers (~15%) detected in never smokers also reveals a likely genetic susceptibility to disease (Chapman, Sun, Ruestow, Cowan, & Madl, 2016; McCarthy, Meza, Jeon,

& Moolgavkar, 2012). In both smokers and never smokers, respiratory co-morbidities including chronic obstructive pulmonary disease (COPD) and interstitial lung diseases (ILDs) are major interacting factors in lung cancer initiation, progression and metastasis adversely affecting survival (Leduc, Antoni, Charloux, Falcoz, & Quoix, 2017).

Approximately 80% of lung cancers are classified as non-small cell lung cancers (NSCLC), while 15–20% are classified as small cell lung cancers (SCLC). NSCLC are further subtyped as adenocarcinomas (ADC), large cell carcinomas (LCC) or squamous cell carcinomas (SCC). The inter- and intra-tumor cellular heterogeneity of lung cancer subtypes and their propensity to metastasize during disease progression is a manifestation of the intrinsic developmental potential of lung cancer initiating cells and their reciprocal interaction with the tumor microenvironment in which they reside.

The concept of the stem cell niche proposed by Schofield (Schofield, 1978) evolved from the analysis and interpretation of irradiated mouse models (Curry & Trentin, 1967; McCulloch, Siminovitch, Till, Russell, & Bernstein, 1965) showing that the fate of transplanted hematopoietic stem cells was not solely determined by their intrinsic developmental potential but also by the permissive or restrictive inductive tissue microenvironment in which they lodged. According to Schofield's model, stem cells anchored in a complex anatomical niche comprising extracellular matrix (ECM) proteins, adhesion molecules, soluble and insoluble factors, and contiguous stromal, hematopoietic and immunomodulatory cells retain their stem cell properties. Stem cells leaving the niche are induced to proliferate, differentiate, and gradually lose their regenerative potential unless they are able to reoccupy an available niche and revert to the stem cell state. This model conveyed an understanding of the stem cell niche as a unique anatomical structure providing factors conducive to the attraction, retention and survival of specific stem cell populations, and the protection and regulation of their regenerative potential. This underpinned a classical, stem cell centric understanding of the organization and regulation of renewing cell populations where a complex disease such as cancer could simply be described as a caricature of normal tissue renewal (Pierce, 1974; Pierce & Speers, 1988). Accordingly, the degree of malignancy of cancer stem cells reflects the developmental potential of their normal counterpart targeted by specific mutations (Kreso & Dick, 2014); and the evolution of tumor heterogeneity during disease progression is a consequence of the genetic instability of mutated stem and progenitor cells (Nowell, 1976). However, recognition that stem cells and their niche function as a trophic unit has focussed attention on the role of critical bidirectional cell-cell and cell-matrix interactions within the niche that regulate organ development as well as adult tissue renewal, regeneration and repair (Clevers, Loh, & Nusse, 2014; Nelson & Bissell, 2006; Schultz, Davidson, Kirsner, Bornstein, & Herman, 2011; Voog & Jones, 2010; Xu, Boudreau, & Bissell, 2009). These highly conserved signaling pathways are utilized by stem cells to reorganize, remodel and repair their niche, while reciprocal signaling mediated by the niche directs the proliferation, differentiation and functionality of stem cells and their progeny.

The defining hallmarks of cancer (Hanahan & Weinberg, 2011) describe the acquired properties of cancer initiating cells and cancer propagating cells that give them a competitive advantage in subverting normal tissue homeostasis. These properties include sustained proliferative signaling, the evasion of growth suppressors, the resistance to cell death, replicative immortality, the induction of angiogenesis, the activation of invasion and metastasis, genomic instability, tumor promotion of inflammation, the reprogramming of energy metabolism, and the evasion of immune destruction. The cumulative genetic, epigenetic and histopathological changes to the stem cell and its niche in the course of cancer initiation, progression and metastasis are a manifestation of the field of injury, and field cancerization (Dakubo, Jakupciak, Birch-Machin, & Parr, 2007; Gower, Steiling, Brothers 2nd, Lenburg, & Spira, 2011; Steiling, Ryan, Brody, & Spira, 2008) where the field of injury reflects the host response to insult or injury characterized by inflammation, oxidative stress, the elaboration of factors, and ECM and tissue remodeling; and field cancerization describes the acquisition of premalignant irreversible genetic changes and mutations in noncancerous cells that eventually lead to their transformation into cancer stem cells.

In the lung, the respiratory epithelium, sub-epithelial mesenchymal cells, neural cells, microvasculature, innate and adaptive immune cells, immunomodulatory cells, and the ECM operate as trophic unit (Evans, Van Winkle, Fanucchi, & Plopper, 1999; Holgate et al., 2000) regulated by highly conserved gene networks and signaling pathways (Metzger, Klein, Martin, & Krasnow, 2008; Morrisey & Hogan, 2010; Warburton, Wuenschell, Flores-Delgado, & Anderson, 1998) that are subverted in lung cancer. That said, deconstructing the interplay between stem cells and their niche in lung cancer initiation and progression is not straight-forward. The lung is a highly complex, intricate organ that is truly a marvel of bioengineering (Dorrello & Vunjak-Novakovic, 2020; Weibel, 2013). The lung is unique in that: (1) it comprises at least 60 functionally differentiated cell types; (2) the respiratory epithelium is maintained by both dedicated stem cells and functional differentiated cells, termed facultative stem cells, that are able to dedifferentiate in response to injury (Jain et al., 2015; Leach & Morrisey, 2018); (3) cyclic mechanical stretch, shear, strain and airflow obstruction modulate lung regeneration and repair (Hsia, 2017); (4) the airway microbiome interacts with immunomodulatory cells and host cells to modulate the interaction between lung stem cells and the niche (Dickson, Erb-Downward, Martinez, & Huffnagle, 2016); and (5) regional differences in factors including pH, temperature, O_2 tension, nutrient availability along the proximal-distal axis influences proliferation and differentiation.

This chapter aims to briefly review how disruption of the reciprocal signaling between lung epithelial stem cells and their niche contributes to lung cancer initiation and progression; and how each element of the tumor microenvironment contributes to, and regulates, the dissemination and fate of metastatic cells.

2. Epithelial stem cell compartments in the adult lung

The respiratory system arises from an outpouching of the anterior foregut that undergoes branching morphogenesis to form the arborizing proximal and distal airways and alveolar bed (Morrisey & Hogan, 2010; Metzger et al., 2008). This process culminates in the generation of a highly vascularized respiratory tree comprising at least 60 functionally differentiated cell types distributed along the proximal-distal lung axis that are essential for maintaining airway patency, mucocilliary clearance and gas exchange (Warburton et al., 1998).

In mice, the trachea and mainstem bronchiolar airway is lined by a pseudostratified epithelium comprising basal cells, secretory cells and ciliated cells. Submucosal glands (SMG) that produce the majority of serous and mucous secretions in the airway are restricted to the sub-epithelial layer of the upper trachea. Pseudostratified epithelium transitions into a simple cuboidal epithelium in the proximal bronchiolar airways that are devoid of basal cells and predominantly lined with secretory club cells and ciliated cells. Rare innervated pulmonary neuroendocrine cells (PNEC) are dispersed throughout proximal and distal bronchiolar airways. Neuroepithelial bodies (NEB) are located at airway branch-points and at the bronchoalveolar duct junction (BADJ) of terminal bronchioles that in turn give rise to alveolar ducts and alveoli lined with cuboidal surfactant-secreting alveolar epithelial Type 2 (AT2) cells and squamous gas-exchanging alveolar epithelial Type 1 (AT1) cells.

The characterization of epithelial stem and progenitor cells and their niches in the adult lung has relied on cell lineage tracing in vivo, and the analysis of the clonal dynamics of three-dimensional multicellular organoid cultures that recapitulate the apical-basal organization, differentiation and functionality of respiratory epithelia in the normal and mutant mouse models from which they are derived. These studies have revealed that each of the anatomical compartments described above is maintained by diverse pools of multipotent and lineage restricted stem and progenitor cells whose activation, recruitment and participation in lung regeneration and repair is determined by the type and severity of injury to the respiratory epithelium and their niche (reviewed in Bertoncello, 2016).

2.1 Developmental origins and regulation of the lung epithelial stem cell hierarchy

The epithelial stem cells of the proximodistal airways and alveolar bed reside in unique niche microenvironments where epithelial-mesenchymal cell crosstalk plays a critical role in their lineage commitment and differentiation by activating highly conserved signaling pathways and factors including FGF-10, BMP4, Shh, Myb, Notch, Wnt/β-catenin, and TGF-β (Table 2) (Morrisey & Hogan, 2010; Warburton et al., 2001). For example, FGF-10 which maintains distal lung epithelial progenitors and prevents their differentiation into proximal airway epithelial cells during development, initiates distal bronchiolar epithelial repair by variant club cells in the adult lung (Volckaert & De Langhe, 2014). BMP4 which regulates proximal-distal patterning in embryonic tracheal epithelium, acts as a brake on basal cell proliferation during lung homeostasis (Tadokoro, Gao, Hong, Hotten, & Hogan, 2016). During development, Notch signaling regulates differentiation of secretory epithelium from proximal endoderm, while Myb is required for multiciliated cell differentiation, whereas in the adult lung, differential expression of c-myb and Notch signaling via the Notch2 intracellular domain directs the differentiation of tracheal basal cells towards a ciliated or secretory cell lineage respectively (Pan et al., 2014; Pardo-Saganta et al., 2015). The sonic hedgehog (SHH) pathway which regulates lung morphogenesis during development, maintains the balance between proliferation and quiescence of lung epithelial stem cells in the adult lung (Peng et al., 2015).

Table 2 The role of key signaling pathways, genes and factors associated with lung epithelial cell proliferation and differentiation during lung development, regeneration and repair.[a]

Gene	Molecular function	Lung development	Adult lung regeneration and repair	Pathophysiology
Wnt	Signaling pathway	Regulates early lung endodermal specification	Expansion of the BASC compartment following injury	Chronic activation associated with increased fibrosis
Notch	Signaling pathway	Differentiation of secretory epithelium from proximal endoderm	Differentiation of basal cells into secretory cells after severe injury	Excess notch signaling promotes mucous metaplasia; Loss of notch signaling induces ciliated cell differentiation
EGFR	Transmembrane receptor tyrosine kinase	$EGFR^{-/-}$ mice are characterized by defective alveolarization and loss of surfactant expression	Proliferation of airway and alveolar stem/progenitor cells	Overexpression associated with epithelial cell hyperplasia and goblet cell metaplasia; Mutations drive lung adenocarcinoma
TTF-1 (Nkx2.1)	Transcription factor	Required for formation of lung parenchyma and differentiation of lung epithelial lineages	Required for normal differentiation of airway and alveolar epithelial cells	Mutations cause lung diseases of varying severity (e.g. alveolar dysgenesis, interstitial lung disease)
Sox-2	Transcription factor	Normal patterning of proximal airway epithelial lineages	Activated after injury to induce differentiation of ciliated, club, basal and goblet cells	Deletion produces airways devoid of normal epithelial cell types
Gata-6	Transcription factor	Regulates lung epithelial gene transcription during development	Maintains distal airway homeostasis, regulating balance between BASC renewal and differentiation via Wnt signaling pathway	Suppression of Gata-6 is associated with lung adenocarcinoma progression

Continued

Table 2 The role of key signaling pathways, genes and factors associated with lung epithelial cell proliferation and differentiation during lung development, regeneration and repair.[a]—cont'd

Gene	Molecular function	Lung development	Adult lung regeneration and repair	Pathophysiology
SPDEF	Transcription factor	Differentiation of goblet cells from basal and non-ciliated columnar epithelial cells	Required for goblet cell differentiation in normal submucosal glands; and following allergen exposure	Deletion blocks goblet cell differentiation in airways and sub-mucosal glands
Myb	Transcription factor	Required for multiciliated cell differentiation	Required for differentiation of ciliated and secretory cells in proximal airways	Myb^{pos} cells are elevated in chronic airway disease; Inactivation leads to failure of proper airway ciliation
p63	Transcription factor	Critical for development of normal tracheobronchial epithelium	Preferentially expressed in basal cell compartment; Plays role in maintenance of lung epithelium	Deletion in mouse results in development of simple columnar epithelium devoid of basal cells; Overexpressed in squamous cell carcinoma
HDAC1/2	Histone deacetylases: catalyse removal of acetyl functional groups	Regulation of Sox2 expression required for development of proximal airways	Regulates lung epithelial regeneration by reactivation of cell cycle progression	Deletion of HDAC1/2 results in defective airway repair; inhibition causes emphysema
SHH	Growth factor	Expressed in epithelium during lung development regulating interaction between epithelial and mesenchymal cells during branching morphogenesis	Promotes segregation between proximal and distal mesenchyme determining distinct airway and alveolar stem cell niches	Upregulation exacerbates lung fibrosis; associated with bronchopulmonary dysplasia

FGF-10	Growth factor	Branching morphogenesis; saccular lung development; tracheal basal cell maintenance and differentiation	Promotes re-epithelialization of injured airways by stimulating lung epithelial stem cell proliferation	Reduced FGF-10 expression is associated with bronchopulmonary dysplasia
HGF	Growth factor	Required for alveologenesis	Promotes lung repair following injury	Mice lacking HGF receptor (Met) have impaired airspace formation
TGF-β	Growth factor	Required for epithelial-mesenchymal interactions in branching morphogenesis and alveologenesis	Orchestrates lung regeneration and repair	Overexpression promotes ECM remodeling and fibrosis, and facilitates lung cancer progression

[a]Revised and updated from (Bertoncello, 2016). Data compiled from (Akram, Patel, Spiteri, & Forsyth, 2016; Boucherat, Morissette, Provencher, Bonnet, & Maltais, 2016; Herriges & Morrisey, 2014; Maeda, Dave, & Whitsett, 2007; Volckaert & De Langhe, 2014; Whitsett, Haitchi, & Maeda, 2011).

Wnt signaling regulates early lung endodermal specification, and canonical Wnt/β-catenin and non-canonical Wnt/β-catenin independent signaling pathways are critical regulators of lung development and adult lung epithelial stem cell regeneration and repair (Baarsma & Konigshoff, 2017; Konigshoff & Eickelberg, 2010; Li, Bellusci, Borok, & Minoo, 2015). Different cell lineages variably express specific Wnt ligands that are able to activate multiple Wnt signaling cascades simultaneously (Baarsma & Konigshoff, 2017; Frank et al., 2016; Zepp et al., 2017), and an imbalance between canonical and non-canonical Wnt signaling can exacerbate disease (Baarsma & Konigshoff, 2017).

TGF-β is the archetypal pleotropic cytokine. It plays a major role in the regulation of branching morphogenesis and alveolarization during development al (Shi, Xu, & Warburton, 2009), and is also a key regulator of both lung mesenchymal and epithelial cell regeneration and repair in the adult lung. TGF-β signaling pathways interact with a myriad of gene networks and pathways that regulate the communication between lung epithelial stem cells and their niche in a context-dependent fashion (Zhang, 2009). In different contexts TGF-β can maintain stem cell quiescence, can be pro- or anti-apoptotic, pro- or anti-proliferative, pro- or anti-differentiative. TGF-β can also stimulate or inhibit differentiated cell functions, or induce opposing transcriptional responses (Ruscetti, Akel, & Bartelmez, 2005). Homeostatic regulation mediated by TGF-β invariably involves the precise balancing of opposing signals in multiple cell lineages leading Shi et al. (2009) to propose the "Goldilocks hypothesis" to describe regulatory signaling by TGF-β in the orchestration of lung development, regeneration and repair: namely, that everything must be done just right!

2.2 Stem and progenitor cells of the proximal airways

Multipotent basal stem cells located in the ducts of SMG in the trachea, and in the pseudostratified epithelium of the proximal airways, are characterized by their variable expression of cell surface markers including transformation related protein 63 (p63), the keratins 5, 8, and 14 (Krt5, Krt8 and Krt14), the nerve growth factor receptor (NGRF), the epithelial cell surface glycoprotein (Trop2), and the integrin alpha chain 6 (α6).

In the trachea, Trop2^{pos}α6^{pos} basal cells regenerate SMG ducts, tubules and the mucus and serous secreting cells of SMG (Hegab, Nickerson, Ha, Darmawan, & Gomperts, 2012). Following severe injury, a normally dormant multipotent population of αSMA^{pos} myoepithelial cells located deep within SMG is activated by canonical Wnt/β-catenin signaling to adopt a basal cell signature profile to regenerate all epithelial cell lineages of SMG and the proximal airways (Lynch et al., 2018).

In proximal airways, $p63^{pos}Krt5^{pos}$ basal cells give rise to $Krt8^{pos}$ progenitor cells that in turn generate bipotent short-lived luminal secretory cell precursors of goblet, club, and ciliated cell lineages (Pardo-Saganta et al., 2015). Injury models reveal significant plasticity in the stem and progenitor cell compartments of the proximal airways with Notch signaling plays a key role in regulating basal cell lineage commitment and differentiation (Carraro & Stripp, 2015; Hogan et al., 2014). After ablation of endogenous basal cells in the proximal airways, fully differentiated secretory cells are also able to dedifferentiate and replenish the basal stem cell compartment (Tata et al., 2013). Differential expression of c-myb and Notch signaling via the Notch2 intracellular domain directs the differentiation of tracheal basal cells towards a ciliated or secretory cell lineage respectively (Pan et al., 2014; Pardo-Saganta et al., 2015)

2.3 Stem and progenitor cells of the distal airways

The distal airways of mice are devoid of basal cells, and $Scb1a1^{pos}$ club cells are the major epithelial cell type. These cells secrete club cell secretory protein (CCSP), otherwise known as uteroglobin, which has potent anti-inflammatory and anti-toxicant activity (Broeckaert, Clippe, Knoops, Hermans, & Bernard, 2000; Zhu et al., 2019). Club cells also function as a large reservoir of transit amplifying progenitor cells giving rise to secretory and ciliated cells (Rawlins et al., 2009). A sub-population of naphthalene-resistant ($CYP2F2^{neg}$) $Scgb1a1^{pos}$ club cells, termed variant club cells, juxtaposed to NEB at airway branch-points have been identified as a critical multipotent stem cell pool essential for bronchiolar epithelial regeneration following severe injury (Giangreco et al., 2009; Hong, Reynolds, Giangreco, Hurley, & Stripp, 2001).

Surfactant protein C positive (SPC^{pos}) bronchioalveolar stem cells (BASC) located at the BADJ in terminal bronchioles are also activated in response to injury to replace both bronchiolar and alveolar epithelial cells (Kim et al., 2005; Nolen-Walston et al., 2008; Salwig et al., 2019). BASC share an $EpCAM^{hi}Sca\text{-}1^{lo}CD24^{lo}$ signature profile (Zacharek et al., 2011) with multipotent $EpCAM^{hi}Sca\text{-}1^{lo}CD24^{lo}\alpha6^{pos}\beta4^{pos}$ epithelial stem/progenitor cells (EpiSPC) isolated from the distal lung that give rise to mixed colonies of multipotent and differentiated airway and alveolar epithelial cells, as well as lineage-restricted colonies of either airway or alveolar epithelial cells in organotypic assays (Chapman et al., 2011; McQualter, Yuen, Williams, & Bertoncello, 2010). Analysis of the distribution of EpCAM^{hi}CD24^{lo} colony-forming cells in Rosa26R-Confetti mice and SFTPC-GFP reporter mice, show that multipotent progenitors similar to multipotent EpiSPC are restricted to the bronchiolar airways while lineage-restricted airway- and alveolar progenitors are predominantly located in the intra-pulmonary conducting airways and the alveolar bed respectively (Chen et al., 2012).

In contrast, PNEC are a rare cell population that originate in the foregut endoderm and comprises less than 1% of adult distal airway epithelium. They are the first epithelial cell type to differentiate in the primitive airways (Cutz, 2015), and migrate to cluster at airway branch-points during development to form NEB by a process termed "slithering" (Kuo & Krasnow, 2015). NEB function as hypoxia-, chemo- and mechanotransduction-sensory cells that regulate immune responses in the adult lung by elaborating neuropeptides (Branchfield et al., 2016; Cutz, 2015). PNEC and NEB proliferate rapidly by hyperplastic growth in response to injury (Peake, Reynolds, Stripp, Stephens, & Pinkerton, 2000; Reynolds, Giangreco, Power, & Stripp, 2000).

2.4 Stem and progenitor cells of the alveolar bed

SPC^{pos}AT2 cells are the stem cells primarily responsible for the maintenance of alveolar epithelium (Barkauskas et al., 2013), although terminally differentiated AT1 cells that comprise 95% of the alveolar epithelium are also able to regenerate AT2 cells during alveolar regrowth following injury (Jain et al., 2015). However, recent lineage tracing studies analyzing alveolar regeneration and repair following H1N1 influenza virus infection have revealed alternative mechanisms and stem cell pools that can be recruited to regenerate alveolar epithelium (Kumar et al., 2011; Vaughan et al., 2015; Zacharias et al., 2018; Zuo et al., 2015).

Following severe H1N1 virus infection, rare Sox2^{pos}Scgb1a1^{neg}Krt5^{neg} peribronchiolar cells (Ray et al., 2016) give rise to Scgb1a1^{neg}SPC^{neg}p63^{pos}Krt5^{pos} progenitor cells termed distal airway stem cell (DASC) (Kumar et al., 2011; Zuo et al., 2015) or lineage negative epithelial cells (LNEP) (Vaughan et al., 2015) that migrate to the inflamed interstitial space of the injured lung to form proliferating cellular pods resembling nascent alveoli that express AT2 and AT1 differentiation markers. Deletion of p63^{pos}Krt5^{pos}

cells prior to H1N1 infection results in failure of alveolar repair (Zuo et al., 2015) showing that these progenitors are essential for the regeneration of alveolar epithelium following severe injury. A more recent study has also identified a $Krt5^{neg}Axin2^{pos}$ subset of AT2 cells termed alveolar epithelial progenitor (AEP) that do not generate $Sox2^{pos}$ or $Krt5^{pos}$ cells but act in concert with $Krt5^{pos}$ cells to regenerate alveolar epithelial cells following H1N1 infection (Zacharias et al., 2018). This latter study suggests that following severe injury, Sox2-derived $Krt5^{pos}$ progenitors rapidly repair the alveolar epithelial barrier while AEP are the major lineage responsible for robust generation of functional AT2 and AT1 cells.

2.5 Lung stem and progenitor cell plasticity

The foregoing analysis of the organization and regulation of lung epithelial stem cell hierarchies in adult mouse models using in vivo lineage tracing techniques and in vitro organoid assays has revealed the plasticity of all epithelial cell compartments along the proximal-distal lung axis (Cole et al., 2010; Hogan et al., 2014; Tata & Rajagopal, 2017).

In the steady state, regional epithelial cell lineages lining the proximal and distal conducting airways and the alveolar bed are maintained by an ordered hierarchy of lineage-restricted regional stem and progenitor cells. Following injury functionally differentiated epithelial cells including basal, club, AT2 and AT1 cells responsible for maintaining epithelial barrier function, mucocilliary clearance, and gas exchange can de-differentiate and temporarily revert to a facultative stem cell state. Or if required in the case of severe injury which compromises the ability of dedicated regional stem and progenitor cells to rapidly restore lung epithelial barrier function, rare multipotent epithelial stem cells are mobilized and recruited to restore epithelial barrier function. All are a target of context dependent field cancerization and lung cancer initiation.

2.6 Human lung stem cell homologs

While the mouse and human lung have much in common, there are significant disparities in anatomical structure, gas-exchange capacity, and cellular composition (Hyde, Hamid, & Irvin, 2009; Knust, Ochs, Gundersen, & Nyengaard, 2009; Rock, Randell, & Hogan, 2010; Weibel, 2009) (Table 3). There is approximately a 5000-fold difference in weight (0.2 gm vs 1.1 Kg), a 200-fold difference in alveolar number (2.3×10^6 vs 480×10^6), and a 15,000-fold difference in alveolar surface area ($85\,cm^2$ vs $130\,m^2$) (Knust et al., 2009; Weibel, 2009).

Table 3 Differences between mouse and human airways.

	Mouse	**Human**
Genome	Highly conserved: 99% genes in common	
Size	0.2 gm	~1.1 Kg
Airway generations	Approximately 13	Approximately 23
Alveolar bed	85 cm^2	130 m^2
Cartilage rings	Only extrapulmonary airways	Extend several airway generations
Submucosal glands	Restricted to upper trachea	Tracheobronchial region
Pseudostratified epithelium (basal cells)	Restricted to trachea	Extends to terminal bronchioles
Cuboidal epithelium	Bronchiolar airways	Restricted to respiratory bronchioles
Terminal bronchioles	End in bronchioalveolar ducts	Transition to respiratory bronchioles
Goblet cells	Rare	Abundant
Inflammation/innate immunity	Poorly reproduced in mice	

In mice, cartilage rings are restricted to the extrapulmonary airways, whereas human cartilaginous airways extend for several generations of the branching airways. In mice, terminal bronchioles end in bronchioalveolar ducts, whereas in humans, terminal bronchioles transition into respiratory bronchioles separated by alveolar outpocketings. Pseudostratified epithelium primarily comprising basal and ciliated cells is restricted to the trachea in mice but extends to the terminal bronchioles in humans. Mucin-secreting goblet cells are abundant in human airways, but relatively rare in mice. Conversely, club cells are abundant in mice but a relatively rare cell population in human conducting airways; and simple cuboidal epithelium devoid of basal cells lines the bronchiolar airways of mice, but is only found in the respiratory bronchioles in humans.

Recent studies have confirmed that basal cells are the dominant stem cell type responsible for maintenance and repair of epithelial cell lineages of the human conducting airways (Butler et al., 2016; Ghosh et al., 2013; Hegab et al., 2012; Ma et al., 2018; Rock et al., 2009). Like their murine counterparts, human club cells are able to dedifferentiate and function as facultative stem cells following insult or injury (Zuo et al., 2018), and progenitors in the distal human lung include AT2 cells as well as homologs of murine AEP (Zacharias et al., 2018) and DASC (Tanaka et al., 2018).

In the proximal airways human basal cell homologs include $NGFR^{pos}CD166^{pos}CD44^{pos}$ tracheal cells that generate $Krt^{pos}Krt5^{pos}$ basal cells and serous and mucous secretory cells (Hegab, Ha, et al., 2012); $Krt5^{pos}Krt14^{pos}$ tissue factor (TF^{pos}) $CD151^{neg}$ tracheobronchial cells that generate $Krt5^{pos}Krt14^{neg}p63^{pos}$ clones (Ghosh et al., 2013); $NGFR^{pos}\alpha6^{pos}$ cells that gives rise to tracheospheres comprising peripheral $Krt14^{pos}p63^{pos}$ basal cells and luminal $Krt8^{pos}$ non-ciliated and ciliated cells (Rock et al., 2009); and $Krt5^{pos}NGFR^{pos}\alpha6^{pos}Trop2^{pos}$ cells that generate tracheospheres comprising ciliated cells (Butler et al., 2016).

Single cell RNA sequencing and flow cytometric analysis has revealed significant heterogeneity in the human basal cell compartment suggesting the existence of a hierarchy of self-renewing multipotent stem cells and transit amplifying cells with more limited self-renewal capacity (Carraro et al., 2020; King et al., 2020). These include a $p63^{pos}Krt5^{pos}Krt14^{pos}$ multipotent basal cell cluster with a strong Wnt signaling signature profile; a proliferating basal cell cluster expressing serpins and transcripts associated with protein folding and mucin production; basal cells transitioning to a secretory phenotype; basal cells expressing high level of AP-1 family members, JUN and FOS; and a basal cell cluster uniquely expressing β-catenin.

In the distal lung, organoid assays have been used to characterize an $EpCAM^{pos}TM4SF1^{pos}HT\text{-}280^{pos}$ human homolog of murine AEP that comprises approximately 29% of the AT2 cell compartment and generates large clones of AT1 and AT2 cells (Zacharias et al., 2018). Yet another study has recently identified a human homolog of DASC that expresses TTF-1 and a truncated isoform of p63 ($\Delta Np63^{pos}TTF\text{-}1^{pos}$) that generate bronchiolar or alveolar cell organoids in culture (Tanaka et al., 2018).

3. Lung cancer stem cells

Lung cancer is the consequence of cumulative genetic mutations and epigenetic changes in lung stem and progenitor cells commonly associated with smoking or repeated exposure to carcinogenic chemicals and toxicants. These mutations typically involve DNA base substitutions, insertions or deletions, or gene truncations, rearrangements, altered copy numbers, and

DNA repair errors in multiple driver and passenger genes (Pon & Marra, 2015; Stratton, Campbell, & Futreal, 2009; Yokota & Kohno, 2004). The context-dependent acquisition of the hallmarks of cancer by transformed lung stem and progenitor cells along the proximal-distal lung axis then drive the generation of different lung tumor types and the development of inter- and intra-tumor heterogeneity (de Bruin et al., 2014; Zhang et al., 2014).

3.1 Non-small cell lung cancer

Eighty percent of lung cancers are classified NSCLC and comprise a heterogeneous set of diseases. Driver mutations include KRAS, EGFR, ALK, ROS1, LKB1, BRAF, PIK3CA, and RET (Chen, Fillmore, Hammerman, Kim, & Wong, 2014). The major driver mutations KRAS, EGFR and ALK, are mutually exclusive, and co-mutations involve a large number of diverse gene partners. Each NSCLC subtype (ADC, SCC, and LCC), and NSCLC detected in smokers compared to never-smokers, is characterized by significant differences in the repertoire of driver mutations and epigenetic changes (Pietanza & Ladanyi, 2012) (Fig. 1). KRAS is the most frequent

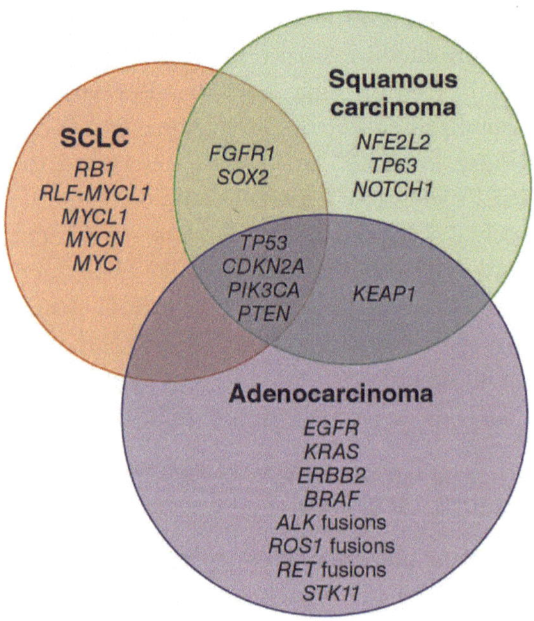

Fig. 1 Frequently shared or restricted cancer-related genes in three lung cancer subtypes. *Figure reproduced from Pietanza, M. C., & Ladanyi, M. (2012). Bringing the genomic landscape of small-cell lung cancer into focus.* Nature Genetics, *44(10), 1074-1075. https://doi.org/10.1038/ng.2415 with permission.*

driver mutation in smoking associated NSCLC, and is frequently associated with co-mutation of other tumor suppressor genes (Ferrer et al., 2018). EGFR mutations are associated with 60% of NSCLC and are more prevalent in never-smokers.

ADC is the most common subtype accounting for approximately 40–50% of NSCLC. ADC are a heterogeneous group of cancers of the distal lung which can be subtyped as alveolar, bronchiolar, mixed or undifferentiated (Table 1) on the basis of the expression of lung epithelial immunohistochemical markers (Tabbo et al., 2018). ADC harboring EGFR mutations have a bronchioalveolar carcinoma phenotype whereas mucinous ADC are more commonly associated with KRAS mutations. Evidence points to AT2 cells as the cell of origin of ADC (Desai, Brownfield, & Krasnow, 2014; Xu et al., 2012) but AT1 cells (Jain et al., 2015), BASC (Kim et al., 2005) and club cells (Spella et al., 2019) have also been implicated as cancer initiating cells. Single cell transcriptional profiling of human primary and metastatic lung ADC has recently revealed the emergence of AFP- and DASC-like cancer stem and progenitor cells that have reactivated transcriptional programs typically associated with lung development or severe injury (Laughney et al., 2020). Further analysis revealed evidence of stem cell plasticity, lineage promiscuity and reduced lineage differentiation in the course of tumor progression and metastasis.

SCC is a cancer of basal cell origin (Ferone et al., 2016). SCC accounts for approximately 20–30% of NSCLC (Bozinovski et al., 2016). The incidence of SCC is strongly associated with exposure to tobacco smoke, and COPD is a common comorbidity. Chronic irritation of the airways leads to progression from metaplasia to dysplasia, to carcinoma in situ, somatic mutations, oxidative stress, field cancerization and cancer where SCC is correlated with the severity of airway obstruction. Dominant driver mutations such as EGFR, ALK, ROS1 and KRAS relatively infrequent in SCC compared to ADC.

LCC is a poorly differentiated NSCLC subtype that accounts for less than 5% of NSCLC (Wang et al., 2019). KRAS activation and changes in the level of PDL1 expression are the most common oncogenic alterations, but EGFR mutations and ALK or ROS1 rearrangements have not been observed (Wang et al., 2019). The expression of AT2- or club cell-associated markers by several variants of LCC point to a bronchioalveolar stem or progenitor cell as the cancer initiating cell. The most recent WHO Classification (Travis et al., 2015) has reclassified LCC tumors as variants of ADC or non-keratinizing SCC.

3.2 Small cell lung cancer

SCLC is a cancer of neuroendocrine cell origin that is invariably associated with heavy smoking (Cardona et al., 2019; Sutherland et al., 2011). Inactivating mutations of tumor suppressor genes, TP53 and RB1, are considered the most important driver mutations in SCLC (Cardona et al., 2019; Pietanza & Ladanyi, 2012). SCLC tumors are also characterized by a large number of additional mutations (Ireland et al., 2020; Rudin et al., 2019) including CDKN2A, and CEBPA, and amplification of FGFR1, Sox2, and Myc family members in heavy smokers, and EGFR, MET, and SMAD4 mutations in never-smokers.

4. Deconstructing the lung cancer stem cell niche

The lung stem cell niche is a highly complex and malleable microenvironment (Schiller et al., 2015). Lung epithelial cell lineages maintain barrier function and ward off airborne pathogens by elaborating chemokines and immunomodulatory cytokines that orchestrate innate and adaptive immune responses. Epithelial cell derived factors regulate mesenchymal and endothelial cell proliferation and differentiation, while autocrine, paracrine and juxtacrine factors elaborated by mesenchymal stromal cells and endothelial cells regulate lung epithelial cell proliferation and differentiation (McQualter & Bertoncello, 2012; Shi et al., 2009). Cytokines, chemokines and inflammatory mediators elaborated by mesenchymal cells and hematopoietic cells during injury orchestrate immune responses which directly and indirectly influence epithelial progenitors, ECM deposition and remodeling (Burgess & Weckmann, 2012; Whitsett & Alenghat, 2015). In addition, the amplitude and periodicity of cyclic stretch (Waters, Roan, & Navajas, 2012), and circadian rhythms in lung inflammatory responses (Sundar, Yao, Sellix, & Rahman, 2015) also affect stem cell fate choices. Inside-out and outside-in integrin signaling regulates epithelial cell proliferation and differentiation as well as cell motility.

4.1 Field of injury and field cancerization

Host responses to carcinogenic insult, and comorbidities such as COPD and ILDs conspire to create fields of injury in the conducting airways and distal lung that generate dysregulated stem cell niche microenvironments that facilitate lung cancer initiation, progression and metastasis (Steiling et al., 2008). Dominant driver mutations that define malignant lung cancer subtypes (e.g. KRAS, EGFR, ALK, TP53) and their associated partner mutations are closely coupled to tumor-promoting signaling pathways, and are often shared by cells that comprise the field of injury and the cancerized field at the periphery of the developing tumor (Curtius, Wright, & Graham, 2018). These mutations endow precancerous cells and cancer initiating cells with a competitive advantage, promoting the sequential development of irreversible pre-neoplastic changes leading to malignancy.

Inter- and intra-lung tumor heterogeneity, and the different patterns of tumor progression and metastasis characteristic of different lung cancer types and subtypes is the consequence of unique interactions between lung cancer stem cells and the different cellular and structural elements of their niches.

4.2 Epithelial cells in the niche

The airway epithelium is the primary defense barrier against airborne pathogens, toxins and pollutants. Epithelial cell lineages are the source of mucins and surfactants synthesized and secreted by submucosal glands (Muc5B), club cells (Muc5AC) and goblet and alveolar cells (Muc1, SPC) along the proximal-distal axis of the lung (Whitsett & Alenghat, 2015). Lung epithelial cells also elaborate a long list of anti-microbial peptides, cytokines, chemokines, proteases and inflammatory mediators including IL-1β, IL-6, IL-13, GM-CSF, and G-CSF (Glaser et al., 2019; Parker & Prince, 2011). Toll-like receptors (TLR) expressed by lung epithelial cells play a critical role in mounting adaptive and innate immune responses to pathogens.

Especially in smokers, lung cancer is typically preceded by chronic injury to lung epithelium leading to the sequential development of epithelial hyperplasia, metaplasia, and mucus hypersecretion that drives inflammation, fibrotic scarring, airflow obstruction, and dysplasia. Recent analysis of clonogenic epithelial cells from COPD patients suggests that these preneoplastic changes are the result of clonal expansion and proliferation of metaplastic E-cadherin^{pos}p63^{pos}Krt5^{pos} variant distal epithelial progenitor cells of small airways and terminal bronchioles (Rao et al., 2020). These include a goblet cell progenitor giving rise to p63^{pos} basal cells and goblet cells co-expressing Scgb1a1, Muc5AC and Muc5b, and squamous cell progenitors giving rise to Krt10^{pos}involucrin (IVL^{pos}) cells. These variant progenitors express a broad array of interleukins, chemokines and inflammatory mediators and profibrotic genes that directly and indirectly influence cancer progression by interacting with other niche components.

Dominant driver mutations and co-mutations in lung cancer stem cells and their progeny are also closely coupled with tumor-promoting inflammatory signaling pathways and elicit unique immune responses (Busch et al., 2016). For example, in KRAS-driven ADC, dysregulated STAT3/NF-κ B signaling enhances the secretion of IL-6 by tumorigenic cells, while KRAS-TP53 tumors generate a T-cell inflamed microenvironment, and KRAS-STK11 tumors secrete pro-inflammatory cytokines (Kitajima, Thummalapalli, & Barbie, 2016). Recruitment of neutrophils as a result of these reactions then promotes ECM remodeling.

Dominant driver mutations in lung cancer stem cells also interact with gene networks to regulate stem cell fate decisions and complex interactions between stem cells and their niche. For example, RAS family members control gene networks that regulate cell survival, proliferation, differentiation, and metabolic reprogramming. In the lung, KRAS activation promotes secretory and ciliated cell differentiation at the expense of squamous cell differentiation (Ogawa et al., 2019).

EGFR is a member of the HER/*erb*B family of receptor tyrosine kinases. EGFR signaling mediated by the RAS-RAF-MEK-MAPK, PI3K-PTEN-AKT, and the STAT signaling pathways drives tumorigenesis by transducing signals from the tumor microenvironment to stimulate proliferation, regulate mitochondrial activity, block apoptosis, promote angiogenesis, alter cell adhesiveness and promote metastasis (Sigismund, Avanzato, & Lanzetti, 2018).

ALK encodes a tyrosine kinase transmembrane protein that binds pleiotrophin and midkine to activate a signaling cascade via canonical pathways including MAPK, PI3K/mTOR, JAK-STAT, and SHH. The consequence of ALK activation and signaling in lung cancer is determined by its fusion partners (e.g. EML4-ALK, TGF-ALK, KIF5B-ALK) regulating diverse cell functions including cell survival, apoptosis, proliferation, migration, metabolism, and cytoskeletal remodeling (Hallberg & Palmer, 2016).

4.3 Mesenchymal-epithelial cell crosstalk

Resident lung mesenchymal cells are a phenotypically and functionally diverse population of fibroblastoid cells that play a multifaceted role in the lung stem cell niche (Riccetti, Gokey, Aronow, & Perl, 2020). Different subtypes distributed throughout the conducting airways of the proximal and distal lung and the alveolar bed are characterized on the basis of their differential expression of non-specific cell lineage markers such as

PDGFRα, PDGFRβ, αSMA, FAP, vimentin, and FSP1, as well as their histomorphology (e.g. myofibroblasts, matrix fibroblasts, lipofibroblasts), their spatial location, and their matrisome and secretome (Gieniec, Butler, Worthley, & Woods, 2019; Zepp et al., 2017).

Different mesenchymal cell subtypes maintain the integrity of the lung biomatrix scaffold, synthesizing and secreting the Type I, III, and V collagens and fibronectin of the ECM, and collagen IV and laminin of basement membranes. Other subtypes are a source of proteases, inflammatory mediators, angiogenic factors, chemokines and immune modulators. Mesenchymal cells are also a source of the many cytokines including EGF, FGF, HGF, IGF, TGF-β and WNTs that maintain and regulate lung epithelial stem and progenitor cells and orchestrate lung regeneration and repair.

Many studies have documented the key role of resident lung mesenchymal cells in maintaining lung basal, club, and alveolar stem cell compartments, and regulating their proliferation and differentiation (Yuan, Volckaert, Chanda, Thannickal, & De Langhe, 2018). These studies show that the specificity of airway and alveolar epithelial stem cell niches is context dependent (Watson & Francavilla, 2018), and conferred by different cohorts of mesenchymal cells distributed along the proximal-distal axis (Lee et al., 2017; Riccetti, Gokey, Aronow, & Perl, 2020; Xie et al., 2018; Zepp et al., 2017). For example, $Lgr6^{pos}\alpha SMA^{pos}$ myofibroblastic cells support the growth of bronchiolar epithelial stem cells, whereas $Lgr5^{pos}$ mesenchymal cells (Lee et al., 2017) or $Pdgfr\alpha^{pos}$ lipofibroblastic cells (Barkauskas et al., 2013; Zepp et al., 2017) support the growth of alveolar progenitor cells. Conversely, Wnt signaling by bronchiolar and alveolar epithelial cells stimulate their respective mesenchymal cell partners (Lee et al., 2017), and activation of TGF-β signaling by mesenchymal cells suppresses epithelial stem and progenitor cell proliferation (McQualter et al., 2013; Ng-Blichfeldt et al., 2019).

Mesenchymal cells and juxtaposed bronchiolar and alveolar epithelial stem cells engage in reciprocal signaling by the secretion of paracrine signaling molecules, the establishment of gap-junction communication (Badri et al., 2011), or by exosome transfer of nucleic acids, miRNA, proteins, lipids and metabolites (Kalluri & LeBleu, 2020). Although many factors elaborated by mesenchymal cells co-operate to regulate lung epithelial stem and progenitor cells, the weight of evidence points to a central role for FGF-10, Notch, canonical and non-canonical Wnt signaling, and TGF-β (Li, Bellusci, et al., 2015; Yuan et al., 2018; Zepp et al., 2017).

FGF-10 is a member of the FGF-7 subfamily of fibroblast growth factors that are exclusively expressed by mesenchyme, and signal primarily via the FGFR2b receptor, and to a lesser extent via FGFR1b receptor (Zinkle & Mohammadi, 2019). During embryonic development in mice, FGF-10 expressed by the lung mesenchyme adjacent to developing lung buds is a critical regulator of airway branching and lung epithelial proliferation and differentiation (Bellusci, Grindley, Emoto, Itoh, & Hogan, 1997; Yuan et al., 2018). Deletion of the FGF-10 or FGFR2b receptor genes results in lung agenesis. In the adult lung, basal stem cells are maintained by FGF10-FGFR2b receptor signaling, and generate their own FGF10-expressing niche mediated by the Yap-Wnt7b signaling pathway (Yuan et al., 2018). FGF-10 expressing lipofibroblastic cells in the alveolar bed regulate AT2 progenitor cells (Yuan et al., 2019). This study also showed that overexpression of FGF-10 following bleomycin-induced lung epithelial cell ablation promoted the dedifferentiation of club cells and their generation of AT2 and AT1 cells.

The role of FGF-10-FGFR2b signaling in human lung development and lung disease is less clear (reviewed by Danopoulos, Shiosaki, & Al Alam, 2019; El Agha, Seeger, & Bellusci, 2017). Unlike the developing mouse lung, where FGF-10 expression is concentrated in the distal mesenchyme and regulates branching morphogenesis, in the developing human lung FGF-10 is diffusely expressed throughout the mesenchyme (Danopoulos et al., 2019). Nor does FGF-10 induce the branching of human fetal lung explants, consistent with a different role for FGF-10 in human lung development. However, although a causal role has not been established, single nucleotide polymorphisms (SNPs) of FGF-10 have been associated with congenital human airway branching defects. FGF-10 haploinsufficiency has been associated with an increased risk of developing COPD. Co-localization of FGF7 and FGFR2 correlates with poor prognosis in ADC, but not SCC, and amplification of the FGFR1 gene also correlates with poor prognosis in NSCLC.

The critical role of Notch signaling in regulating the proximodistal specification of lung epithelial cell lineages during development and in the adult lung was briefly discussed earlier in this chapter, and reviewed by others (Carraro & Stripp, 2015; Hussain et al., 2017). Notch signaling also plays a key role the regulation of lung myofibroblast differentiation, the specification of lung vascular endothelium, and in epithelial-mesenchymal transition (Hu & Phan, 2016; Wieland et al., 2017). Dysregulated Notch signaling is a significant factor in lung cancer initiation, tumor progression and metastasis in both NSCLC (Yuan et al., 2014) and SCLC (Leonetti et al., 2019). Context-dependent Notch ligand-receptor interactions within the lung cancer stem cell niche can be tumor suppressive or pro-tumorigenic (Lim et al., 2017) depending on the cellular composition of the niche and the driver mutations expressed by cancer stem cells, resulting in inter- and intra-tumor heterogeneity.

Canonical and non-canonical Wnt signaling pathways play a critical role in the regulation of lung development, the maintenance of lung homeostasis, and in lung pathogenesis (Baarsma & Konigshoff, 2017; Konigshoff & Eickelberg, 2010; Li, Bellusci, et al., 2015). Specific Wnt ligands (e.g. Wnt3a) activate canonical Wnt/β-catenin signaling while others (e.g. Wnt 5a) activate non-canonical Wnt/β-catenin independent signaling pathways. In the distal lung, both mesenchymal cells and AT2 stem cells are characterized by the expression of the Axin-related protein, Axin2, which regulates the stability of β-catenin. Spatial mapping has shown that $Axin2^{bright}Pdgfr\alpha^{pos}$ mesenchymal alveolar niche (MANC) cells closely juxtaposed to AT2 cells are a critical component of the alveolar niche promoting AT2 self-renewal and differentiation into AT1 cells by paracrine secretion of IL-6 and FGF7 (Zepp et al., 2017). A recent study has proposed that the alveolar stem cell niche comprises a single Wnt-expressing mesenchymal cell attached to a single $Axin2^{pos}$AT2 stem cell (Nabhan, Brownfield, Harbury, Krasnow, & Desai, 2018). According to this model, activation of canonical Wnt/β-catenin signaling in $Axin2^{pos}$ AT2 stem cells by mesenchymal cell derived Wnt5a maintains the stemness of tethered $Axin2^{pos}$ AT2 stem cells while daughter cells leaving the niche progressively differentiate into AT1 cells and lose their regenerative potential. Autocrine mesenchymal Wnt5a signaling maintains the alveolar niche, while activation of Wnt signaling in $Axin2^{dim}Pdgfr\beta^{pos}Acta2^{pos}$ alveolar myofibrogenic cells promotes fibrotic remodeling of the alveolar niche following injury.

Using a different experimental design analyzing the proliferation and differentiation of AT2 stem cells in precision cut lung slices and in organoid culture, others have shown that Wnt5a and Wnt5b repress AT2 cell proliferation by activation of non-canonical Wnt/β-catenin independent signaling (Wu et al., 2019). These different outcomes may be explained by the existence of sub-populations of $Axin2^{pos}$AT2 cells with differing responsiveness to Wnt ligands (Zepp et al., 2017), and the expression of multiple Wnt ligands by mesenchymal cells in the alveolar niche (Nabhan et al., 2018). Consequently, stimulation or suppression of AT2 cell proliferation in different contexts will be determined by the opposing effects of activation of multiple canonical and non-canonical Wnt signaling cascades acting in synergy with other cytokines elaborated in the niche.

This is particularly relevant in understanding the regulation of cancer stem cells in NSCLC. Activation of Wnt signaling promotes tumor progression in KRAS and BRAF mutant lung tumors in mice (Pacheco-Pinedo et al., 2011), and elevated Wnt gene expression is correlated with poor survival in human lung ADC (Tammela et al., 2017). The latter study showed that multiple Wnt ligands and their receptors are robustly expressed in ADC, and that paracrine signaling cells form a niche that drives cancer cell proliferation and tumor progression.

4.4 Cancer associated fibroblasts and TGF-β

Cancer associated fibroblasts (CAF) are a permanently activated multi-functional stromal cell component of the cancer stem cell niche that plays a major role in tumor progression and metastasis by remodeling the ECM, promoting angiogenesis, interacting with immune cells, secreting tumor-promoting chemokines, cytokines and inflammatory mediators, and proteins that control lung cancer cell survival and proliferation (Chan, Shaked, & Tsai, 2019; Ishii, Ochiai, & Neri, 2016).

The origin of CAF is uncertain. They are thought to be derived from lung-resident mesenchymal cells, or stromal fibroblasts, or bone marrow-derived cells recruited to the tumor microenvironment, or by trans-differentiation of lung epithelial cells undergoing EMT (Yoshida, 2020). Various mechanisms have been suggested, including activation by tumor cell secreted factors and inflammatory cytokines, or reprogramming by miRNAs delivered by exosomes from adjacent cancer cells (Gascard & Tlsty, 2016; Sahai et al., 2020). TGF-β is considered a major orchestrator of the cancer stem cell niche (Erdogan & Webb, 2017; Morrison, Parvani, & Schiemann, 2013).

Epithelial cells, macrophages, and the ECM are the major sources of latent TGF-β in the lung (Saito, Horie, & Nagase, 2018; Shi et al., 2009). Once activated by proteolysis, thrombospondin-1, low pH, reactive oxygen species, or integrins $\alpha_V\beta6$ and $\alpha_V\beta8$, TGF-β acts alone or in synergy with other factors to regulate epithelial-mesenchymal interactions, signaling via the canonical TGF-β/Smad pathway, or other signaling cascades co-opting a diverse array of kinases including p38, JNK, MAPK, ERK1, ERK2, or mTOR.

CAF are a source of factors and exosomes that promote tumor growth, a secretome that drives angiogenesis, conditions the microenvironment at metastatic sites, and generate chemokines, immunosuppressive of T cell responses. Sustained release of TGF-β by lung cancer cells drives the secretion of cytokines, pro-inflammatory mediators, chemokines and proteases that further drive tumor progression and metastasis. Paradoxically, TGF-β acts both as a tumor suppressor, inducing cell cycle arrest, promoting apoptosis (Calon, Tauriello, & Batlle, 2014) and downregulating the secretion of cytokines such as FGF-10, FGF-7 and HGF and skewing Wnt/β-catenin signaling (McQualter et al., 2013; Ng-Blichfeldt et al., 2019); and also as a tumor promoter driving EMT and stimulating metastasis (Morrison et al., 2013; Neel, Humbert, & Lebrun, 2012).

As a tumor promoter, TGF-β downregulates E-Cadherin expression in epithelial cancer cells and induces the expression of mesenchymal cell surface proteins such as N-Cadherin, vimentin, fibronectin, and tenascin-C that enable lung cancer cells to migrate and invade the microvasculature. TGF-β stimulates angiogenesis by inducing secretion of VEGF; and potentiates the chemoattraction of cancer cells to distant organs; inhibits expression of angiopoetin-1 to increase blood vessel permeability; and stimulates the secretion of immunomodulatory factors that allow cancer cells to escape immunosurveillance.

While EMT programs drive epithelial cancer stem cells to disseminate and metastasize to distant sites, MET programs drive those metastasized cells to revert from a mesenchymal to epithelial phenotype prerequisite to reinitiating programs that drive the formation of secondary tumors (Karacosta et al., 2019; Morrison et al., 2013). MET involves the withdrawal of TGF-β signaling, but the participation of other members of the TGF-β superfamily.

4.5 Microvascular endothelial cells

The primary role of the lung microvasculature is to provide an efficient gas-exchange mechanism for oxygenation of the blood (Wagner, 2015). However, the specialized endothelial cells in different compartments of the lung also play an important role in the lung epithelial stem cell niche, synthesizing and secreting factors that regulate lung epithelial stem cell proliferation and differentiation. A number of studies have identified significant organotypic lung endothelial cell factors and molecular pathways that are activated to stimulate bronchiolar and alveolar epithelial stem cell regeneration and restore homeostasis following injury. Lee et al. (2014) showed that co-cultured lung endothelial cells supported the clonogenic proliferation and differentiation of BASC in a 3D organoid culture assay (Lee et al., 2014), and that endothelial-BASC crosstalk involved the triggering of nuclear factor of activated T cell c1 (NFATc1) dependent activation of thrombospondin-1 (TSP1). Employing a mouse pneumonectomy model and 3D co-culture assay, Ding et al. (2011) showed that partial pneumonectomy stimulated the production of MMP14 by lung endothelial cells. This leads to cleavage of EGF-like fragments that stimulate the proliferation of alveolar stem cells to repair the alveolar epithelium (Ding et al., 2011). Repair of the alveolar bed requires reciprocal signaling between endothelial and alveolar epithelial cells. Niethamer et al., have recently identified a unique subpopulation of lung endothelial progenitor cells termed $Car4^{high}$ endothelial cells that are juxtaposed and potentially primed to interact with alveolar progenitor cells (Niethamer et al., 2020).

Lung endothelial cells also elaborate ECM proteins, autocrine, paracrine and angiocrine factors that mediate the crosstalk between endothelium, myofibroblasts, immunomodulatory cells and epithelial cells to maintain the integrity of the lung stem cell niche. These include notch signaling factors, VEGF, FGF and chemokines. Consequently, the microvasculature plays an important role in orchestrating the interaction between lung cancer stem cells and the niche in cancer progression, and especially in metastasis.

Lung capillaries also constitutively express adhesion molecules that selectively retard different types of hematopoietic cells slowing their transit time through the microvasculature, and regulating their trafficking and activation as they transit the pulmonary circulation (Aird, 2007; Jambusaria et al., 2020).

4.6 Innate and adaptive immune cells

Injury or damage to the epithelium of the conducting airways and the alveolar bed activates an immune response mediated by the release of cytokines, chemokines, mucins and surfactants that mobilize and activate cells of the innate and adaptive immune system (Beamer & Shepherd, 2013; Moldoveanu et al., 2009). The ensuing cascade of growth factors, pro-inflammatory and anti-inflammatory cytokines, chemokines, and proteolytic enzymes released by these cells then act in concert on all components of the niche to affect lung regeneration and repair. Persistent activation of innate and adaptive immune responses, or incomplete resolution of acute inflammation results in dysregulated cell proliferation and degradation and remodeling of the ECM. The way in which the interaction between tumor cells and the immune system impinges on the niche has a profound effect on the regulation of lung cancer stem cells, lung cancer progression and metastasis (Fig. 2).

Innate lymphoid cells (ILC) that provide a first line of defense against pathogens in the lung are early mediators of inflammation following impairment of epithelial barrier function. ILC are highly enriched for genes and factors that regulate inflammation, lung tissue regeneration and repair, and the maintenance of homeostasis (Globinska & Kowalski, 2017). When activated, ILC secrete a range of cytokines including INFγ and TNFα (Group 1); amphiregulin, GM-CSF, IL-3, IL-4, IL-5, IL-8, IL-9, and IL-13 (Group 2); IL-2, IL-17 and IL-22 (Group 3)

Lung endothelium is characterized by the upregulated expression of genes related to immune function including leukocyte cell-cell adhesion, T cell activation, and leukocyte migration (Jambusaria et al., 2020). The lung microvasculature is a major site of neutrophil sequestration, with the reservoir of marginated neutrophils comprising approximately 40% of the circulating neutrophil pool. As a major source of proteases, oxidases, VEGF and matrix metalloproteinases (MMP9), this marginated neutrophil population plays a critical role in orchestrating pro-and anti-inflammatory-activity in the lung, releasing and activating ECM-bound factors, remodeling the ECM and degrading elastin (Bozinovski et al., 2016; Hartl et al., 2018; Treffers, Hiemstra, Kuijpers, van den Berg, & Matlung, 2016). Neutrophil elastase promotes the detachment and metastasis of tumor cells by degrading E-cadherin, while neutrophil proteases promote angiogenesis by cleaving the heparin-binding domain of ECM-bound VEGF (Treffers et al., 2016). A high neutrophil to lymphocyte ratio is a poor prognostic index for both NSCLC and SCLC (Yin et al., 2015).

In addition to their role as an effector cell of the innate immune system, macrophages are master regulators of organ development, tissue patterning and remodeling, and regeneration and repair (Stefater, Ren, Lang, & Duffield, 2011). Their ability to synthesize and secrete almost every effector molecule involved in these processes, including Interleukins, colony-stimulating factors, FGFs, EGF, PDGFs, VEGF, TGF-β, Wnt ligands, chemokines, angiogenic factors, and proteases (Arango Duque & Descoteaux, 2014) enables them to play a unique role in maintaining and regulating lung epithelial stem cell proliferation and differentiation.

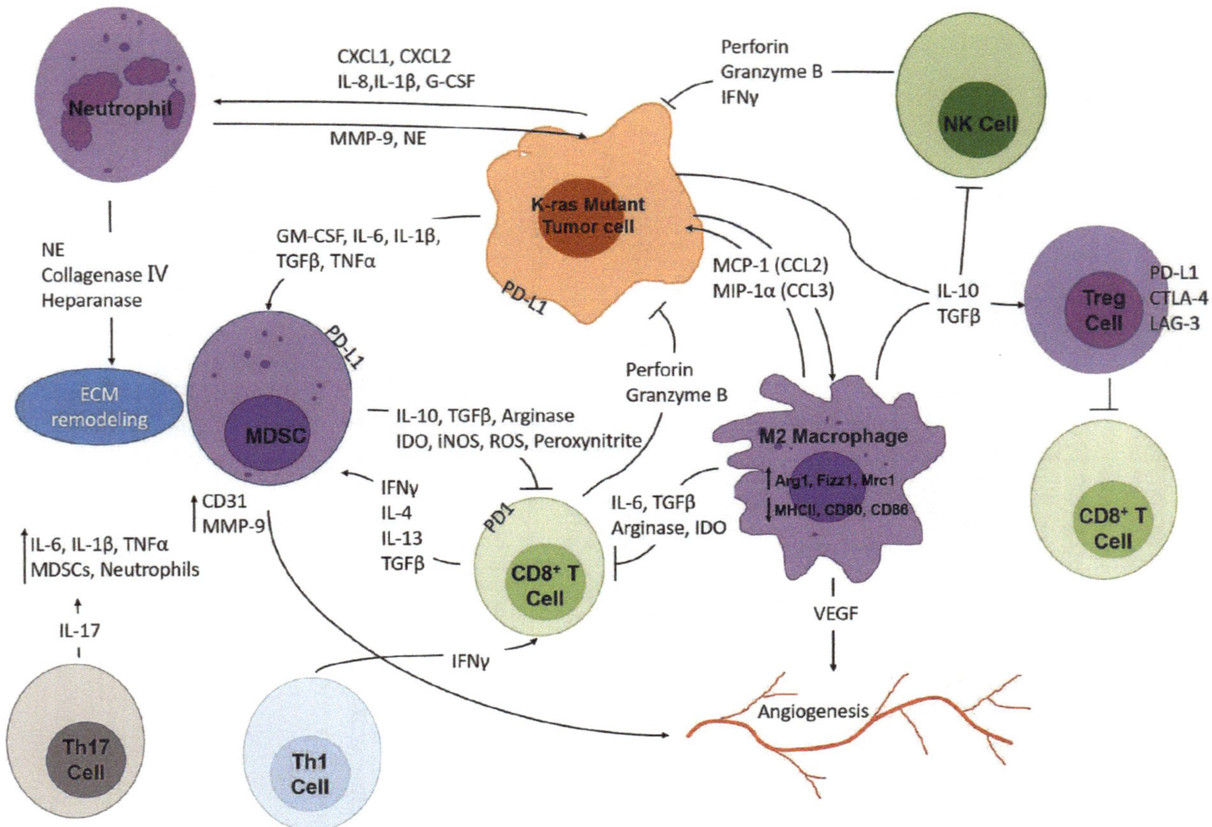

Fig. 2 The interaction between lung cancer cells, innate and adaptive immune cells, and myeloid derived suppressor cells (MDSC), induces the secretion of chemokines, cytokines and soluble factors that promote ECM remodeling, angiogenesis, to generate an immunosuppressive, pro-tumorigenic microenvironment that supports tumor progression and metastasis. *Reproduced from Deng, S., Clowers, M. J., Velasco, W. V., Ramos-Castaneda, M., & Moghaddam, S. J. (2019). Understanding the complexity of the tumor microenvironment in K-ras mutant lung cancer: Finding an alternative path to prevention and treatment.* Frontiers in Oncology, 9, *1556. https://doi.org/10.3389/fonc.2019.01556 with permission.*

Classically activated (M1) macrophages mount an inflammatory response to tissue perturbation or injury, secreting high levels of pro-inflammatory mediators, whereas alternatively activated (M2) macrophages act to attenuate inflammation by secreting anti-inflammatory mediators that stimulate tissue regeneration. Imbalance in the M1/M2 ratio results in dysregulated tissue regeneration and remodeling.

Tumor-associated macrophages (TAM) are a key cellular component of the lung cancer stem cell niche (Laviron & Boissonnas, 2019; Mantovani & Sica, 2010), and high infiltration of TAM is a poor prognostic index in NSCLC (Sumitomo et al., 2019). The consensus view is that TAM originate from circulating bone marrow derived monocytes recruited to the tumor microenvironment and are reprogrammed by chemokines and cytokines elaborated by tumor cells including CCL2, CSF1, TNFα, IL-1, IL-10, VEGF and TGF-β (Laviron & Boissonnas, 2019). However in the lung, evidence also supports derivation of TAM from resident yolk-sac derived pro-inflammatory (M2) macrophages. Together with CAF, they are considered the major orchestrators of interactions within the niche promoting the acquisition and persistence of the hallmarks of cancer (Mantovani & Sica, 2010).

TAM influence virtually all stages of lung cancer progression, dissemination and metastasis. Cytokines elaborated by TAM promote cancer cell proliferation, tumor angiogenesis, epithelial-mesenchymal transition (EMT), and the recruitment and activation of immunomodulatory cells (Biswas & Mantovani, 2010; Sarode, Schaefer, Grimminger, Seeger, & Savai, 2020). TAM have justifiably been described as "the incessant builders and destroyers of the cancer stroma" (Liguori, Solinas, Germano, Mantovani, & Allavena, 2011). Their synthesis and secretion of proteolytic enzymes results in initiation of a vicious cycle of ECM degradation and remodeling that cleaves and activates a plethora of ECM-bound bioactive peptides, factors and cytokines that amplify the crosstalk between TAM, CAF, the innate and adaptive immune system and tumor cells.

4.7 Immunoediting and the evolution of tumor heterogeneity

Spatial and temporal heterogeneity in the pattern of evolution and growth of primary and metastatic lung cancer clones and sub-clones is a fundamental feature of tumor progression driven by the interaction between cancer stem cells and their niche (de Bruin et al., 2014; Joshi et al., 2019). This is the consequence of many factors, but immunosurveillance and immunoediting play a major role (Kunimasa & Goto, 2020).

In lung cancer, different tumor types characterized by different driver and passenger mutations express different repertoires of neoantigens and elicit distinct immune responses that affect tumor progression (Joshi et al., 2019; Lavin et al., 2017; Rosenthal et al., 2019). As a consequence individual tumor clones can exhibit dramatic differences in immune cell infiltration, immune cell clustering, and the ratio of immune cells to tumor cells. For example, mouse lung ADC expressing KRAS and EGFR driver mutations are characterized by differential expansion of B cells, $CD8^{pos}T$ cells, Tregs, and IL-17A-producing T cells (Busch et al., 2016). The immune profiling of early human ADC lesions also reveals a reduced $CD8^{pos}T$ effector/Treg ratio, impaired NK cell function, and the presence of immunosuppressive, pro-tumorigenic macrophages (Lavin et al., 2017).

The process of immunoediting comprises three phases (elimination, equilibrium, and escape) involving crosstalk between cancer stem cells and the immune system (Bhatia & Kumar, 2011; Kunimasa & Goto, 2020). During the elimination phase, innate and adaptive immune cells recruited to the developing tumor by the release of inflammatory cytokines and chemokines, co-operate to activate cytotoxic lymphocytes and eliminate cancer cells expressing immunogenic neoantigens. During the equilibrium phase the immune system continues to apply selective pressure on tumor development and growth. Surviving cancer cells enter dormancy, while variants participate niche remodeling triggering the release of INF-γ and interleukins by NK cells and macrophages which activate anti-proliferative, pro-apoptotic and angiostatic mechanisms that keep their proliferation and expansion in check (Bhatia & Kumar, 2011). During the escape phase, genomic instability and epigenetic mechanisms conspire with microenvironmental factors to favor the uncontrolled proliferation and metastasis of variant clones able to evade immunosurveillance (Joshi et al., 2019; Kunimasa & Goto, 2020; Rosenthal et al., 2019; Yuan, 2016).

The evolutionary trajectory, spatial heterogeneity, and pattern of clonal diversification in human lung ADC has been described in a number of multiregional, longitudinal RNA-sequencing of the TRACERx (**TRA**cking non-small cell lung **C**ancer **E**volution through therapy [**Rx**]) cohort (Jamal-Hanjani et al., 2014; Jamal-Hanjani et al., 2017; Joshi et al., 2019; Rosenthal et al., 2019), and confirmed by others (de Bruin et al., 2014; Zhang et al., 2019). Briefly, these studies show that early canonical driver mutations (e.g. EGFR, MET, BRAF) tended to be clonally dominant and ubiquitously expressed across multiple sites of disease. Clonal diversification is characterized by the expression of driver mutations that are dominant in some regions and absent in others, while mutations of tumor suppressor genes, and genes regulating functions such as DNA repair, chromatin remodeling and histone methylation emerge late in cancer progression. Recent integration of exome and RNA sequencing data with spatial histological mapping of the distribution pattern of cancer cells, lymphocytes and stromal cells confirms the role of the niche in constraining tumor growth, and in driving genetic diversification during the immunoediting process (AbdulJabbar et al., 2020).

4.8 The lung microbiome

The lung microbiome has largely been ignored as a critical component of the lung stem cell niche. The conducting airways and alveolar bed are host to a constant low level of microbes, of which bacteriodetes and firmicutes are the most abundant phyla (Dickson et al., 2016; Dickson et al., 2018). In the steady state, regional differences in microbial communities along the proximal-distal lung axis is maintained by the balance between migration and elimination of commensal microbes, and the interaction between microbiota, the respiratory epithelium and immune cells regulates the baseline concentration of key inflammatory cytokines. Following injury or in exacerbations of lung disease, imbalances in the microflora of the lung (dysbiosis) trigger innate and adaptive immune responses that result in a cascade of pro-inflammatory factors, leading to disruption of epithelial regeneration.

Several studies show that there is a strong correlation between the composition of the lung microbiome and the response to injury in mouse models (Dolma et al., 2020; Jin et al., 2019; O'Dwyer et al., 2019). Neonatal germ-free mice did not exhibit hyperoxia-induced lung damage unlike conventional mice, because of their inability to mount a robust pro-inflammatory response in the absence of a lung microbiota (Dolma et al., 2020). In another study, the absence of lung microbiota in germ-free adult mice conferred a survival advantage compared to conventional mice following bleomycin-induced lung fibrosis (O'Dwyer et al., 2019). Yet another study showed that the absence of lung microbiota protected germ-free mice from development of lung cancer induced by KRAS mutation and loss of p53 by preventing the expansion of tumor associated IL-17 producing $\gamma\delta$-T cells that would normally promote neutrophil infiltration and tumor progression (Jin et al., 2019). Likewise, changes in the composition of the lung microbiome also affects the initiation and progression of human lung cancer and the response to therapy (reviewed in Sommariva et al., 2020).

4.9 ECM Scaffold and mechanotransduction

The ECM of the lung comprises the thin sheet of basement membrane that separates the respiratory epithelium from underlying connective tissues that provides an adhesive substrate for epithelial cell anchorage, and the interstitial fibrillary meshwork of the biomatrix scaffold which endows the lung with its biomechanical properties (Burgess & Weckmann, 2012; Burgstaller et al., 2017). The ECM also serves as a reservoir of insoluble proteins, bound and latent growth factors activated by proteolytic release that together form an instructive stem cell niche (Brizzi, Tarone, & Defilippi, 2012; Taipale & Keski-Oja, 1997). Bi-directional cell-ECM integrin-mediated signaling plays a key role in maintaining stem cells and specifying stem cell fate (Bissell, Hall, & Parry, 1982; Schultz et al., 2011). Mina Bissell termed this process "dynamic reciprocity" (Bissell et al., 1982) whereby the ECM exerts physical and chemical influences on the geometry and biochemistry of the cell *"so as to alter the pattern of gene expression by changing the association of the cytoskeleton with the mRNA and the interaction of the chromatin with the nuclear matrix"* and vice versa.

Integrin-mediated mechanotransduction is especially important in the lung where mechanical traction, elastic recoil, shear and airway resistance are significant variables. Mechanical distortion of cells mediated by the amplitude and periodicity of cyclic stretch profoundly affect lung cell proliferation, differentiation and function (Gjorevski & Nelson, 2009; Torday, Sanchez-Esteban, & Rubin, 1998; Waters et al., 2012). Progressive proteolytic degradation, remodeling and stiffening of ECM symptomatic of chronic lung disease changes the integrin repertoire of ECM facilitating conversion of mesenchymal cells to CAF in the cancerized field, altering the polarity and favoring the migration and metastasis of lung cancer stem cells (Burgstaller et al., 2017; Gotte & Kovalszky, 2018; Zhou et al., 2018).

5. Organotropism and the metastatic niche

The dissemination of lung cancer cells from the primary tumor to colonize distant organs is a complex multistep process (Paolillo & Schinelli, 2019; Wood, Pernemalm, Crosbie, & Whetton, 2014). Loss of polarity and the induction of epithelial-mesenchymal transition (EMT) primes cancer cells for mobilization and migration from the primary tumor. Degradation of the basement membrane supports cancer cell invasion of the surrounding stroma. The intravasation of cancer cells into blood vessels and lymphatics leads to their circulation and subsequent entrapment in the microvasculature where they lodge in pre-metastatic niches conducive to their survival. Eventually these migrating cancer cells rework their microenvironment to form a metastatic niche that supports the growth of secondary tumors. Each of these steps involves the coordinated and orchestrated interaction of lung cancer stem cells, their progeny and diverse stromal cell lineages. Exosomes are an important mediator of cell-cell communication, and are implicated in virtually all pathways defining the hallmarks of cancer in the lung (Hu, Meiners, Lukas, Stathopoulos, & Chen, 2020; Zheng et al., 2018). Exosomes elaborated by lung cancer stem cells regulate the behavior of CAF, endothelial cells and immune cells in the niche. Conversely, CAF-derived exosomes, together with TAM-derived and neutrophil-derived exosomes (Genschmer et al., 2019) influence behavior of cancer stem cells, while also playing a role in angiogenic and ECM remodeling.

EMT reprogramming is characterized by downregulation of the cell-cell adhesion protein E-cadherin (E-Cad), which leads to loss of polarity, intercellular contact and epithelial morphology, and increased motility and invasiveness (Wood et al., 2014; Yousefi et al., 2017). This process is mediated by diverse signaling pathways including the TGF-β, Wnt, Notch and HIF signaling pathways that activate transcriptional repressor genes including Twist, Slug, Snail 1 and 2, Zeb 1, Sip 1, and ZEB. Downregulation of E-Cad is associated with upregulation of mesenchymal markers (e.g. N-cadherin, vimentin), the elaboration of proteolytic enzymes, MMPs and proangiogenic factors that enhance the motility and migration of lung cancer cells. CAF and TAM are the major cell types contributing to degradation of the basement membrane and promoting angiogenesis enabling the migration and intravasation of cancer cells.

As long ago as 1889, Paget observed that certain cancers preferentially metastasized to specific organs, and proposed that the growth of secondary tumors was contingent on circulating cancer cells lodging in "congenial soil" rather than random entrapment in the microvasculature (Paget, 1889). However, these alternatives are not mutually exclusive (Chambers, Groom, & MacDonald, 2002; Yousefi et al., 2017). Cancer cells can migrate as individual cells, or aggregates of cancer cells with or without attached CAF that provide survival cues resembling the microenvironment of the primary tumor (Gieniec et al., 2019; Popper, 2016). The metastatic site will then be determined by a combination of factors including the pattern of blood flow from the primary tumor, and the entrapment of cells or cell aggregates in lymphatic and microvascular capillary beds by size constraints, as well as attraction to the preferential site favoring engraftment.

Lung cancer displays a marked propensity for metastasis to the brain, accounting for approximately 50% of all metastases (Yousefi et al., 2017). Another comprehensive analysis of 159,000 lung cancer cases showed that brain, bone, liver and distant lymph nodes accounted for 94% of metastatic lesions (Wang et al., 2020). This study showed that brain and bone were the two leading sites for metastatic spread in NSCLC, and revealed significant differences in the frequency and pattern of metastatic spread in NSCLC subtypes and SCLC. It also revealed that SCLC had the highest frequency of metastasis, especially to the liver.

Given that different driver mutations affect lung cancer stem cell fate, and the interactions between lung cancer stem cells and their niche, it is not surprising that driver mutations also affect the pattern of organotropic metastasis. One study compared metastatic spread of non-squamous, stage 4 NSCLC in 290 patients with EGFR, KRAS, ALK and triple negative mutations (Doebele et al., 2012). All cohorts were equally predisposed to brain or bone metastasis, although tumors with ALK rearrangements or EGFR mutations were more predisposed to liver metastasis when compared to those with KRAS mutations or triple negatives. Another study of 543 patients showed that the incidence of lung and brain metastasis was higher in NSCLC with EGFR mutations compared to EGFR wildtype tumors (Hsu et al., 2017). Predisposition to metastasize to liver was associated with exon 21 EGFR mutation (Hsu et al., 2017) whereas predisposition to metastasize to brain was associated with exon 19 EGFR mutation (Li, Sun, et al., 2015). On the other hand, NSCLC tumors with fusion mutations (ROS1, ALK and RET) had a higher propensity for lymph node metastasis and vascular invasion compared to those with upstream (EGFR, HER2, and MET) or downstream receptor (KRAS and BRAF) mutations (Liu et al., 2019). Others have noted that a functional polymorphism of the EGFR promoter that induces EGFR activation and overexpression (-216G/T (rs712829)) is associated with the pleural metastasis of ADC (Guo et al., 2013).

The brain organotropism of NSCLC stem cells relies on bi-directional crosstalk between metastasizing cells and the brain metastatic niche initiated by tumor cell derived exosomes (Hoshino et al., 2015; Rodrigues et al., 2019). Uptake of exosomal factors such as the Wnt-related, hyaluronan-binding protein (CEMIP) by brain endothelial and microglial cells upregulates pro-inflammatory cytokines that promote the microvascular remodeling and the preparation of a pre-metastatic niche (Rodrigues et al., 2019). Lung cancer-derived brain metastases then form by co-opting the existing brain microvasculature where adhering metastasizing cells grow as sheets in close apposition to astrocytes that secrete proliferation enhancing cytokines such as IL-6, TGF-β and IGF (Gao et al., 2019; Winkler, 2017). Paradoxically, reactive astrocytes can suppress metastatic cells by elaborating plasmin which induces FasL-dependent apoptosis, and by inactivating L1CAM used by metastatic cells to adhere and guide their spread along capillaries. Lung metastatic cells are shielded from both of these activities by elaborating anti-plasmin serpins NS and B2 (Valiente et al., 2014).

The bone organotropism of metastasizing lung cancer stem cells is thought to be initiated by the secretion of pro-inflammatory osteoclastogenic factors such as IL-11 and PTHrP by metastasizing cells. This in turn activates bone resorption and the proteolytic activation of TGF-β which drives a feed-forward mechanism stimulating the release of osteolytic factors, as well as RANK ligand and Osteoprotegerin (OPG) that inhibits bone-forming osteoblasts (Juarez & Guise, 2011). It is interesting to note that a recent study has found that expression of the formin-like protein (FMNL1) is upregulated in NSCLC bone metastatic tissue (Yang, Liao, & Xue, 2019). Formin-related proteins play a role in regulating cell polarity, cytoskeletal organization, cell motility, EMT and TGF-β signaling (Rana, Aloisio, Choi, & Barber, 2018). Knockdown of FMNL1 in human lung cancer cell lines inhibits their migration and invasion in transwell culture, and suppresses tumor growth and bone metastasis in a nude mouse model (Yang et al., 2019). Others have also implicated chemokine signaling pathways in mediating the bone organotropism of metastasizing NSCLC cells (Ajona et al., 2018; Guo et al., 2016; Wood et al., 2014).

6. Conclusion

Much of our knowledge of the role of the different components of the lung cancer stem cell niche has come from reductionist in vivo and in vitro studies in mouse models, and translational studies of human lung cancers. Lineage tracing studies in genetically engineered gain-of-function and loss-of-function mouse models have also been invaluable in teasing out the role of specific cell lineages, genes, and signaling pathways in regulating cross-talk between stem cells and their niche (Cheon & Orsulic, 2011; Meuwissen & Berns, 2005). However these approaches do have limitations in faithfully recapitulating the human condition, and a number of traits of inbred mouse strains compromise their ability to model lung cancer stem cell—niche interactions.

Different mouse strains exhibit different susceptibilities to lung cancer, and fail to recapitulate the inter- and intra-tumor heterogeneity observed in human lung cancer (Popper, 2015; Safari & Meuwissen, 2015). Inbred mouse strains are also characterized by genetically determined differences in lung ECM composition (Antunes et al., 2009; Wang et al., 2015), airway tree structure (Thiesse et al., 2010), immune surveillance and inflammatory responses (Mestas & Hughes, 2004), and susceptibility to lung injury and fibrosis (Johnston et al., 1998; Walkin et al., 2013). The very much longer telomeres of murine cells also endow them with a replicative capacity unmatched by their human counterparts (Calado & Dumitriu, 2013) potentially amplifying the tumorigenic potential of otherwise inconsequential microenvironmental perturbations.

On the other hand, the more recent application of high throughput "omic" platforms to profile human lung cancers at the single cell level has revealed an unimagined level of complexity in the evolution of lung cancer and the cancer stem cell niche (Chen et al., 2020; Gentles et al., 2020; Gillette et al., 2020; Xu et al., 2020). The mining of such databases by applying machine learning algorithms does have the potential to unmask the role of previously ignored genes in cancer progression (Zhang et al., 2020). There is cause for optimism that the interrogation of high throughput datasets will ultimately reveal the mysteries of bidirectional communication between lung cancer stem cells and their niches. However, it remains a daunting challenge best articulated by Peter Quesenberry some years ago when describing the interaction of an estimated 40 cytokines regulating hematopoietic stem cells (Quesenberry, 1993). A not dissimilar set of circumstances applicable to lung cancer stem cells and their niches.

I quote verbatim:

> "....at least 40 different cytokines act on the hemolymphopoietic system. The number of possible combinations *(with order being important)* is 2.2178877^{48}. This is the simplest case. These cytokines may stimulate or inhibit or in certain cases do both, dependent on dose or target cell constitution. In most cases they appear to act synergistically or additively. Many of these actions may induce juxtacrine, paracrine, holocrine, endocrine, autocrine or protocrine loops involving a wide variety of cytokines. The action of these cytokines also probably depends on the physical and spatial orientation of soluble, extracellular bound or membrane associated cytokines, [their] proteolytic activation or degradation, and the temporal sequence in which they are presented to their target cells. ..."

The contribution of variables including cell-cell and cell matrix interactions, tissue oxygenation, metabolism and pH should also be considered. Changes wrought as a consequence of genetic and epigenetic mutations and co-morbidities in the course of tumor initiation and progression add even further complexity characteristic of inter- and intra-tumor heterogeneity.

In summary, how the interaction between lung cancer stem cells and their niche differs from their normal counterparts leading to the development of tumor heterogeneity, and influencing the course of lung cancer progression and metastasis, can be described in much the same way as families portrayed in the opening sentence of Tolstoy's novel "Anna Karenina":

> "All happy families are alike, but every unhappy family is unhappy in its own way"

References

AbdulJabbar, K., Raza, S. E. A., Rosenthal, R., Jamal-Hanjani, M., Veeriah, S., Akarca, A., et al. (2020). Geospatial immune variability illuminates differential evolution of lung adenocarcinoma. *Nature Medicine*, *26*(7), 1054–1062. https://doi.org/10.1038/s41591-020-0900-x.

Aird, W. C. (2007). Phenotypic heterogeneity of the endothelium: II. Representative vascular beds. *Circulation Research*, *100*(2), 174–190. https://doi.org/10.1161/01.RES.0000255690.03436.ae.

Ajona, D., Zandueta, C., Corrales, L., Moreno, H., Pajares, M. J., Ortiz-Espinosa, S., et al. (2018). Blockade of the complement C5a/C5aR1 axis impairs lung cancer bone metastasis by CXCL16-mediated effects. *American Journal of Respiratory and Critical Care Medicine*, *197*(9), 1164–1176. https://doi.org/10.1164/rccm.201703-0660OC.

Akram, K. M., Patel, N., Spiteri, M. A., & Forsyth, N. R. (2016). Lung regeneration: Endogenous and exogenous stem cell mediated therapeutic approaches. *International Journal of Molecular Sciences*, *17*(1), 128. https://doi.org/10.3390/ijms17010128.

Antunes, M. A., Abreu, S. C., Damaceno-Rodrigues, N. R., Parra, E. R., Capelozzi, V. L., Pinart, M., et al. (2009). Different strains of mice present distinct lung tissue mechanics and extracellular matrix composition in a model of chronic allergic asthma. *Respiratory Physiology & Neurobiology*, *165*(2–3), 202–207. https://doi.org/10.1016/j.resp.2008.12.003.

Arango Duque, G., & Descoteaux, A. (2014). Macrophage cytokines: Involvement in immunity and infectious diseases. *Frontiers in Immunology*, *5*, 491. https://doi.org/10.3389/fimmu.2014.00491.

Baarsma, H. A., & Konigshoff, M. (2017). "WNT-er is coming": WNT signalling in chronic lung diseases. *Thorax*, *72*(8), 746–759. https://doi.org/10.1136/thoraxjnl-2016-209753.

Badri, L., Walker, N. M., Ohtsuka, T., Wang, Z., Delmar, M., Flint, A., et al. (2011). Epithelial interactions and local engraftment of lung-resident mesenchymal stem cells. *American Journal of Respiratory Cell and Molecular Biology*, *45*(4), 809–816. https://doi.org/10.1165/rcmb.2010-0446OC.

Barkauskas, C. E., Cronce, M. J., Rackley, C. R., Bowie, E. J., Keene, D. R., Stripp, B. R., et al. (2013). Type 2 alveolar cells are stem cells in adult lung. *Journal of Clinical Investigation*, *123*(7), 3025–3036. https://doi.org/10.1172/JCI68782.

Barta, J. A., Powell, C. A., & Wisnivesky, J. P. (2019). Global epidemiology of lung cancer. *Annals of Global Health*, *85*(1). https://doi.org/10.5334/aogh.2419.

Beamer, C. A., & Shepherd, D. M. (2013). Role of the aryl hydrocarbon receptor (AhR) in lung inflammation. *Seminars in Immunopathology, 35*(6), 693–704. https://doi.org/10.1007/s00281-013-0391-7.

Bellusci, S., Grindley, J., Emoto, H., Itoh, N., & Hogan, B. L. (1997). Fibroblast growth factor 10 (FGF10) and branching morphogenesis in the embryonic mouse lung. *Development, 124*(23), 4867–4878.

Bertoncello, I. (2016). Properties of adult lung stem and progenitor cells. *Journal of Cellular Physiology, 231*(12), 2582–2589. https://doi.org/10.1002/jcp.25404.

Bhatia, A., & Kumar, Y. (2011). Cancer-immune equilibrium: Questions unanswered. *Cancer Microenvironment, 4*(2), 209–217. https://doi.org/10.1007/s12307-011-0065-8.

Bissell, M. J., Hall, H. G., & Parry, G. (1982). How does the extracellular matrix direct gene expression? *Journal of Theoretical Biology, 99*(1), 31–68. https://doi.org/10.1016/0022-5193(82)90388-5.

Biswas, S. K., & Mantovani, A. (2010). Macrophage plasticity and interaction with lymphocyte subsets: Cancer as a paradigm. *Nature Immunology, 11*(10), 889–896. https://doi.org/10.1038/ni.1937.

Boucherat, O., Morissette, M. C., Provencher, S., Bonnet, S., & Maltais, F. (2016). Bridging lung development with chronic obstructive pulmonary disease. Relevance of developmental pathways in chronic obstructive pulmonary disease pathogenesis. *American Journal of Respiratory and Critical Care Medicine, 193*(4), 362–375. https://doi.org/10.1164/rccm.201508-1518PP.

Bozinovski, S., Vlahos, R., Anthony, D., McQualter, J., Anderson, G., Irving, L., et al. (2016). COPD and squamous cell lung cancer: Aberrant inflammation and immunity is the common link. *British Journal of Pharmacology, 173*(4), 635–648. https://doi.org/10.1111/bph.13198.

Branchfield, K., Nantie, L., Verheyden, J. M., Sui, P., Wienhold, M. D., & Sun, X. (2016). Pulmonary neuroendocrine cells function as airway sensors to control lung immune response. *Science, 351*(6274), 707–710. https://doi.org/10.1126/science.aad7969.

Bray, F., Ferlay, J., Soerjomataram, I., Siegel, R. L., Torre, L. A., & Jemal, A. (2018). Global cancer statistics 2018: GLOBOCAN estimates of incidence and mortality worldwide for 36 cancers in 185 countries. *CA: A Cancer Journal for Clinicians, 68*(6), 394–424. https://doi.org/10.3322/caac.21492.

Brizzi, M. F., Tarone, G., & Defilippi, P. (2012). Extracellular matrix, integrins, and growth factors as tailors of the stem cell niche. *Current Opinion in Cell Biology, 24*(5), 645–651. https://doi.org/10.1016/j.ceb.2012.07.001.

Broeckaert, F., Clippe, A., Knoops, B., Hermans, C., & Bernard, A. (2000). Clara cell secretory protein (CC16): Features as a peripheral lung biomarker. *Annals of the New York Academy of Sciences, 923*, 68–77. https://doi:10.1111/j.1749-6632.2000.tb05520.x.

Burgess, J. K., & Weckmann, M. (2012). Matrikines and the lungs. *Pharmacology and Therapeutics, 134*(3), 317–337. https://doi.org/S0163-7258(12)00042-3 [pii] https://doi.org/10.1016/j.pharmthera.2012.02.002.

Burgstaller, G., Oehrle, B., Gerckens, M., White, E. S., Schiller, H. B., & Eickelberg, O. (2017). The instructive extracellular matrix of the lung: Basic composition and alterations in chronic lung disease. *European Respiratory Journal, 50*(1), 1601805. https://doi.org/10.1183/13993003.01805-2016.

Busch, S. E., Hanke, M. L., Kargl, J., Metz, H. E., MacPherson, D., & Houghton, A. M. (2016). Lung cancer subtypes generate unique immune responses. *Journal of Immunology, 197*(11), 4493–4503. https://doi.org/10.4049/jimmunol.1600576.

Butler, C. R., Hynds, R. E., Gowers, K. H., Lee Ddo, H., Brown, J. M., Crowley, C., et al. (2016). Rapid expansion of human epithelial stem cells suitable for airway tissue engineering. *American Journal of Respiratory and Critical Care Medicine, 194*(2), 156–168. https://doi.org/10.1164/rccm.201507-1414OC.

Calado, R. T., & Dumitriu, B. (2013). Telomere dynamics in mice and humans. *Seminars in Hematology*, *50*(2), 165–174. https://doi.org/10.1053/j.seminhematol.2013.03.030.

Calon, A., Tauriello, D. V., & Batlle, E. (2014). TGF-beta in CAF-mediated tumor growth and metastasis. *Seminars in Cancer Biology*, *25*, 15–22. https://doi.org/10.1016/j.semcancer.2013.12.008.

Cardona, A. F., Rojas, L., Zatarain-Barron, Z. L., Ruiz-Patino, A., Ricaurte, L., Corrales, L., et al. (2019). Multigene mutation profiling and clinical characteristics of small-cell lung cancer in never-smokers vs. heavy smokers (Geno1.3-CLICaP). *Frontiers in Oncology*, *9*, 254. https://doi.org/10.3389/fonc.2019.00254.

Carraro, G., Mulay, A., Yao, C., Mizuno, T., Konda, B., Petrov, M., et al. (2020). Single cell reconstruction of human basal cell diversity in normal and IPF lung. *American Journal of Respiratory and Critical Care Medicine*. [EPub ahead of Print]. https://doi.org/10.1164/rccm.201904-0792OC.

Carraro, G., & Stripp, B. R. (2015). A new notch for lung stem cells. *Cell Stem Cell*, *16*(2), 107–109. https://doi.org/10.1016/j.stem.2015.01.012.

Chambers, A. F., Groom, A. C., & MacDonald, I. C. (2002). Dissemination and growth of cancer cells in metastatic sites. *Nature Reviews Cancer*, *2*(8), 563–572. https://doi.org/10.1038/nrc865.

Chan, T. S., Shaked, Y., & Tsai, K. K. (2019). Targeting the interplay between cancer fibroblasts, mesenchymal stem cells, and cancer stem cells in desmoplastic cancers. *Frontiers in Oncology*, *9*, 688. https://doi.org/10.3389/fonc.2019.00688.

Chapman, H. A., Li, X., Alexander, J. P., Brumwell, A., Lorizio, W., Tan, K., et al. (2011). Integrin alpha6beta4 identifies an adult distal lung epithelial population with regenerative potential in mice. *Journal of Clinical Investigation*, *121*(7), 2855–2862. https://doi.org/57673 [pii] https://doi.org/10.1172/JCI57673.

Chapman, A. M., Sun, K. Y., Ruestow, P., Cowan, D. M., & Madl, A. K. (2016). Lung cancer mutation profile of EGFR, ALK, and KRAS: Meta-analysis and comparison of never and ever smokers. *Lung Cancer*, *102*, 122–134. https://doi.org/10.1016/j.lungcan.2016.10.010.

Chen, Z., Fillmore, C. M., Hammerman, P. S., Kim, C. F., & Wong, K. K. (2014). Non-small-cell lung cancers: A heterogeneous set of diseases. *Nature Reviews Cancer*, *14*(8), 535–546. https://doi.org/10.1038/nrc3775.

Chen, H., Matsumoto, K., Brockway, B. L., Rackley, C. R., Liang, J., Lee, J. H., et al. (2012). Airway epithelial progenitors are region specific and show differential responses to bleomycin-induced lung injury. *Stem Cells*, *30*(9), 1948–1960. https://doi.org/10.1002/stem.1150.

Chen, Y. J., Roumeliotis, T. I., Chang, Y. H., Chen, C. T., Han, C. L., Lin, M. H., et al. (2020). Proteogenomics of non-smoking lung cancer in east asia delineates molecular signatures of pathogenesis and progression. *Cell*, *182*(1), 226–244. e217. https://doi.org/10.1016/j.cell.2020.06.012.

Cheon, D. J., & Orsulic, S. (2011). Mouse models of cancer. *Annual Review of Pathology: Mechanisms of Disease*, *6*, 95–119. https://doi.org/10.1146/annurev.pathol.3.121806.154244.

Clevers, H., Loh, K. M., & Nusse, R. (2014). Stem cell signaling. An integral program for tissue renewal and regeneration: Wnt signaling and stem cell control. *Science*, *346*(6205), 1248012. https://doi.org/10.1126/science.1248012.

Cole, B. B., Smith, R. W., Jenkins, K. M., Graham, B. B., Reynolds, P. R., & Reynolds, S. D. (2010). Tracheal basal cells: A facultative progenitor cell pool. *American Journal of Pathology*, *177*(1), 362–376. https://doi.org/S0002-9440(10)60092-2 [pii] https://doi.org/10.2353/ajpath.2010.090870.

Curry, J. L., & Trentin, J. J. (1967). Hemopoietic spleen colony studies. I. Growth and differentiation. *Developmental Biology*, *15*(5), 395–413. https://doi.org/10.1016/0012-1606(67)90034-6.

Curtius, K., Wright, N. A., & Graham, T. A. (2018). An evolutionary perspective on field cancerization. *Nature Reviews Cancer*, *18*(1), 19–32. https://doi.org/10.1038/nrc.2017.102.
Cutz, E. (2015). Hyperplasia of pulmonary neuroendocrine cells in infancy and childhood. *Seminars in Diagnostic Pathology*, *32*(6), 420–437. https://doi.org/10.1053/j.semdp.2015.08.001.
Dakubo, G. D., Jakupciak, J. P., Birch-Machin, M. A., & Parr, R. L. (2007). Clinical implications and utility of field cancerization. *Cancer Cell International*, 7, 2. https://doi.org/10.1186/1475-2867-7-2.
Danopoulos, S., Shiosaki, J., & Al Alam, D. (2019). FGF signaling in lung development and disease: Human versus mouse. *Frontiers in Genetics*, *10*, 170. https://doi.org/10.3389/fgene.2019.00170.
Danopoulos, S., Thornton, M. E., Grubbs, B. H., Frey, M. R., Warburton, D., Bellusci, S., et al. (2019). Discordant roles for FGF ligands in lung branching morphogenesis between human and mouse. *Journal of Pathology*, *247*(2), 254–265. https://doi.org/10.1002/path.5188.
de Bruin, E. C., McGranahan, N., Mitter, R., Salm, M., Wedge, D. C., Yates, L., et al. (2014). Spatial and temporal diversity in genomic instability processes defines lung cancer evolution. *Science*, *346*(6206), 251–256. https://doi.org/10.1126/science.1253462.
Desai, T. J., Brownfield, D. G., & Krasnow, M. A. (2014). Alveolar progenitor and stem cells in lung development, renewal and cancer. *Nature*, *507*(7491), 190–194. https://doi.org/10.1038/nature12930.
Dickson, R. P., Erb-Downward, J. R., Falkowski, N. R., Hunter, E. M., Ashley, S. L., & Huffnagle, G. B. (2018). The lung microbiota of healthy mice are highly variable, cluster by environment, and reflect variation in baseline lung innate immunity. *American Journal of Respiratory and Critical Care Medicine*, *198*(4), 497–508. https://doi.org/10.1164/rccm.201711-2180OC.
Dickson, R. P., Erb-Downward, J. R., Martinez, F. J., & Huffnagle, G. B. (2016). The microbiome and the respiratory tract. *Annual Review of Physiology*, *78*, 481–504. https://doi.org/10.1146/annurev-physiol-021115-105238.
Ding, B. S., Nolan, D. J., Guo, P., Babazadeh, A. O., Cao, Z., Rosenwaks, Z., et al. (2011). Endothelial-derived angiocrine signals induce and sustain regenerative lung alveolarization. *Cell*, *147*(3), 539–553. https://doi.org/S0092-8674(11)01194-9 [pii] https://doi.org/10.1016/j.cell.2011.10.003.
Doebele, R. C., Lu, X., Sumey, C., Maxson, D. A., Weickhardt, A. J., Oton, A. B., et al. (2012). Oncogene status predicts patterns of metastatic spread in treatment-naive non-small cell lung cancer. *Cancer*, *118*(18), 4502–4511. https://doi.org/10.1002/cncr.27409.
Dolma, K., Freeman, A. E., Rezonzew, G., Payne, G. A., Xu, X., Jilling, T., et al. (2020). Effects of hyperoxia on alveolar and pulmonary vascular development in germ-free mice. *American Journal of Physiology. Lung Cellular and Molecular Physiology*, *318*(2), L421–L428. https://doi.org/10.1152/ajplung.00316.2019.
Dorrello, N. V., & Vunjak-Novakovic, G. (2020). Bioengineering of pulmonary epithelium with preservation of the vascular niche. *Frontiers in Bioengeneering and Biotechnology*, *8*, 269. https://doi.org/10.3389/fbioe.2020.00269.
El Agha, E., Seeger, W., & Bellusci, S. (2017). Therapeutic and pathological roles of fibroblast growth factors in pulmonary diseases. *Developmental Dynamics*, *246*(4), 235–244. https://doi.org/10.1002/dvdy.24468.
Erdogan, B., & Webb, D. J. (2017). Cancer-associated fibroblasts modulate growth factor signaling and extracellular matrix remodeling to regulate tumor metastasis. *Biochemical Society Transactions*, *45*(1), 229–236. https://doi.org/10.1042/BST20160387.

Evans, M. J., Van Winkle, L. S., Fanucchi, M. V., & Plopper, C. G. (1999). The attenuated fibroblast sheath of the respiratory tract epithelial-mesenchymal trophic unit. *American Journal of Respiratory Cell and Molecular Biology, 21*, 655–657. https://doi:10.1165/ajrcmb.21.6.3807.

Ferone, G., Song, J. Y., Sutherland, K. D., Bhaskaran, R., Monkhorst, K., Lambooij, J. P., et al. (2016). SOX2 is the determining oncogenic switch in promoting lung squamous cell carcinoma from different cells of origin. *Cancer Cell, 30*(4), 519–532. https://doi.org/10.1016/j.ccell.2016.09.001.

Ferrer, I., Zugazagoitia, J., Herbertz, S., John, W., Paz-Ares, L., & Schmid-Bindert, G. (2018). KRAS-mutant non-small cell lung cancer: From biology to therapy. *Lung Cancer, 124*, 53–64. https://doi.org/10.1016/j.lungcan.2018.07.013.

Frank, D. B., Peng, T., Zepp, J. A., Snitow, M., Vincent, T. L., Penkala, I. J., et al. (2016). Emergence of a wave of Wnt signaling that regulates lung alveologenesis by controlling epithelial self-renewal and differentiation. *Cell Reports, 17*(9), 2312–2325. https://doi.org/10.1016/j.celrep.2016.11.001.

Gao, Y., Bado, I., Wang, H., Zhang, W., Rosen, J. M., & Zhang, X. H. (2019). Metastasis organotropism: Redefining the congenial soil. *Developmental Cell, 49*(3), 375–391. https://doi.org/10.1016/j.devcel.2019.04.012.

Gascard, P., & Tlsty, T. D. (2016). Carcinoma-associated fibroblasts: Orchestrating the composition of malignancy. *Genes and Development, 30*(9), 1002–1019. https://doi.org/10.1101/gad.279737.116.

Genschmer, K. R., Russell, D. W., Lal, C., Szul, T., Bratcher, P. E., Noerager, B. D., et al. (2019). Activated PMN exosomes: Pathogenic entities causing matrix destruction and disease in the lung. *Cell, 176*(1–2), 113–126. e115. https://doi.org/10.1016/j.cell.2018.12.002.

Gentles, A. J., Hui, A. B., Feng, W., Azizi, A., Nair, R. V., Bouchard, G., et al. (2020). A human lung tumor microenvironment interactome identifies clinically relevant cell-type cross-talk. *Genome Biology, 21*(1), 107. https://doi.org/10.1186/s13059-020-02019-x.

Ghosh, M., Ahmad, S., Jian, A., Li, B., Smith, R. W., Helm, K. M., et al. (2013). Human tracheobronchial basal cells. Normal versus remodeling/repairing phenotypes in vivo and in vitro. *American Journal of Respiratory Cell and Molecular Biology, 49*(6), 1127–1134. https://doi.org/10.1165/rcmb.2013-0049OC.

Giangreco, A., Arwert, E. N., Rosewell, I. R., Snyder, J., Watt, F. M., & Stripp, B. R. (2009). Stem cells are dispensable for lung homeostasis but restore airways after injury. *Proceedings of the National Academy of Sciences of the United States of America, 106*(23), 9286–9291. https://doi:10.1073/pnas.0900668106.

Gieniec, K. A., Butler, L. M., Worthley, D. L., & Woods, S. L. (2019). Cancer-associated fibroblasts-heroes or villains? *British Journal of Cancer, 121*(4), 293–302. https://doi.org/10.1038/s41416-019-0509-3.

Gillette, M. A., Satpathy, S., Cao, S., Dhanasekaran, S. M., Vasaikar, S. V., Krug, K., et al. (2020). Proteogenomic characterization reveals therapeutic vulnerabilities in lung adenocarcinoma. *Cell, 182*(1), 200–225. e235. https://doi.org/10.1016/j.cell.2020.06.013.

Gjorevski, N., & Nelson, C. M. (2009). Bidirectional extracellular matrix signaling during tissue morphogenesis. *Cytokine & Growth Factor Reviews, 20*(5–6), 459–465. https://doi:10.1073/pnas.0901269106.

Glaser, L., Coulter, P. J., Shields, M., Touzelet, O., Power, U. F., & Broadbent, L. (2019). Airway epithelial derived cytokines and chemokines and their role in the immune response to respiratory syncytial virus infection. *Pathogens, 8*(3), 106. https://doi.org/10.3390/pathogens8030106.

Globinska, A., & Kowalski, M. L. (2017). Innate lymphoid cells: The role in respiratory infections and lung tissue damage. *Expert Review of Clinical Immunology*, *13*(10), 991–999. https://doi.org/10.1080/1744666X.2017.1366314.

Gotte, M., & Kovalszky, I. (2018). Extracellular matrix functions in lung cancer. *Matrix Biology*, *73*, 105–121. https://doi.org/10.1016/j.matbio.2018.02.018.

Gower, A. C., Steiling, K., Brothers, J. F., 2nd, Lenburg, M. E., & Spira, A. (2011). Transcriptomic studies of the airway field of injury associated with smoking-related lung disease. *Proceedings of the American Thoracic Society*, *8*(2), 173–179. https://doi.org/10.1513/pats.201011-066MS.

Guo, F., Wang, Y., Liu, J., Mok, S. C., Xue, F., & Zhang, W. (2016). CXCL12/CXCR4: A symbiotic bridge linking cancer cells and their stromal neighbors in oncogenic communication networks. *Oncogene*, *35*(7), 816–826. https://doi.org/10.1038/onc.2015.139.

Guo, H., Xing, Y., Liu, R., Chen, S., Bian, X., Wang, F., et al. (2013). 216G/T (rs712829), a functional variant of the EGFR promoter, is associated with the pleural metastasis of lung adenocarcinoma. *Oncology Letters*, *6*(3), 693–698. https://doi.org/10.3892/ol.2013.1442.

Hallberg, B., & Palmer, R. H. (2016). The role of the ALK receptor in cancer biology. *Annals of Oncology*, *27*(Suppl. 3). iii4–iii15. https://doi.org/10.1093/annonc/mdw301.

Hanahan, D., & Weinberg, R. A. (2011). Hallmarks of cancer: The next generation. *Cell*, *144*(5), 646–674. https://doi.org/10.1016/j.cell.2011.02.013.

Hartl, D., Tirouvanziam, R., Laval, J., Greene, C. M., Habiel, D., Sharma, L., et al. (2018). Innate immunity of the lung: From basic mechanisms to translational medicine. *Journal of Innate Immunology*, *10*(5-6), 487–501. https://doi.org/10.1159/000487057.

Hegab, A. E., Ha, V. L., Darmawan, D. O., Gilbert, J. L., Ooi, A. T., Attiga, Y. S., et al. (2012). Isolation and in vitro characterization of basal and submucosal gland duct stem/progenitor cells from human proximal airways. *Stem Cells Translational Medicine*, *1*(10), 719–724. https://doi.org/sctm.2012-0056 [pii]. https://doi.org/10.5966/sctm.2012-0056.

Hegab, A. E., Nickerson, D. W., Ha, V. L., Darmawan, D. O., & Gomperts, B. N. (2012). Repair and regeneration of tracheal surface epithelium and submucosal glands in a mouse model of hypoxic-ischemic injury. *Respirology*, *17*(7), 1101–1113. https://doi.org/10.1111/j.1440-1843.2012.02204.x.

Herriges, M., & Morrisey, E. E. (2014). Lung development: Orchestrating the generation and regeneration of a complex organ. *Development*, *141*(3), 502–513. https://doi.org/10.1242/dev.098186.

Hogan, B. L., Barkauskas, C. E., Chapman, H. A., Epstein, J. A., Jain, R., Hsia, C. C., et al. (2014). Repair and regeneration of the respiratory system: Complexity, plasticity, and mechanisms of lung stem cell function. *Cell Stem Cell*, *15*(2), 123–138. https://doi.org/10.1016/j.stem.2014.07.012.

Holgate, S. T., Davies, D. E., Lackie, P. M., Wilson, S. J., Puddicombe, S. M., & Lordan, J. L. (2000). Epithelial-mesenchymal interactions in the pathogenesis of asthma. *Journal of Allergy and Clinical Immunology*, *105*(2 Pt 1), 193–204. https://doi.org/S0091674900606255 [pii].

Hong, K. U., Reynolds, S. D., Giangreco, A., Hurley, C. M., & Stripp, B. R. (2001). Clara cell secretory protein-expressing cells of the airway neuroepithelial body microenvironment include a label-retaining subset and are critical for epithelial renewal after progenitor cell depletion. *American Journal of Respiratory Cell and Molecular Biology*, *24*(6), 671–681. https://doi.org/10.1165/ajrcmb.24.6.4498.

Hoshino, A., Costa-Silva, B., Shen, T. L., Rodrigues, G., Hashimoto, A., Tesic Mark, M., et al. (2015). Tumour exosome integrins determine organotropic metastasis. *Nature*, *527*(7578), 329–335. https://doi.org/10.1038/nature15756.

Hsia, C. C. (2017). Comparative analysis of the mechanical signals in lung development and compensatory growth. *Cell and Tissue Research*, *367*(3), 687–705. https://doi.org/10.1007/s00441-016-2558-8.

Hsu, F., De Caluwe, A., Anderson, D., Nichol, A., Toriumi, T., & Ho, C. (2017). Patterns of spread and prognostic implications of lung cancer metastasis in an era of driver mutations. *Current Oncology*, *24*(4), 228–233. https://doi.org/10.3747/co.24.3496.

Hu, C., Meiners, S., Lukas, C., Stathopoulos, G. T., & Chen, J. (2020). Role of exosomal microRNAs in lung cancer biology and clinical applications. *Cell Proliferation*, *53*(6), e12828. https://doi.org/10.1111/cpr.12828.

Hu, B., & Phan, S. H. (2016). Notch in fibrosis and as a target of anti-fibrotic therapy. *Pharmacological Research*, *108*, 57–64. https://doi.org/10.1016/j.phrs.2016.04.010.

Hussain, M., Xu, C., Ahmad, M., Yang, Y., Lu, M., Wu, X., et al. (2017). Notch signaling: Linking embryonic lung development and asthmatic airway remodeling. *Molecular Pharmacology*, *92*(6), 676–693. https://doi.org/10.1124/mol.117.110254.

Hyde, D. M., Hamid, Q., & Irvin, C. G. (2009). Anatomy, pathology, and physiology of the tracheobronchial tree: Emphasis on the distal airways. *Journal of Allergy and Clinical Immunology*, *124*(6 Suppl), S72–S77. https://doi.org/10.1016/j.jaci.2009.08.048.

Ireland, A. S., Micinski, A. M., Kastner, D. W., Guo, B., Wait, S. J., Spainhower, K. B., et al. (2020). MYC drives temporal evolution of small cell lung cancer subtypes by reprogramming neuroendocrine fate. *Cancer Cell*, *38*(1), 60–78. e12. https://doi.org/10.1016/j.ccell.2020.05.001.

Ishii, G., Ochiai, A., & Neri, S. (2016). Phenotypic and functional heterogeneity of cancer-associated fibroblast within the tumor microenvironment. *Advances in Drug Delivery Reviews*, *99*(Pt. B), 186–196. https://doi.org/10.1016/j.addr.2015.07.007.

Jain, R., Barkauskas, C. E., Takeda, N., Bowie, E. J., Aghajanian, H., Wang, Q., et al. (2015). Plasticity of Hopx(+) type I alveolar cells to regenerate type II cells in the lung. *Nature Communications*, *6*, 6727. https://doi.org/10.1038/ncomms7727.

Jamal-Hanjani, M., Hackshaw, A., Ngai, Y., Shaw, J., Dive, C., Quezada, S., et al. (2014). Tracking genomic cancer evolution for precision medicine: The lung TRACERx study. *PLoS Biology*, *12*(7), e1001906. https://doi.org/10.1371/journal.pbio.1001906.

Jamal-Hanjani, M., Wilson, G. A., McGranahan, N., Birkbak, N. J., Watkins, T. B. K., Veeriah, S., et al. (2017). Tracking the evolution of non-small-cell lung cancer. *New England Journal of Medicine*, *376*(22), 2109–2121. https://doi.org/10.1056/NEJMoa1616288.

Jambusaria, A., Hong, Z., Zhang, L., Srivastava, S., Jana, A., Toth, P. T., et al. (2020). Endothelial heterogeneity across distinct vascular beds during homeostasis and inflammation. *eLife*, *9*, e51413. https://doi.org/10.7554/eLife.51413.

Jin, C., Lagoudas, G. K., Zhao, C., Bullman, S., Bhutkar, A., Hu, B., et al. (2019). Commensal microbiota promote lung cancer development via gammadelta T cells. *Cell*, *176*(5), 998–1013. e1016. https://doi.org/10.1016/j.cell.2018.12.040.

Johnston, C. J., Stripp, B. R., Piedbeouf, B., Wright, T. W., Mango, G. W., Reed, C. K., et al. (1998). Inflammatory and epithelial responses in mouse strains that differ in sensitivity to hyperoxic injury. *Experimental Lung Research*, *24*(2), 189–202. https://doi.org/10.3109/01902149809099582.

Joshi, K., de Massy, M. R., Ismail, M., Reading, J. L., Uddin, I., Woolston, A., et al. (2019). Spatial heterogeneity of the T cell receptor repertoire reflects the mutational landscape in lung cancer. *Nature Medicine*, *25*(10), 1549–1559. https://doi.org/10.1038/s41591-019-0592-2.

Juarez, P., & Guise, T. A. (2011). TGF-beta in cancer and bone: Implications for treatment of bone metastases. *Bone*, *48*(1), 23–29. https://doi.org/10.1016/j.bone.2010.08.004.

Kalluri, R., & LeBleu, V. S. (2020). The biology, function, and biomedical applications of exosomes. *Science*, *367*(6478), eaau6977. https://doi.org/10.1126/science.aau6977.

Karacosta, L. G., Anchang, B., Ignatiadis, N., Kimmey, S. C., Benson, J. A., Shrager, J. B., et al. (2019). Mapping lung cancer epithelial-mesenchymal transition states and trajectories with single-cell resolution. *Nature Communications*, *10*(1), 5587. https://doi.org/10.1038/s41467-019-13441-6.
Kim, C. F., Jackson, E. L., Woolfenden, A. E., Lawrence, S., Babar, I., Vogel, S., et al. (2005). Identification of bronchioalveolar stem cells in normal lung and lung cancer. *Cell*, *121*(6), 823–835. https://doi:10.1016/j.cell.2005.03.032.
King, N. E., Suzuki, S., Barilla, C., Hawkins, F. J., Randell, S. H., Reynolds, S. D., et al. (2020). Correction of airway stem cells: Genome editing approaches for the treatment of cystic fibrosis. *Human Gene Therapy*, *31*(17-18), 956–972. https://doi.org/10.1089/hum.2020.160.
Kitajima, S., Thummalapalli, R., & Barbie, D. A. (2016). Inflammation as a driver and vulnerability of KRAS mediated oncogenesis. *Seminars in Cell and Developmental Biology*, *58*, 127–135. https://doi.org/10.1016/j.semcdb.2016.06.009.
Knust, J., Ochs, M., Gundersen, H. J., & Nyengaard, J. R. (2009). Stereological estimates of alveolar number and size and capillary length and surface area in mice lungs. *Anatomical Record*, *292*(1), 113–122. https://doi.org/10.1002/ar.20747.
Konigshoff, M., & Eickelberg, O. (2010). WNT signaling in lung disease: A failure or a regeneration signal? *American Journal of Respiratory Cell and Molecular Biology*, *42*(1), 21–31.
Kreso, A., & Dick, J. E. (2014). Evolution of the cancer stem cell model. *Cell Stem Cell*, *14*(3), 275–291. https://doi.org/10.1016/j.stem.2014.02.006.
Kumar, P. A., Hu, Y., Yamamoto, Y., Hoe, N. B., Wei, T. S., Mu, D., et al. (2011). Distal airway stem cells yield alveoli in vitro and during lung regeneration following H1N1 influenza infection. *Cell*, *147*(3), 525–538. https://doi.org/S0092-8674(11)01173-1 [pii] https://doi.org/10.1016/j.cell.2011.10.001.
Kunimasa, K., & Goto, T. (2020). Immunosurveillance and immunoediting of Lung Cancer: Current perspectives and challenges. *International Journal of Molecular Sciences*, *21*(2), 597. https://doi.org/10.3390/ijms21020597.
Kuo, C. S., & Krasnow, M. A. (2015). Formation of a neurosensory organ by epithelial cell slithering. *Cell*, *163*(2), 394–405. https://doi.org/10.1016/j.cell.2015.09.021.
Laughney, A. M., Hu, J., Campbell, N. R., Bakhoum, S. F., Setty, M., Lavallee, V. P., et al. (2020). Regenerative lineages and immune-mediated pruning in lung cancer metastasis. *Nature Medicine*, *26*(2), 259–269. https://doi.org/10.1038/s41591-019-0750-6.
Lavin, Y., Kobayashi, S., Leader, A., Amir, E. D., Elefant, N., Bigenwald, C., et al. (2017). Innate immune landscape in early lung adenocarcinoma by paired single-cell analyses. *Cell*, *169*(4), 750–765. e717. https://doi.org/10.1016/j.cell.2017.04.014.
Laviron, M., & Boissonnas, A. (2019). Ontogeny of tumor-associated macrophages. *Frontiers in Immunology*, *10*, 1799. https://doi.org/10.3389/fimmu.2019.01799.
Leach, J. P., & Morrisey, E. E. (2018). Repairing the lungs one breath at a time: How dedicated or facultative are you? *Genes and Development*, *32*(23-24), 1461–1471. https://doi.org/10.1101/gad.319418.118.
Leduc, C., Antoni, D., Charloux, A., Falcoz, P. E., & Quoix, E. (2017). Comorbidities in the management of patients with lung cancer. *European Respiratory Journal*, *49*(3). https://doi.org/10.1183/13993003.01721-2016.
Lee, J. H., Bhang, D. H., Beede, A., Huang, T. L., Stripp, B. R., Bloch, K. D., et al. (2014). Lung stem cell differentiation in mice directed by endothelial cells via a BMP4-NFATc1-thrombospondin-1 axis. *Cell*, *156*(3), 440–455. https://doi.org/10.1016/j.cell.2013.12.039.
Lee, J. H., Tammela, T., Hofree, M., Choi, J., Marjanovic, N. D., Han, S., et al. (2017). Anatomically and functionally distinct lung mesenchymal populations marked by Lgr5 and Lgr6. *Cell*, *170*(6), 1149–1163. e1112. https://doi.org/10.1016/j.cell.2017.07.028.

Leonetti, A., Facchinetti, F., Minari, R., Cortellini, A., Rolfo, C. D., Giovannetti, E., et al. (2019). Notch pathway in small-cell lung cancer: From preclinical evidence to therapeutic challenges. *Cellular Oncology*, *42*(3), 261–273. https://doi.org/10.1007/s13402-019-00441-3.

Li, C., Bellusci, S., Borok, Z., & Minoo, P. (2015). Non-canonical WNT signalling in the lung. *Journal of Biochemistry*, *158*(5), 355–365. https://doi.org/10.1093/jb/mvv081.

Li, B., Sun, S. Z., Yang, M., Shi, J. L., Xu, W., Wang, X. F., et al. (2015). The correlation between EGFR mutation status and the risk of brain metastasis in patients with lung adenocarcinoma. *Journal of Neurooncology*, *124*(1), 79–85. https://doi.org/10.1007/s11060-015-1776-3.

Liguori, M., Solinas, G., Germano, G., Mantovani, A., & Allavena, P. (2011). Tumor-associated macrophages as incessant builders and destroyers of the cancer stroma. *Cancers*, *3*(4), 3740–3761. https://doi.org/10.3390/cancers3043740.

Lim, J. S., Ibaseta, A., Fischer, M. M., Cancilla, B., O'Young, G., Cristea, S., et al. (2017). Intratumoural heterogeneity generated by Notch signalling promotes small-cell lung cancer. *Nature*, *545*(7654), 360–364. https://doi.org/10.1038/nature22323.

Liu, Z., Liang, H., Lin, J., Cai, X., Pan, Z., Liu, J., et al. (2019). The incidence of lymph node metastasis in patients with different oncogenic driver mutations among T1 non-small-cell lung cancer. *Lung Cancer*, *134*, 218–224. https://doi.org/10.1016/j.lungcan.2019.06.026.

Lynch, T. J., Anderson, P. J., Rotti, P. G., Tyler, S. R., Crooke, A. K., Choi, S. H., et al. (2018). Submucosal gland myoepithelial cells are reserve stem cells that can regenerate mouse tracheal epithelium. *Cell Stem Cell*, *22*(5), 779. https://doi.org/10.1016/j.stem.2018.04.007.

Ma, Q., Ma, Y., Dai, X., Ren, T., Fu, Y., Liu, W., et al. (2018). Regeneration of functional alveoli by adult human SOX9(+) airway basal cell transplantation. *Protein & Cell*, *9*(3), 267–282. https://doi.org/10.1007/s13238-018-0506-y.

Maeda, Y., Dave, V., & Whitsett, J. A. (2007). Transcriptional control of lung morphogenesis. *Physiological Reviews*, *87*(1), 219–244. https://doi:10.1152/physrev.00028.2006.

Mantovani, A., & Sica, A. (2010). Macrophages, innate immunity and cancer: Balance, tolerance, and diversity. *Current Opinion in Immunology*, *22*(2), 231–237. https://doi.org/10.1016/j.coi.2010.01.009.

McCarthy, W. J., Meza, R., Jeon, J., & Moolgavkar, S. H. (2012). Chapter 6: Lung cancer in never smokers: Epidemiology and risk prediction models. *Risk Analysis*, *32*(Suppl. 1), S69–S84. https://doi.org/10.1111/j.1539-6924.2012.01768.x.

McCulloch, E. A., Siminovitch, L., Till, J. E., Russell, E. S., & Bernstein, S. E. (1965). The cellular basis of the genetically determined hemopoietic defect in anemic mice of genotype Sl-Sld. *Blood*, *26*(4), 399–410.

McQualter, J. L., & Bertoncello, I. (2012). Concise review: Deconstructing the lung to reveal its regenerative potential. *Stem Cells*, *30*(5), 811–816. https://doi.org/10.1002/stem.1055.

McQualter, J. L., McCarty, R. C., Van der Velden, J., O'Donoghue, R. J., Asselin-Labat, M. L., Bozinovski, S., et al. (2013). TGF-beta signaling in stromal cells acts upstream of FGF-10 to regulate epithelial stem cell growth in the adult lung. *Stem Cell Research*, *11*(3), 1222–1233. https://doi.org/10.1016/j.scr.2013.08.007.

McQualter, J. L., Yuen, K., Williams, B., & Bertoncello, I. (2010). Evidence of an epithelial stem/progenitor cell hierarchy in the adult mouse lung. *Proceedings of the National Academy of Sciences of the United States of America*, *107*(4), 1414–1419. https://doi.org/10.1073/pnas.0909207107.

Mestas, J., & Hughes, C. C. (2004). Of mice and not men: Differences between mouse and human immunology. *Journal of Immunology*, *172*(5), 2731–2738. https://doi:10.4049/jimmunol.172.5.2731.

Metzger, R. J., Klein, O. D., Martin, G. R., & Krasnow, M. A. (2008). The branching programme of mouse lung development. *Nature, 453*(7196), 745–750. https://doi.org/10.1038/nature07005.

Meuwissen, R., & Berns, A. (2005). Mouse models for human lung cancer. *Genes and Development, 19*(6), 643–664. https://doi:10.1101/gad.1284505.

Moldoveanu, B., Otmishi, P., Jani, P., Walker, J., Sarmiento, X., Guardiola, J., et al. (2009). Inflammatory mechanisms in the lung. *Journal of Inflammation Research, 2*, 1–11.

Morrisey, E. E., & Hogan, B. L. (2010). Preparing for the first breath: Genetic and cellular mechanisms in lung development. *Developmental Cell, 18*(1), 8–23.4. https://doi:10.1016/j.devcel.2009.12.010.

Morrison, C. D., Parvani, J. G., & Schiemann, W. P. (2013). The relevance of the TGF-beta paradox to EMT-MET programs. *Cancer Letters, 341*(1), 30–40. https://doi.org/10.1016/j.canlet.2013.02.048.

Nabhan, A. N., Brownfield, D. G., Harbury, P. B., Krasnow, M. A., & Desai, T. J. (2018). Single-cell Wnt signaling niches maintain stemness of alveolar type 2 cells. *Science, 359*(6380), 1118–1123. https://doi.org/10.1126/science.aam6603.

Neel, J.-C., Humbert, L., & Lebrun, J.-J. (2012). The dual role of TGFβ in human cancer: From tumor suppression to cancer metastasis. *ISRN Molecular Biology, 2012*, 381428. https://doi.org/10.5402/2012/381428.

Nelson, C. M., & Bissell, M. J. (2006). Of extracellular matrix, scaffolds, and signaling: Tissue architecture regulates development, homeostasis, and cancer. *Annual Review of Cell and Developmental Biology, 22*, 287–309. https://doi:10.1146/annurev.cellbio.22.010305.104315.

Ng-Blichfeldt, J. P., de Jong, T., Kortekaas, R. K., Wu, X., Lindner, M., Guryev, V., et al. (2019). TGF-beta activation impairs fibroblast ability to support adult lung epithelial progenitor cell organoid formation. *American Journal of Physiology. Lung Cellular and Molecular Physiology, 317*(1), L14–L28. https://doi.org/10.1152/ajplung.00400.2018.

Niethamer, T. K., Stabler, C. T., Leach, J. P., Zepp, J. A., Morley, M. P., Babu, A., et al. (2020). Defining the role of pulmonary endothelial cell heterogeneity in the response to acute lung injury. *eLife, 9*, e53072. https://doi.org/10.7554/eLife.53072.

Nolen-Walston, R. D., Kim, C. F., Mazan, M. R., Ingenito, E. P., Gruntman, A. M., Tsai, L., et al. (2008). Cellular kinetics and modeling of bronchioalveolar stem cell response during lung regeneration. *American Journal of Physiology. Lung Cellular and Molecular Physiology, 294*(6), L1158–L1165. https://doi:10.1152/ajplung.00298.2007.

Nowell, P. C. (1976). The clonal evolution of tumor cell populations. *Science, 194*, 23–28. https://doi:10.1126/science.959840.

O'Dwyer, D. N., Ashley, S. L., Gurczynski, S. J., Xia, M., Wilke, C., Falkowski, N. R., et al. (2019). Lung microbiota contribute to pulmonary inflammation and disease progression in pulmonary fibrosis. *American Journal of Respiratory and Critical Care Medicine, 199*(9), 1127–1138. https://doi.org/10.1164/rccm.201809-1650OC.

Ogawa, F., Walters, M. S., Shafquat, A., O'Beirne, S. L., Kaner, R. J., Mezey, J. G., et al. (2019). Role of KRAS in regulating normal human airway basal cell differentiation. *Respiratory Research, 20*(1), 181. https://doi.org/10.1186/s12931-019-1129-4.

Pacheco-Pinedo, E. C., Durham, A. C., Stewart, K. M., Goss, A. M., Lu, M. M., Demayo, F. J., et al. (2011). Wnt/beta-catenin signaling accelerates mouse lung tumorigenesis by imposing an embryonic distal progenitor phenotype on lung epithelium. *Journal of Clinical Investigation, 121*(5), 1935–1945. https://doi.org/10.1172/JCI44871.

Paget, S. (1889). The distribution of secondary growths in cancer of the breast. *Lancet, 1*, 571–573.

Pan, J. H., Adair-Kirk, T. L., Patel, A. C., Huang, T., Yozamp, N. S., Xu, J., et al. (2014). Myb permits multilineage airway epithelial cell differentiation. *Stem Cells, 32*(12), 3245–3256. https://doi.org/10.1002/stem.1814.

Paolillo, M., & Schinelli, S. (2019). Extracellular matrix alterations in metastatic processes. *International Journal of Molecular Sciences*, *20*(19). https://doi.org/10.3390/ijms20194947.
Pardo-Saganta, A., Law, B. M., Tata, P. R., Villoria, J., Saez, B., Mou, H., et al. (2015). Injury induces direct lineage segregation of functionally distinct airway basal stem/progenitor cell subpopulations. *Cell Stem Cell*, *16*(2), 184–197. https://doi.org/10.1016/j.stem.2015.01.002.
Parker, D., & Prince, A. (2011). Innate immunity in the respiratory epithelium. *American Journal of Respiratory Cell and Molecular Biology*, *45*(2), 189–201. https://doi.org/10.1165/rcmb.2011-0011RT.
Peake, J. L., Reynolds, S. D., Stripp, B. R., Stephens, K. E., & Pinkerton, K. E. (2000). Alteration of pulmonary neuroendocrine cells during epithelial repair of naphthalene-induced airway injury. *American Journal of Pathology*, *156*(1), 279–286. https://doi.org/10.1016/S0002-9440(10)64728-1.
Peng, T., Frank, D. B., Kadzik, R. S., Morley, M. P., Rathi, K. S., Wang, T., et al. (2015). Hedgehog actively maintains adult lung quiescence and regulates repair and regeneration. *Nature*, *526*(7574), 578–582. https://doi.org/10.1038/nature14984.
Pierce, G. B. (1974). Neoplasms, differentiations and mutations. *American Journal of Pathology*, *77*(1), 103–118.
Pierce, G. B., & Speers, W. C. (1988). Tumors as caricatures of the process of tissue renewal: Prospects for therapy by directing differentiation. *Cancer Research*, *48*(8), 1996–2004.
Pietanza, M. C., & Ladanyi, M. (2012). Bringing the genomic landscape of small-cell lung cancer into focus. *Nature Genetics*, *44*(10), 1074–1075. https://doi.org/10.1038/ng.2415.
Pon, J. R., & Marra, M. A. (2015). Driver and passenger mutations in cancer. *Annual Review of Pathology*, *10*, 25–50. https://doi.org/10.1146/annurev-pathol-012414-040312.
Popper, H. H. (2015). Lung adenocarcinomas: Comparison between mice and men. *Methods in Molecular Biology*, *1267*, 19–43. https://doi.org/10.1007/978-1-4939-2297-0_2.
Popper, H. H. (2016). Progression and metastasis of lung cancer. *Cancer and Metastasis Reviews*, *35*(1), 75–91. https://doi.org/10.1007/s10555-016-9618-0.
Quesenberry, P. J. (1993). Too much of a good thing. Reductionism run amok [editorial]. *Experimental Hematology*, *21*, 193–194.
Rana, M. K., Aloisio, F. M., Choi, C., & Barber, D. L. (2018). Formin-dependent TGF-beta signaling for epithelial to mesenchymal transition. *Molecular Biology of the Cell*, *29*(12), 1465–1475. https://doi.org/10.1091/mbc.E17-05-0325.
Rao, W., Wang, S., Duleba, M., Niroula, S., Goller, K., Xie, J., et al. (2020). Regenerative metaplastic clones in COPD lung drive inflammation and fibrosis. *Cell*, *181*(4), 848–864. e818 https://doi.org/10.1016/j.cell.2020.03.047.
Rawlins, E. L., Okubo, T., Xue, Y., Brass, D. M., Auten, R. L., Hasegawa, H., et al. (2009). The role of Scgb1a1 + Clara cells in the long-term maintenance and repair of lung airway, but not alveolar, epithelium. *Cell Stem Cell*, *4*(6), 525–534. https://doi:10.1016/j.stem.2009.04.002.
Ray, S., Chiba, N., Yao, C., Guan, X., McConnell, A. M., Brockway, B., et al. (2016). Rare SOX2(+) airway progenitor cells generate KRT5(+) cells that repopulate damaged alveolar parenchyma following influenza virus infection. *Stem Cell Reports*, 7(5), 817–825. https://doi.org/10.1016/j.stemcr.2016.09.010.
Reynolds, S. D., Giangreco, A., Power, J. H., & Stripp, B. R. (2000). Neuroepithelial bodies of pulmonary airways serve as a reservoir of progenitor cells capable of epithelial regeneration. *American Journal of Pathology*, *156*(1), 269–278.
Riccetti, M., Gokey, J. J., Aronow, B., & Perl, A. T. (2020). The elephant in the lung: Integrating lineage-tracing, molecular markers, and single cell sequencing data to identify distinct fibroblast populations during lung development and regeneration. *Matrix Biology*, *91-92*, 51–74. https://doi.org/10.1016/j.matbio.2020.05.002.

Rock, J. R., Onaitis, M. W., Rawlins, E. L., Lu, Y., Clark, C. P., Xue, Y., et al. (2009). Basal cells as stem cells of the mouse trachea and human airway epithelium. *Proceedings of the National Academy of Sciences of the United States of America*, *106*(31), 12771–12775.

Rock, J. R., Randell, S. H., & Hogan, B. L. (2010). Airway basal stem cells: A perspective on their roles in epithelial homeostasis and remodeling. *Disease Models & Mechanisms*, *3*(9–10), 545–556. https://doi.org/10.1242/dmm.006031.

Rodrigues, G., Hoshino, A., Kenific, C. M., Matei, I. R., Steiner, L., Freitas, D., et al. (2019). Tumour exosomal CEMIP protein promotes cancer cell colonization in brain metastasis. *Nature Cell Biology*, *21*(11), 1403–1412. https://doi.org/10.1038/s41556-019-0404-4.

Rodriguez-Canales, J., Parra-Cuentas, E., & Wistuba, I. I. (2016). Diagnosis and molecular classification of lung cancer. *Cancer Treatment and Research*, *170*, 25–46. https://doi.org/10.1007/978-3-319-40389-2_2.

Rosenthal, R., Cadieux, E. L., Salgado, R., Bakir, M. A., Moore, D. A., Hiley, C. T., et al. (2019). Neoantigen-directed immune escape in lung cancer evolution. *Nature*, *567*(7749), 479–485. https://doi.org/10.1038/s41586-019-1032-7.

Rudin, C. M., Poirier, J. T., Byers, L. A., Dive, C., Dowlati, A., George, J., et al. (2019). Molecular subtypes of small cell lung cancer: A synthesis of human and mouse model data. *Nature Reviews Cancer*, *19*(5), 289–297. https://doi.org/10.1038/s41568-019-0133-9.

Ruscetti, F. W., Akel, S., & Bartelmez, S. H. (2005). Autocrine transforming growth factor-beta regulation of hematopoiesis: Many outcomes that depend on the context. *Oncogene*, *24*(37), 5751–5763. https://doi.org/10.1038/sj.onc.1208921.

Safari, R., & Meuwissen, R. (2015). Practical use of advanced mouse models for lung cancer. *Methods in Molecular Biology*, *1267*, 93–124. https://doi.org/10.1007/978-1-4939-2297-0_5.

Sahai, E., Astsaturov, I., Cukierman, E., DeNardo, D. G., Egeblad, M., Evans, R. M., et al. (2020). A framework for advancing our understanding of cancer-associated fibroblasts. *Nature Reviews Cancer*, *20*(3), 174–186. https://doi.org/10.1038/s41568-019-0238-1.

Saito, A., Horie, M., & Nagase, T. (2018). TGF-beta signaling in lung health and disease. *International Journal of Molecular Sciences*, *19*(8). https://doi.org/10.3390/ijms19082460.

Salwig, I., Spitznagel, B., Vazquez-Armendariz, A. I., Khalooghi, K., Guenther, S., Herold, S., et al. (2019). Bronchioalveolar stem cells are a main source for regeneration of distal lung epithelia in vivo. *EMBO Journal*, *38*(12), e102099. https://doi.org/10.15252/embj.2019102099.

Sarode, P., Schaefer, M. B., Grimminger, F., Seeger, W., & Savai, R. (2020). Macrophage and tumor cell cross-talk is fundamental for lung tumor progression: We need to talk. *Frontiers in Oncology*, *10*, 324. https://doi.org/10.3389/fonc.2020.00324.

Schiller, H. B., Fernandez, I. E., Burgstaller, G., Schaab, C., Scheltema, R. A., Schwarzmayr, T., et al. (2015). Time- and compartment-resolved proteome profiling of the extracellular niche in lung injury and repair. *Molecular Systems Biology*, *11*(7), 819. https://doi.org/10.15252/msb.20156123.

Schofield, R. (1978). The relationship between the spleen colony-forming cell and the haemopoietic stem cell. *Blood Cells*, *4*(1-2), 7–25.

Schultz, G. S., Davidson, J. M., Kirsner, R. S., Bornstein, P., & Herman, I. M. (2011). Dynamic reciprocity in the wound microenvironment. *Wound Repair and Regeneration*, *19*(2), 134–148. https://doi.org/10.1111/j.1524-475X.2011.00673.x.

Shi, W., Xu, J., & Warburton, D. (2009). Development, repair and fibrosis: What is common and why it matters. *Respirology*, *14*(5), 656–665.

Sigismund, S., Avanzato, D., & Lanzetti, L. (2018). Emerging functions of the EGFR in cancer. *Molecular Oncology*, *12*(1), 3–20. https://doi.org/10.1002/1878-0261.12155.

Sommariva, M., Le Noci, V., Bianchi, F., Camelliti, S., Balsari, A., Tagliabue, E., et al. (2020). The lung microbiota: Role in maintaining pulmonary immune homeostasis and its implications in cancer development and therapy. *Cell and Molecular Life Sciences*, 77(14), 2739–2749. https://doi.org/10.1007/s00018-020-03452-8.

Spella, M., Lilis, I., Pepe, M. A., Chen, Y., Armaka, M., Lamort, A. S., et al. (2019). Club cells form lung adenocarcinomas and maintain the alveoli of adult mice. *eLife*, *8*, e45571. https://doi.org/10.7554/eLife.45571.

Stefater, J. A., 3rd, Ren, S., Lang, R. A., & Duffield, J. S. (2011). Metchnikoff's policemen: Macrophages in development, homeostasis and regeneration. *Trends in Molecular Medicine*, *17*(12), 743–752. https://doi.org/10.1016/j.molmed.2011.07.009.

Steiling, K., Ryan, J., Brody, J. S., & Spira, A. (2008). The field of tissue injury in the lung and airway. *Cancer Prevention Research*, *1*(6), 396–403. https://doi.org/10.1158/1940-6207.CAPR-08-0174.

Stratton, M. R., Campbell, P. J., & Futreal, P. A. (2009). The cancer genome. *Nature*, *458*(7239), 719–724. https://doi.org/10.1038/nature07943.

Sumitomo, R., Hirai, T., Fujita, M., Murakami, H., Otake, Y., & Huang, C. L. (2019). M2 tumor-associated macrophages promote tumor progression in non-small-cell lung cancer. *Experimental and Therapeutic Medicine*, *18*(6), 4490–4498. https://doi.org/10.3892/etm.2019.8068.

Sundar, I. K., Yao, H., Sellix, M. T., & Rahman, I. (2015). Circadian molecular clock in lung pathophysiology. *American Journal of Physiology. Lung Cellular and Molecular Physiology*, *309*(10), L1056–L1075. https://doi.org/10.1152/ajplung.00152.2015

Sutherland, K. D., Proost, N., Brouns, I., Adriaensen, D., Song, J. Y., & Berns, A. (2011). Cell of origin of small cell lung cancer: Inactivation of Trp53 and Rb1 in distinct cell types of adult mouse lung. *Cancer Cell*, *19*(6), 754–764. https://doi.org/S1535-6108(11)00167-X [pii] https://doi.org/10.1016/j.ccr.2011.04.019.

Tabbo, F., Nottegar, A., Guerrera, F., Migliore, E., Luchini, C., Maletta, F., et al. (2018). Cell of origin markers identify different prognostic subgroups of lung adenocarcinoma. *Human Pathology*, *75*, 167–178. https://doi.org/10.1016/j.humpath.2018.01.017.

Tadokoro, T., Gao, X., Hong, C. C., Hotten, D., & Hogan, B. L. (2016). BMP signaling and cellular dynamics during regeneration of airway epithelium from basal progenitors. *Development*, *143*(5), 764–773. https://doi.org/10.1242/dev.126656.

Taipale, J., & Keski-Oja, J. (1997). Growth factors in the extracellular matrix. *FASEB Journal*, *11*(1), 51–59. https://doi.org/10.1096/fasebj.11.1.9034166.

Tammela, T., Sanchez-Rivera, F. J., Cetinbas, N. M., Wu, K., Joshi, N. S., Helenius, K., et al. (2017). A Wnt-producing niche drives proliferative potential and progression in lung adenocarcinoma. *Nature*, *545*(7654), 355–359. https://doi.org/10.1038/nature22334.

Tanaka, Y., Yamaguchi, M., Hirai, S., Sumi, T., Tada, M., Saito, A., et al. (2018). Characterization of distal airway stem-like cells expressing N-terminally truncated p63 and thyroid transcription factor-1 in the human lung. *Experimental Cell Research*, *372*(2), 141–149. https://doi.org/10.1016/j.yexcr.2018.09.020.

Tata, P. R., Mou, H., Pardo-Saganta, A., Zhao, R., Prabhu, M., Law, B. M., et al. (2013). Dedifferentiation of committed epithelial cells into stem cells in vivo. *Nature*, *503*(7475), 218–223. https://doi.org/10.1038/nature12777.

Tata, P. R., & Rajagopal, J. (2017). Plasticity in the lung: Making and breaking cell identity. *Development*, *144*(5), 755–766. https://doi.org/10.1242/dev.143784.

Thiesse, J., Namati, E., Sieren, J. C., Smith, A. R., Reinhardt, J. M., Hoffman, E. A., et al. (2010). Lung structure phenotype variation in inbred mouse strains revealed through in vivo micro-CT imaging. *Journal of Applied Physiology*, *109*(6), 1960–1968. https://doi.org/10.1152/japplphysiol.01322.2009.

Torday, J. S., Sanchez-Esteban, J., & Rubin, L. P. (1998). Paracrine mediators of mechanotransduction in lung development. *American Journal of the Medical Sciences, 316*(3), 205–208.

Travis, W. D., Brambilla, E., Nicholson, A. G., Yatabe, Y., Austin, J. H. M., Beasley, M. B., et al. (2015). The 2015 World Health Organization classification of lung tumors: Impact of genetic, clinical and radiologic advances since the 2004 classification. *Journal of Thoracic Oncology, 10*(9), 1243–1260. https://doi.org/10.1097/JTO.0000000000000630.

Treffers, L. W., Hiemstra, I. H., Kuijpers, T. W., van den Berg, T. K., & Matlung, H. L. (2016). Neutrophils in cancer. *Immunology Reviews, 273*(1), 312–328. https://doi.org/10.1111/imr.12444.

Valiente, M., Obenauf, A. C., Jin, X., Chen, Q., Zhang, X. H., Lee, D. J., et al. (2014). Serpins promote cancer cell survival and vascular co-option in brain metastasis. *Cell, 156*(5), 1002–1016. https://doi.org/10.1016/j.cell.2014.01.040.

Vaughan, A. E., Brumwell, A. N., Xi, Y., Gotts, J. E., Brownfield, D. G., Treutlein, B., et al. (2015). Lineage-negative progenitors mobilize to regenerate lung epithelium after major injury. *Nature, 517*(7536), 621–625. https://doi.org/10.1038/nature14112.

Volckaert, T., & De Langhe, S. (2014). Lung epithelial stem cells and their niches: Fgf10 takes center stage. *Fibrogenesis & Tissue Repair*, 7, 8. https://doi.org/10.1186/1755-1536-7-8.

Voog, J., & Jones, D. L. (2010). Stem cells and the niche: A dynamic duo. *Cell Stem Cell, 6*(2), 103–115.

Wagner, P. D. (2015). The physiological basis of pulmonary gas exchange: Implications for clinical interpretation of arterial blood gases. *European Respiratory Journal, 45*(1), 227–243. https://doi.org/10.1183/09031936.00039214.

Walkin, L., Herrick, S. E., Summers, A., Brenchley, P. E., Hoff, C. M., Korstanje, R., et al. (2013). The role of mouse strain differences in the susceptibility to fibrosis: A systematic review. *Fibrogenesis & Tissue Repair, 6*(1), 18. https://doi.org/10.1186/1755-1536-6-18.

Wang, L., Liu, H., Jiao, Y., Wang, E., Clark, S. H., Postlethwaite, A. E., et al. (2015). Differences between mice and humans in regulation and the molecular network of collagen, Type III, Alpha-1 at the gene expression level: Obstacles that translational research must overcome. *International Journal of Molecular Sciences, 16*(7), 15031–15056. https://doi.org/10.3390/ijms160715031.

Wang, F., Lu, J. B., Wu, X. Y., Feng, Y. F., Shao, Q., An, X., et al. (2019). Clinical genetic features and related survival implications in patients with surgically resected large-cell lung cancer. *Cancer Management and Research, 11*, 5489–5499. https://doi.org/10.2147/CMAR.S200263.

Wang, X., Wang, Z., Pan, J., Lu, Z. Y., Xu, D., Zhang, H. J., et al. (2020). Patterns of extrathoracic metastases in different histological types of lung cancer. *Frontiers in Oncology, 10*, 715. https://doi.org/10.3389/fonc.2020.00715.

Warburton, D., Tefft, D., Mailleux, A., Bellusci, S., Thiery, J. P., Zhao, J., et al. (2001). Do lung remodeling, repair, and regeneration recapitulate respiratory ontogeny? *American Journal of Respiratory and Critical Care Medicine, 164*(10 Pt. 2), S59–S62.

Warburton, D., Wuenschell, C., Flores-Delgado, G., & Anderson, K. (1998). Commitment and differentiation of lung cell lineages. *Biochemistry and Cell Biology, 76*, 971–995.

Waters, C. M., Roan, E., & Navajas, D. (2012). Mechanobiology in lung epithelial cells: Measurements, perturbations, and responses. *Comparative Physiology, 2*(1), 1–29. https://doi.org/10.1002/cphy.c100090.

Watson, J., & Francavilla, C. (2018). Regulation of FGF10 signaling in development and disease. *Frontiers in Genetics, 9*, 500. https://doi.org/10.3389/fgene.2018.00500.

Weibel, E. R. (2009). What makes a good lung? *Swiss Medical Weekly, 139*(27–28), 375–386. https://doi.org/10.4414/smw.2009.12270. [pii] smw-12270.

Weibel, E. R. (2013). It takes more than cells to make a good lung. *American Journal of Respiratory and Critical Care Medicine*, *187*(4), 342–346. https://doi.org/10.1164/rccm.201212-2260OE.

Whitsett, J. A., & Alenghat, T. (2015). Respiratory epithelial cells orchestrate pulmonary innate immunity. *Nature Immunology*, *16*(1), 27–35. https://doi.org/10.1038/ni.3045.

Whitsett, J. A., Haitchi, H. M., & Maeda, Y. (2011). Intersections between pulmonary development and disease. *American Journal of Respiratory and Critical Care Medicine*, *184*(4), 401–406. https://doi.org/10.1164/rccm.201103-0495PP.

Wieland, E., Rodriguez-Vita, J., Liebler, S. S., Mogler, C., Moll, I., Herberich, S. E., et al. (2017). Endothelial notch1 activity facilitates metastasis. *Cancer Cell*, *31*(3), 355–367. https://doi.org/10.1016/j.ccell.2017.01.007.

Winkler, F. (2017). Hostile takeover: How tumours hijack pre-existing vascular environments to thrive. *Journal of Pathology*, *242*(3), 267–272. https://doi.org/10.1002/path.4904.

Wood, S. L., Pernemalm, M., Crosbie, P. A., & Whetton, A. D. (2014). The role of the tumor-microenvironment in lung cancer-metastasis and its relationship to potential therapeutic targets. *Cancer Treatment Reviews*, *40*(4), 558–566. https://doi.org/10.1016/j.ctrv.2013.10.001.

Wu, X., van Dijk, E. M., Ng-Blichfeldt, J. P., Bos, I. S. T., Ciminieri, C., Konigshoff, M., et al. (2019). Mesenchymal WNT-5A/5B signaling represses lung alveolar epithelial progenitors. *Cells*, *8*(10), 1147. https://doi.org/10.3390/cells8101147.

Xie, T., Wang, Y., Deng, N., Huang, G., Taghavifar, F., Geng, Y., et al. (2018). Single cell deconvolution of fibroblast heterogeneity in mouse pulmonary fibrosis. *Cell Reports*, *22*(13), 3625–3640. https://doi.org/10.1016/j.celrep.2018.03.010.

Xu, R., Boudreau, A., & Bissell, M. J. (2009). Tissue architecture and function: Dynamic reciprocity via extra- and intra-cellular matrices. *Cancer Metastasis Reviews*, *28*(1-2), 167–176. https://doi.org/10.1007/s10555-008-9178-z.

Xu, X., Rock, J. R., Lu, Y., Futtner, C., Schwab, B., Guinney, J., et al. (2012). Evidence for type II cells as cells of origin of K-Ras-induced distal lung adenocarcinoma. *Proceedings of the National Academy of Sciences of the United States of America*, *109*(13), 4910–4915. https://doi.org/10.1073/pnas.1112499109.

Xu, J. Y., Zhang, C., Wang, X., Zhai, L., Ma, Y., Mao, Y., et al. (2020). Integrative proteomic characterization of human lung adenocarcinoma. *Cell*, *182*(1), 245–261. e217. https://doi.org/10.1016/j.cell.2020.05.043.

Yang, X. Y., Liao, J. J., & Xue, W. R. (2019). FMNL1 down-regulation suppresses bone metastasis through reducing TGF-beta1 expression in non-small cell lung cancer (NSCLC). *Biomedicine & Pharmacotherapy*, *117*, 109126. https://doi.org/10.1016/j.biopha.2019.109126.

Yin, Y., Wang, J., Wang, X., Gu, L., Pei, H., Kuai, S., et al. (2015). Prognostic value of the neutrophil to lymphocyte ratio in lung cancer: A meta-analysis. *Clinics (São Paulo, Brazil)*, *70*(7), 524–530. https://doi.org/10.6061/clinics/2015(07)10.

Yokota, J., & Kohno, T. (2004). Molecular footprints of human lung cancer progression. *Cancer Science*, *95*(3), 197–204. https://doi.org/10.1111/j.1349-7006.2004.tb02203.x.

Yoshida, G. J. (2020). Regulation of heterogeneous cancer-associated fibroblasts: The molecular pathology of activated signaling pathways. *Journal of Experimental & Clinical Cancer Research*, *39*(1), 112. https://doi.org/10.1186/s13046-020-01611-0.

Yousefi, M., Bahrami, T., Salmaninejad, A., Nosrati, R., Ghaffari, P., & Ghaffari, S. H. (2017). Lung cancer-associated brain metastasis: Molecular mechanisms and therapeutic options. *Cellular Oncology*, *40*(5), 419–441. https://doi.org/10.1007/s13402-017-0345-5.

Yuan, Y. (2016). Spatial heterogeneity in the tumor microenvironment. *Cold Spring Harbor Perspectives in Medicine*, *6*(8), a026583. https://doi.org/10.1101/cshperspect.a026583.

Yuan, T., Volckaert, T., Chanda, D., Thannickal, V. J., & De Langhe, S. P. (2018). Fgf10 signaling in lung development, homeostasis, disease, and repair after injury. *Frontiers in Genetics*, *9*, 418. https://doi.org/10.3389/fgene.2018.00418.

Yuan, T., Volckaert, T., Redente, E. F., Hopkins, S., Klinkhammer, K., Wasnick, R., et al. (2019). FGF10-FGFR2B signaling generates basal cells and drives alveolar epithelial regeneration by bronchial epithelial stem cells after lung injury. *Stem Cell Reports*, *12*(5), 1041–1055. https://doi.org/10.1016/j.stemcr.2019.04.003.

Yuan, X., Wu, H., Han, N., Xu, H., Chu, Q., Yu, S., et al. (2014). Notch signaling and EMT in non-small cell lung cancer: Biological significance and therapeutic application. *Journal of Hematology & Oncology*, 7, 87. https://doi.org/10.1186/s13045-014-0087-z.

Zacharek, S. J., Fillmore, C. M., Lau, A. N., Gludish, D. W., Chou, A., Ho, J. W., et al. (2011). Lung stem cell self-renewal relies on BMI1-dependent control of expression at imprinted loci. *Cell Stem Cell*, *9*(3), 272–281. https://doi.org/S1934-5909(11)00337-7 [pii] https://doi.org/10.1016/j.stem.2011.07.007.

Zacharias, W. J., Frank, D. B., Zepp, J. A., Morley, M. P., Alkhaleel, F. A., Kong, J., et al. (2018). Regeneration of the lung alveolus by an evolutionarily conserved epithelial progenitor. *Nature*, *555*(7695), 251–255. https://doi.org/10.1038/nature25786.

Zepp, J. A., Zacharias, W. J., Frank, D. B., Cavanaugh, C. A., Zhou, S., Morley, M. P., et al. (2017). Distinct mesenchymal lineages and niches promote epithelial self-renewal and myofibrogenesis in the lung. *Cell*, *170*(6), 1134–1148. e1110. https://doi.org/10.1016/j.cell.2017.07.034.

Zhang, Y. E. (2009). Non-Smad pathways in TGF-beta signaling. *Cell Research*, *19*(1), 128–139. https://doi.org/10.1038/cr.2008.328.

Zhang, J., Fujimoto, J., Zhang, J., Wedge, D. C., Song, X., Zhang, J., et al. (2014). Intratumor heterogeneity in localized lung adenocarcinomas delineated by multiregion sequencing. *Science*, *346*(6206), 256–259. https://doi.org/10.1126/science.1256930.

Zhang, Y., Tseng, J. T., Lien, I. C., Li, F., Wu, W., & Li, H. (2020). mRNAsi index: Machine learning in mining lung adenocarcinoma stem cell biomarkers. *Genes (Basel)*, *11*(3), 257. https://doi.org/10.3390/genes11030257.

Zhang, C., Zhang, J., Xu, F. P., Wang, Y. G., Xie, Z., Su, J., et al. (2019). Genomic landscape and immune microenvironment features of preinvasive and early invasive lung adenocarcinoma. *Journal of Thoracic Oncology*, *14*(11), 1912–1923. https://doi.org/10.1016/j.jtho.2019.07.031.

Zheng, H., Zhan, Y., Liu, S., Lu, J., Luo, J., Feng, J., et al. (2018). The roles of tumor-derived exosomes in non-small cell lung cancer and their clinical implications. *Journal of Experimental & Clinical Cancer Research*, *37*(1), 226. https://doi.org/10.1186/s13046-018-0901-5.

Zhou, Y., Horowitz, J. C., Naba, A., Ambalavanan, N., Atabai, K., Balestrini, J., et al. (2018). Extracellular matrix in lung development, homeostasis and disease. *Matrix Biology*, *73*, 77–104. https://doi.org/10.1016/j.matbio.2018.03.005.

Zhu, L., An, L., Ran, D., Lizarraga, R., Bondy, C., Zhou, X., et al. (2019). The club cell marker SCGB1A1 downstream of FOXA2 is reduced in asthma. *American Journal of Respiratory Cell and Molecular Biology*, *60*(6), 695–704. https://doi.org/10.1165/rcmb.2018-0199OC.

Zinkle, A., & Mohammadi, M. (2019). Structural biology of the FGF7 subfamily. *Frontiers in Genetics*, *10*, 102. https://doi.org/10.3389/fgene.2019.00102.

Zuo, W. L., Shenoy, S. A., Li, S., O'Beirne, S. L., Strulovici-Barel, Y., Leopold, P. L., et al. (2018). Ontogeny and biology of human small airway epithelial club cells. *American Journal of Respiratory and Critical Care Medicine*, *198*(11), 1375–1388. https://doi.org/10.1164/rccm.201710-2107OC.

Zuo, W., Zhang, T., Wu, D. Z., Guan, S. P., Liew, A. A., Yamamoto, Y., et al. (2015). p63(+)Krt5(+) distal airway stem cells are essential for lung regeneration. *Nature*, *517*(7536), 616–620. https://doi.org/10.1038/nature13903.

CHAPTER FIVE

The prostate cancer stem cell niche: Genetic drivers and therapeutic approaches

Giacomo Canesin[*,†]

Department of Surgery, Cancer Research Institute, Beth Israel Deaconess Medical Center and Harvard Medical School, Boston, MA, United States

*Corresponding author: e-mail address: giacomo.canesin@gmail.com

Contents

1. Introduction

Prostate cancer (PCa) is the most commonly diagnosed malignancy in men and the second cause of cancer-related death in western countries (Siegel, Miller, & Jemal, 2019). From a clinical perspective, PCa is a very heterogeneous disease, as it can range from indolent to very aggressive

† Current address: Vor Biopharma, Cambridge, MA, United States.

Advances in Stem Cells and their Niches, Volume 5
ISSN 2468-5097
https://doi.org/10.1016/bs.asn.2020.09.001

and it can rapidly progress to metastatic and treatment-refractory (DeMarzo, Nelson, Isaacs, & Epstein, 2003; Watson, Arora, & Sawyers, 2015). The malignant transformation of the prostate is a multistep process, usually starting with prostatic intraepithelial neoplasia (PIN) that evolves into a localized cancer and advanced prostate adenocarcinoma with local invasion, to finally culminate in metastatic PCa (Fig. 1) (Shen & Abate-Shen, 2010; Wang, Zhao, Spring, & DePinho, 2018). Despite the high long-term survival in localized prostate cancer, conventional therapies fail to eradicate the disease, resulting in tumor relapse and in patients developing metastatic castration-resistant PCa (CRPC), which remains largely incurable (Rice, Malhotra, & Stoyanova, 2019). A significant role in this context is played by prostate cancer stem cells (PCSCs), a small population of therapy-resistant quiescent cells that reside within the tumor and are able to differentiate into mature tumor cells, inducing tumor relapse and metastasis (Castillo, Valenzuela, Huidobro, Contreras, & Castellon, 2014). In this chapter, I will describe the role of PCSCs in prostate cancer development, focusing on the genetic drivers of the stem cell niche and the therapeutic approaches that are currently under investigation.

2. The prostate gland: Structure and general overview

The prostate is a walnut sized gland located around the urethra at the base of the bladder, it has secretory functions and its activity is mainly regulated by male sex hormones (DeMarzo et al., 2003; Maitland, Frame, Polson, Lewis, & Collins, 2011). The normal, mature human prostate is highly organized at the cellular level, showing a two-layered epithelium with three phenotypically distinct cell types: basal, luminal and neuroendocrine cells (Fig. 2) (Shen & Abate-Shen, 2010; van Leenders & Schalken, 2003). Luminal cells constitute the major cell type of the prostatic epithelium and represent the exocrine compartment of the prostate, secreting prostate specific antigen (PSA) and prostatic acid phosphatase (PAP) into the lumen of the gland (Berry, Maitland, & Collins, 2008). These cells are terminally differentiated, depend on androgens for their survival and express high levels of cytokeratin-8 (CK-8), cytokeratin-18 (CK-18) and the androgen receptor (AR) (Collins & Maitland, 2006). Basal cells are located between the luminal cells and the basement membrane: they express low to undetectable levels of AR (and hence are androgen-independent) but they express higher levels of cytokeratin-5 (CK-5), cytokeratin-14 (CK-14),

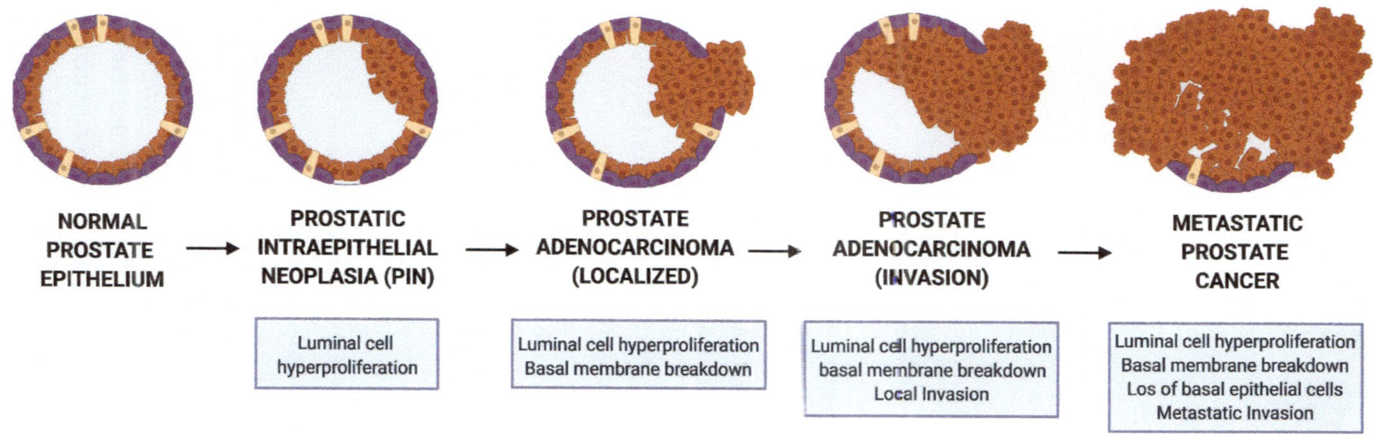

Fig. 1 Stages of progression of human prostate cancer. Prostate cancer initiates as *in situ* carcinoma called prostatic intraepithelial neoplasia (PIN), characterized by a hyper-proliferation of luminal cells. The disease evolves into a localized adenocarcinoma with the breakdown of the basal membrane; it further progresses into invasive carcinoma and eventually culminates into metastatic prostate cancer, with metastasis to distant organs.

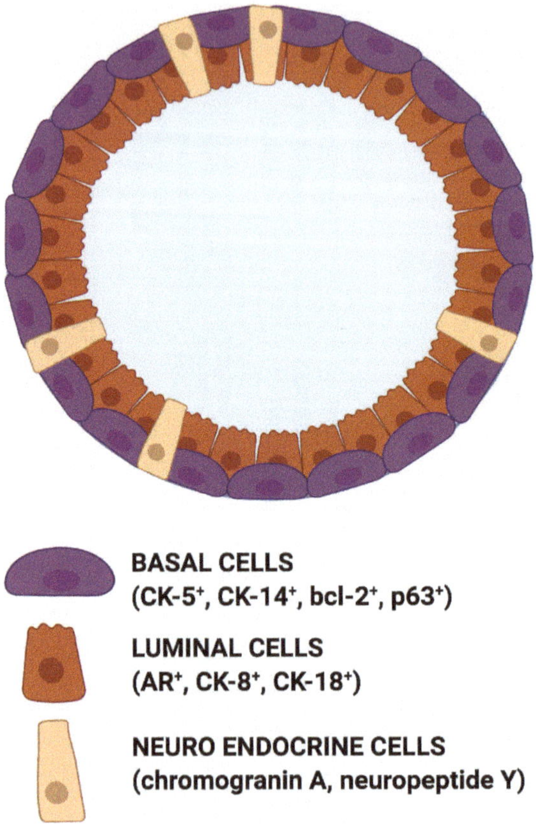

Fig. 2 Schematic representation of the architecture of the human normal prostate epithelium. The prostate epithelium consists of a basal layer of relatively undifferentiated basal cells and a luminal layer of terminally differentiated secretory luminal cells. A small population of scattered neuroendocrine cells with endocrine/paracrine functions is also present.

the anti-apoptotic protein BCL2 and the protein P63, a homologue of the tumor-suppressor P53 (Bonkhoff & Remberger, 1996; Collins & Maitland, 2006). Finally, neuroendocrine cells are terminally differentiated cells residing within the secretory epithelium: they contain serotonin, thyroid-stimulating hormone and chromogranin A, they can express neuropeptide Y but do not express AR and are androgen-insensitive (Bonkhoff & Remberger, 1996; Hudson, O'Hare, Watt, & Masters, 2000). The existence of a stem cell population within the prostate (prostate stem cells, PSCs) was first postulated in the 1980s, when studies showed that the basal cell compartment contains cells capable of regenerating the prostate epithelium after transitional androgen deprivation (English, Santen, & Isaacs, 1987). Later studies based on cytokeratin expression patterns analysis confirmed that PSCs are mainly located in niches within the basal layer and are slow-cycling or quiescent cells that can be recruited into active proliferation in response to different physiological stimuli (Hudson, 2004; Sawicki & Rothman, 2002). Further studies have described PSCs as a heterogeneous population containing sub-populations of self-renewing cells, with the ability to reconstitute either the whole prostate epithelium or with the capacity to differentiate into definitive luminal or basal cells only (Rizzo, Attard, & Hudson, 2005; Schmelz et al., 2005). Additionally, more recent lineage-tracing studies in the mouse prostate suggest that both basal and luminal cells contain lineage-restricted stem/progenitor cells, indicating that the basal layer is not the only niche for stem and progenitor cells within the normal prostate epithelium (Li & Shen, 2019; Zhang, Zhao, Li, Kirk, & Tang, 2018).

3. Prostate cancer and prostate cancer stem cells

Most prostate tumors are adenocarcinomas with a predominant luminal phenotype: for this reason PCa was thought to be primarily a disease of the luminal secretory epithelium, but studies have now observed both luminal (Choi, Zhang, Zhang, Ittmann, & Xin, 2012; Wang et al., 2009, 2013; Yoo et al., 2016) and basal (Goldstein, Huang, Guo, Garraway, & Witte, 2010; Lawson et al., 2010; Lawson, Xin, Lukacs, Cheng, & Witte, 2007) phenotypes in PCa. Additionally, prostate adenocarcinomas with neuroendocrine features (NEPC) have been described, with an aggressive phenotype resulting in metastasis to bones, brain, liver and bladder (Vlachostergios, Puca, & Beltran, 2017). The heterogeneity of PCa in terms of cell composition is further enhanced by the presence of stem-like tumor cells or prostate cancer stem cells (PCSCs), a small population of tumor-initiating cells displaying phenotypic and functional features typical of normal PSCs (Di Zazzo et al., 2016). PCSCs only represent a small fraction of all prostate tumor cells (approximately 0.1%) but they retain the ability to initiate tumor, they are able to self-renew *in vitro* and to reproduce the cellular composition of the original tumor (Collins, Habib, Maitland, & Neal, 2001). Furthermore, PCSCs isolated from patients undergoing radical prostatectomy show high clonogenic and invasive capacity (Collins, Berry, Hyde, Stower, & Maitland, 2005; Lin et al., 2016). Several studies have identified putative PCSCs based on the expression of specific surface markers, including CD44, $\alpha_2\beta_1$ integrin, CD133 and aldehyde dehydrogenase-1 (ALDH1)

(Gu, Yuan, Wills, & Kasper, 2007; Patrawala, Calhoun-Davis, Schneider-Broussard, & Tang, 2007; van den Hoogen et al., 2010; Williams, Motiani, Giridhar, & Kasper, 2013; Williamson et al., 2012).

CD44 is a glycoprotein involved in cell-cell interaction, cell adhesion and migration (Adamowicz, Pakravan, Bakhshinejad, Drewa, & Babashah, 2017). Studies have demonstrated that $CD44^+$ clones from the PC3 PCa cell line are able to initiate serially transplantable tumors after subcutaneous injection *in vivo* (Li, Chen, Calhoun-Davis, Claypool, & Tang, 2008) indicating a possible role for CD44 as a PCSC marker. Further data showed that $CD44^+$ cells harvested from human PCa reconstitute the original tumor *in vivo* and can differentiate into all the epithelial cell lineages of the prostate including luminal, basal and neuroendocrine cells (Gu et al., 2007). Importantly, $CD44^+$ cells possess a higher proliferative, clonogenic, tumorigenic and metastatic potential compared to $CD44^-$ cells, and the expression of CD44 is associated with increased invasion, adhesion, chemoresistance, radioresistance and epithelial-mesenchymal transition (EMT) induction, all features of PCSCs (Ni et al., 2014; Patrawala et al., 2006, 2007). In addition, data from preclinical models show that CD44 expression correlates with larger tumor burden, increased metastasis and a more immunosuppressive microenvironment (Wu, Huang, Chen, & Chen, 2019). From a clinical point of view, high CD44 expression levels were associated with biochemical recurrence and distant metastasis in PCa patients, indicating that CD44 may be a marker of poor prognosis and a potential therapeutic target (Shang et al., 2015; Wu et al., 2015).

CD133, also known as prominin-1, is a well-established human hematopoietic stem cell marker and is described as a stem cell marker in both normal and malignant prostate (Grosse-Gehling et al., 2013). Initial studies revealed that CD133 expression was restricted to a subpopulation of $\alpha_2\beta_1$-integrin$^+$ prostate basal cells with stemness properties (Richardson et al., 2004). Compared to $CD133^-$ cells, $CD133^+$ cells showed greater colony-forming ability *in vitro* and were able to form a fully differentiated prostatic epithelium with associated secretory activity when grafted *in vivo* (Richardson et al., 2004). Subsequent studies demonstrated that $CD133^+$ cells have higher proliferative potential, are able to generate spheres *in vitro* and have increased tumorigenicity ability after implantation *in vivo* (Kelly et al., 2010; Pellacani, Oldridge, Collins, & Maitland, 2013).

ALDH1 is a member of the aldehyde dehydrogenase family, a group of enzymes involved in intracellular retinoic acid production (Yoshida, Hsu, & Davé, 1992). ALDH enzymes have important functions in epithelial homeostasis and, as a result, their deregulation has been implicated in cell proliferation, differentiation, drug resistance and response to oxidative stress in multiple cancers (Tomita, Tanaka, Tanaka, & Hara, 2016; Vassalli, 2019). High ALDH activity has been increasingly used to identify normal cell populations enriched in stem and progenitor cells as well as cell populations from cancer tissues enriched in tumor-initiating stem-like cells (Gentry et al., 2007; Ginestier et al., 2007; Huang et al., 2009). ALDH1$^+$ cells isolated from human PCa cell lines showed important CSCs properties, including high *in vitro* tumorigenicity, *in vivo* tumor initiation and self-renewal capacities, and were able to initiate transplantable tumors that gave rise to a heterogeneous population of cancer cells (Li et al., 2010). Additionally, subpopulations of human PCa cells with high ALDH1 activity were identified as metastasis-initiating cells in preclinical models of orthotopic growth and metastasis (van den Hoogen et al., 2010). Immunohistochemical studies on clinical specimens showed that ALDH1$^+$ cells can be detected in the basal layer of normal prostate glands, while in cancer specimens they are found within both secretory type cancer epithelia and neuroendocrine tumor cells (Li et al., 2010). Importantly, high ALDH1 expression significantly correlates with high Gleason Score, primary tumor stage, regional lymph node involvement and patient's poor survival, indicating the potential role of this protein in PCa tumorigenesis (Kalantari et al., 2017; Matsika et al., 2015).

PCSCs are also characterized by increased gene expression of intracellular stem cell reprogramming factors and transcriptional regulators, such as *OCT3/4*, *SOX2*, *KLF4*, *NANOG* and *MYC* (Civenni et al., 2013; Peitzsch et al., 2016; Rajasekhar, Studer, Gerald, Socci, & Scher, 2011). These genes are defined as pluripotency-determining factors, as they play a critical role in maintaining the self-renewal ability and pluripotency of mouse embryonic stem cells (Mitsui et al., 2003; Okita, Ichisaka, & Yamanaka, 2007). However, mounting evidence suggests that these factors are up-regulated in human malignancies and contribute to carcinogenesis by regulating the stem cell niche (Yasumizu et al., 2020; Zhang et al., 2019). For example, *NANOG* expression in PCSCs promotes tumor development through an up-regulation of CD133 and ALDH1, resulting in resistance to androgen deprivation (Jeter et al., 2011). The expression of *OCT3/4* and *SOX2* is significantly increased in PCa tissue compared to normal prostate or benign prostate hyperplasia tissue and cells expressing high *OCT3/4* and *SOX2* levels show high tumorigenicity *in vivo* (Bae et al., 2010). Additionally, a study identified a rare population of prostate cells expressing *OCT4A*, a splice variant of the *OCT3/4* gene (Sotomayor, Godoy,

Smith, & Huss, 2009). The expression of *OCT4A* increases in patients with high Gleason Score and OCT4A$^+$ cells are non-proliferative, undifferentiated cells co-expressing chromogranin A and synaptophysin, both markers of neuroendocrine differentiation (Sotomayor et al., 2009).

PCSCs are characterized by a high level of plasticity acquired during EMT, which contributes to metastatic progression and treatment resistance (Montanari et al., 2017). Through EMT, polarized epithelial cells lose their apical-basal polarity and the expression of epithelial markers (β-catenin, E-cadherin) in favor of mesenchymal markers (vimentin, N-cadherin, fibronectin, α-smooth muscle actin), becoming motile mesenchymal cells with enhanced migratory/invasive abilities and increased apoptosis resistance (Antony, Thiery, & Huang, 2019; Ye & Weinberg, 2015). This process is typically controlled by specific transcription factors, such as SNAIL, TWIST and ZEB, which induce the transcriptional repression of E-cadherin (Kalluri & Weinberg, 2009; Thiery, 2002). Numerous studies have shown that PCSCs display increased EMT markers and exhibit enhanced migration/invasion ability (Mani et al., 2008; Ruscetti, Quach, Dadashian, Mulholland, & Wu, 2015). Additionally, the expression of the stem cell marker CD44 is increased in PCa cells following androgen deprivation therapy (ADT), which leads to accumulation of de-differentiated CD44$^+$ PCSCs able to initiate EMT and maintain a mesenchymal phenotype (Shang et al., 2015). Several signaling pathways (including AR, STAT3 and Wnt/β-catenin) have been shown to sustain PCSCs growth and self-renewal through regulation of EMT, further confirming the important role played by stem/progenitor cells in PCa progression (see below).

4. Genetic drivers of stemness in prostate cancer

Identifying the genetic drivers of PCa stemness and understanding the genetic pathways that regulate the establishment, expansion and maintenance of the PCSCs pool is fundamental for the development of more effective therapeutic strategies, in order to promote PCSCs exhaustion and prevent treatment resistance and disease recurrence. In this section I will describe the role of canonical pathways identified as drivers of stemness in the context of PCa.

4.1 AR signaling pathway

The normal development and maintenance of the prostate is dependent on androgens, acting through AR (Heinlein & Chang, 2004). The AR has three main domains: a C-terminal ligand binding domain (LBD), an N-terminal transcriptional activation domain and a central DNA-binding domain consisting of two zinc finger motifs and a hinge domain (Setlur & Rubin, 2005). Binding of androgens induces a conformational change that leads to receptor dimerization and the subsequent nuclear translocation of the hormone-receptor complex (Setlur & Rubin, 2005). Once in the nucleus, this complex binds to specific androgen response elements in the upstream promoter of target genes, leading to transcriptional activation and modulation of a transcription complex that determines the expression of androgen-regulated genes (Cano, Godoy, Escamilla, Dhir, & Onate, 2007; Shang, Myers, & Brown, 2002). While AR mediates male sexual differentiation, sperm production and the development of the normal prostatic gland, the AR signaling pathway is also critical to all stages of PCa, from development to progression to castration resistance following ADT (Holzbeierlein et al., 2004). In this context, a study by Nelson et al. in 2002 identified specific androgen-regulated genes and gene networks that regulate cellular responses to androgens, showing that this pathway is implicated in cell proliferation, differentiation, apoptosis, metabolism and secretory activities (Nelson et al., 2002). Since then, several other studies have shown the implication of AR in the regulation of the PCSC niche. Park and colleagues showed the existence of a feedback loop between AR and the family of ZEB transcription factors that mediate EMT in both normal and cancerous prostate, contributing to the formation of PCSCs and to skeletal invasion (Park, Gaur, Lengyel, & Peter, 2008). A different study demonstrated that AR binds directly to the *NANOG* promoter and up-regulates its mRNA and protein levels, with an overall effect of increasing the self-renewal and stem-like properties of PCa cells (Kregel, Szmulewitz, & Vander Griend, 2014). Additionally, androgens and AR have been shown to regulate alternative splicing of the *CD44* transcript, thus contributing to pro-tumorigenic and metastatic phenotypes (Clark et al., 2008; Mongan et al., 2011; Rajan et al., 2008). Hence, these studies point at androgens and AR as key regulators of stemness-related signaling pathways in PCa, indicating that they may contribute to the generation of new subpopulations of PCSCs during tumor progression.

4.2 IL-6 and JAK/STAT3 signaling pathway

Interleukin-6 (IL-6) is a potent inflammatory cytokine recognized as a major regulator of PCa progression (Twillie et al., 1995). Binding of IL-6 to its receptor (IL-6r) triggers the activation of the Janus kinase (JAK) by autophosphorylation, which in turn leads to the activation of Signal Transducer and Activator of Transcription 3 (STAT3) by phosphorylation at tyrosine 705 (Y705) (Canesin, Krzyzanowska, Hellsten, & Bjartell, 2020). As a consequence, pSTAT3 forms stable homodimers that translocate into the nucleus and bind specific DNA sequences, resulting in the transcription of genes that regulate cell proliferation, differentiation, apoptosis and inflammation (Cocchiola et al., 2019; Heinrich et al., 2003). For example, STAT3 regulates the expression of hypoxia-inducible factor (*HIF*)-*1α*, matrix metalloproteinase (*MMP*)-*2* and -*9*, *BCL-XL*, *BCL-2* and vascular endothelial growth factor (*VEGF*), all genes that directly influence PCa growth and progression (Nguyen, Li, & Tewari, 2014). Several studies have also demonstrated a pivotal role for the IL-6/STAT3 axis in the regulation of the PCSC niche. STAT3 is known to be constitutively active in the PCa stem-like population *in vitro* and in stem-like/progenitor cells derived from human PCa patients (Kroon et al., 2013; Talukdar et al., 2019). Gene expression profiling of PCSCs from primary cultures revealed a significant activation of STAT3, indicating that STAT3 inhibition could be a valid strategy to selectively target PCSCs (Birnie et al., 2008; Galoczova, Coates, & Vojtesek, 2018; Han et al., 2014). Furthermore, a recent study revealed that IL-6 levels were significantly correlated with the levels of the stem cell marker CD44 and were associated with larger tumor burden and tumor growth, while blocking the IL-6/STAT3 signaling attenuated the expression of CD44, stemness properties, and aggressive tumor behavior both *in vitro* and *in vivo* (Wu et al., 2019). Finally, studies have shown that therapy-induced increase in IL-6 expression can lead to STAT3 activation, which in turn induces the expression of proteins with stem cell-like function (SOX2, CD44, NANOG) and the development of a PCSCc phenotype (Handle et al., 2016; Schroeder et al., 2014).

4.3 The canonical Wnt/β-catenin pathway

In the canonical pathway, the binding of Wnt ligands to Frizzled receptors and co-receptors leads to stabilization and nuclear translocation of β-catenin, which in turns acts as transcriptional activator of pro-tumorigenic genes (Murillo-Garzón & Kypta, 2017). Both the expression and the localization of β-catenin are frequently altered in advanced PCa and studies have shown that this signaling pathway can directly promote self-renewal of PCSCs

(Bisson & Prowse, 2009; Ma et al., 2016; Zhang et al., 2017). The nuclear translocation of β-catenin is known to increase the expression of PCSC-associated genes, including *CD44*, *OCT4*, *SOX2*, *NANOG*, *KLF4*, *MYC* and *MMP-2* (Song et al., 2018). The aberrant activation of the Wnt/β-catenin signaling pathway also correlates with EMT features and positively regulates both proliferation and invasiveness of PCSCs (Jiang et al., 2007). Additionally, this pathway is also known to maintain the self-renewal ability of $CD44^+$ PCSCs and β-catenin nuclear expression can serve as an important indicator for prognosis in PCa patients (Zhang et al., 2019). Furthermore, a link between the canonical Wnt/β-catenin pathway and the progression of androgen-independent disease and bone metastasis has been shown (Lin et al., 2016; Yardy & Brewster, 2005). Of note, studies have shown that inhibition of Wnt/β-catenin signaling impaired the growth of $CD44^+$ PCSC *in vitro*, indicating that targeting this signaling axis may enhance therapy efficacy in treating PCa (Chen et al., 2019; Hsieh et al., 2013).

4.4 Notch signaling pathway

The Notch signaling pathway is controlled by a set of receptors (Notch1-4) and ligands (DLL1, DLL3, DLL4, Jagged-1 and Jagged-2): upon ligand binding, specific proteolytic enzymes cleave the cytoplasmic domain of the receptor, leading to the release of the Notch intracellular domain, which translocates to the nucleus to activate the transcription of target genes (Li et al., 2017). Since the pathway regulates PSCs maintenance and differentiation, its activation in the context of PCa can contribute to PCSCs expansion (Kwon et al., 2014; O'Brien & Marignol, 2017; Wang, Shou, Wong, French, & Gao, 2004). For example, HIF1α can activate Notch1, leading to increased expression of *OCT4* and ultimately enabling PCSCs to survive and proliferate in a hypoxic microenvironment (Feitelson et al., 2015). Additionally, Notch1 is known to induce EMT in PCa cells, promoting sphere formation and resulting in increased cell migration and invasion (Zhang et al., 2017). A study showed that PCSCs with high levels of active Notch1 also express elevated levels of *ABCC1*, the gene that encodes for the multidrug resistance-associated protein-1 (MRP1). Specifically, Notch1 binds to *ABCC1* promoter to induce its expression, resulting in enhanced chemoresistance of PCSCs (Liu et al., 2014). Interestingly, subpopulations of cells overexpressing Notch were found in

PCa tissues and were related to elevated tumor aggressiveness, poor patient prognosis, potent tumor-initiating capacity and acquired resistance to chemotherapy (Domingo-Domenech et al., 2012).

4.5 The hedgehog signaling pathway

The Hedgehog pathway is critical during embryogenesis, controlling cell renewal and survival in normal stem cells (Takebe et al., 2015). During adulthood the pathway is significantly less active, but studies have shown it plays a crucial role in regulating the cancer stem cell niche in several human cancers, including PCa (Gonnissen, Isebaert, & Haustermans, 2013). The Hedgehog signaling pathway is activated by the binding of specific ligands to the membrane receptors Patched-1 (Ptch1), Patched-2 (Ptch-2) and Smoothened (SMO): in the presence of these ligands, these receptors promote the nuclear translocation of the transcription factor Gli-1, which triggers the expression of specific target genes involved in cancer cell proliferation, survival and metastasis (Karhadkar et al., 2004). Prostate tumors often exhibit abnormally activated Hedgehog signaling, and high Gli-1 expression was found to be significantly associated with advanced tumor stage, positive lymph node metastasis, advanced clinical stage and worse clinical outcome (Lv, Yang, Ma, & Xuan, 2018). By using a mouse model with persistent activation of the Hedgehog pathway, Chang and colleagues demonstrated that this signaling pathway plays important roles not only in transforming normal prostate basal/stem cells into PCSCs, but also in the progression of PCSCs into metastatic tumor cells (Chang et al., 2011). Additionally, Hedgehog pathway promotes the expansion of PCSCs by regulating EMT and by inducing the transcription of the multidrug resistance transporter proteins ABCB1 and ABCG2, indicating that targeting this pathway may result in the depletion of PCSCs (Gonnissen et al., 2013; Luo et al., 2013; Nanta et al., 2013).

5. The PCSC NICHE environment

Just like normal stem cells, CSCs reside in a specialized niche providing a microenvironment that maintains the balance between quiescence and self-renewal and controls differentiation (Pan et al., 2015). This microenvironment includes different cell types, from fibroblasts to immune cells, that are responsible for abnormal signaling pathway activation in CSCs, inducing EMT and leading to cancer cell dissemination and invasion (Borovski, De Sousa E Melo, Vermeulen, & Medema, 2011). Additionally, studies have shown that the tumor microenvironment protects CSCs from drug-induced apoptosis and is thus responsible for drug resistance, indicating that targeting these interactions could be necessary to prevent cancer progression and relapse (Clayton & Mousa, 2011; Yadav & Desai, 2019). In this section I will describe the role of myeloid cells and fibroblasts in the regulation of the PCSC niche.

5.1 Myeloid cells: MDSCs and TAMs

Myeloid cells are the most abundant nucleated hematopoietic cells in the human body and are involved in both innate and adaptive immune response (Sawant & Ponnazhagan, 2013). Specifically, several studies have shown the importance of Myeloid-Derived Suppressor Cells (MDSCs) and Tumor-Associated Macrophages (TAMs) as fundamental regulators of cancer progression, metastasis, and therapy resistance.

MDSCs represent a population of immature myeloid cells, due to a partial block of the differentiation program in the myeloid lineage (Kumar, Patel, Tcyganov, & Gabrilovich, 2016; Talmadge & Gabrilovich, 2013). Overall, these cells are known to promote angiogenesis, tumor cell invasion and metastasis, and they possess potent immunosuppressive activity (Condamine, Ramachandran, Youn, & Gabrilovich, 2015; Gabrilovich & Nagaraj, 2009). MDSC levels in the blood were found to correlate with circulating PSA and with negative prognostic markers levels in patients with mCRPC (Brusa et al., 2013; Idorn, Køllgaard, Kongsted, Sengeløv, & Thor Straten, 2014; Vuk-Pavlović et al., 2010). Several studies have identified a role for MDSCs in PCa development and progression. For example, a mouse model of indolent *PTEN*-null PCa was used to show that MDSCs promote tumor initiation and progression and are involved in senescence evasion and chemoresistance (Di Mitri et al., 2014; Garcia et al., 2014). Similarly, Wang et al. used a prostate adenocarcinoma model driven by loss of *PTEN* and *SMAD4* to show that the Hippo-YAP pathway regulates the expression of CXCL5, a cancer-secreted chemokine that in turn attracts MDSCs to the tumor microenvironment (TME) (Wang et al., 2016). CXCL5 promotes PCa and its expression correlates with disease progression and metastases (Begley et al., 2008). Its role in recruiting MDSCs could thus explain the fact that the metastatic stroma shows higher number of infiltrating MDSCs compared to the stroma of the primary tumors (Wen et al., 2020). Of note, blocking the CXCL5 signaling elicits significant antitumor responses and prolongs survival, which highlights the critical role that MDSCs play in PCa progression (Wang et al., 2016).

A recent study showed that the cytokine IL-30, produced by both PCa cells and lymph node-infiltrating myeloid cells, can directly support the viability, self-renewal and migratory abilities of PCSCs, also increasing their tumorigenicity (Sorrentino et al., 2018). Using a PCa orthotopic immunocompetent mouse model, the authors observed that IL-30 acts as an autocrine and a paracrine growth factor, activating pathways that boost PCSCs viability and reshape the TME to create an "*ad hoc*" pre-metastatic niche in the lymph nodes and bone marrow that helps the systemic dissemination of PCSCs (Sorrentino et al., 2018). On one side, IL-30 acts primarily *via* STAT1 and STAT3 signaling induction, promoting a pro-inflammatory and angiogenic milieu where IL-1α, IL-1β and TNFα are highly expressed, which may directly affect tumor cell survival as well as regulate immune cell recruitment and pro-tumoral functions (Balkwill, 2009). On the other side, high levels of IL-30 boost the local production and release of other cytokines that are known to drive cancer cell dissemination and metastatic colonization, including CXCL13, CXCR5, CXCR4 and CXCL12 (Massagué & Obenauf, 2016; Meijer, Zeelenberg, Sipos, & Roos, 2006). Interestingly, IL-30 expression in PCa patients correlates with advanced grade and stage of the disease, indicating that targeting this driver could be an effective strategy to hinder PCa progression or recurrence (Di Meo et al., 2014).

TAMs are usually classified into tumor-suppressive M1 macrophages or tumor-promoting M2 macrophages (Zhou et al., 2020). This classification is based on the differential expression of transcription factors and surface molecules as well as on differences in their cytokine profiles and metabolism (Murray, 2017; Noy & Pollard, 2014). TAMs are known to promote tumor progression, migration and metastasis, while depletion of TAMs has been shown to suppress tumor growth in murine tumor models (Maolake et al., 2017; Qian & Pollard, 2010; Ries et al., 2014; Wu et al., 2014). Of note, several studies have evaluated the clinical relevance of TAMs in PCa progression, also suggesting a link between TAMs and disease recurrence (Fujii et al., 2013; Gannon et al., 2009; Gollapudi et al., 2013; Lanciotti et al., 2014; Nonomura et al., 2011).

In a preclinical study, surgical castration of mice bearing PCa resulted in increased number of inflammatory cells recruited to the tumor: while B cells and natural killer (NK) cells were detected, TAMs were the major immune cell type in the tumor and their number increased after castration (Ammirante, Luo, Grivennikov, Nedospasov, & Karin, 2010). Clinical studies suggest that TAMs exert a negative impact on treatment response, as their infiltration is increased in tumor biopsies taken from patients who had received ADT and this increase in TAMs correlates with time to tumor progression (Nonomura et al., 2011). Mechanistically, ADT by chemical or physical castration induces tumor cells to express macrophage colony-stimulating factor-1 (M-CSF1), that recruits and modulates macrophages, causing a significant increase in TAMs infiltration (Escamilla et al., 2015). Furthermore, ADT stimulates the release of M2-skewing cytokines such as IL-10 and IL-13, hence polarizing macrophages toward an alternatively activated, pro-tumorigenic state (Escamilla et al., 2015). In turn, M2-TAMs are known to express high levels of the *HIF1α*, *VEGF*, *PDGF*, *MMP-9* and *ARG-1* genes, which can promote treatment resistance by enhancing angiogenesis, tissue-remodeling, and immune suppression (Bingle, Brown, & Lewis, 2002; Sica, Schioppa, Mantovani, & Allavena, 2006). In several solid tumors, CSCs are able to recruit monocytes and educate them to become TAMs, which results in remodeling the stem cell niche and contributing to tumor growth, metastasis, and chemoresistance (Lu et al., 2014; Raggi et al., 2017; Zhou et al., 2015). A recent study showed that PCSCs rely on autophagy-related gene 7 (*ATG7*) to sustain their self-renewal abilities: ATG7 regulates β-catenin, which binds to the *OCT4* promoter and facilitates its transcription, promoting tumor initiation and drug resistance (Huang et al., 2018). Additionally, this study reported that PCSCs can remodel their specific niche by educating monocytes/macrophages toward TAMs: this results in a reciprocal crosstalk between PCSCs and TAMs that facilitates ADT resistance and cancer progression *via* IL-6/STAT3 (Huang et al., 2018). Of note, the combined targeting of PCSCs and TAMs by inhibiting both ATG7/*OCT4* and IL-6 signaling was effective in improving ADT resistance, indicating that targeting PCSCs and their niche may be a more powerful strategy than targeting PCSCs alone (Huang et al., 2018).

5.2 Cancer-associated fibroblasts (CAFs)

Cancer-associated fibroblasts (CAFs) are a key component of the tumor microenvironment and can influence the recruitment of immune cells to the tumor and regulate cell growth and migration in several cancers (Camps et al., 1990). In the context of PCa, it has been known for many years that CAFs can accelerate tumor growth and progression *in vitro* and *in vivo* (Olumi et al., 2000). However, only more recent works have shown that CAFs play a role in regulating the PCSC niche. Compared to normal human prostate fibroblasts (NPFs), CAFs were shown to significantly enhance spheroid formation and promote the proliferation of PCSCs-derived glands in co-culture experiments (Liao, Adisetiyo, Liang, & Roy-Burman, 2010). CAFs are commonly reported to express a wide range of pro-inflammatory cytokines and chemokines that influence the behavior of PCa cells in a sort of regulatory loop, resulting in increased stemness features (Kiskowski et al., 2011). An elegant study by Orr et al. applied next-generation Tag profiling to human CAFs and NPFs, in order to identify molecules specifically expressed in the prostatic stroma (Orr et al., 2012). The study identified more than 600 transcripts that were highly enriched in CAFs, including *CXCL12* and *ZEB*, which have been related to EMT induction and PCa stemness (Cackowski et al., 2018; Orr et al., 2012; Wang et al., 2018). Another study showed that prostate CAFs secrete high levels of Annexin A1 (AnxA1), a phospholipid-binding protein and regulator of glucocorticoid-induced inflammatory signaling (Geary et al., 2014). This study revealed that CAFs-secreted AnxA1 can both induce a de-differentiation process leading to generation of stem-like cells from a subpopulation of cancer epithelial cells and stimulate the proliferation of these cancer stem-like cells (Geary et al., 2014). Remarkably, AnxA1 protein levels are significantly altered in highly metastatic PCa patients (Varambally et al., 2005) and studies have revealed that this protein controls the expression of *CD44*, *CD133*, *NANOG*, *SNAIL* and *STAT3* in PCa cells, inducing EMT and promoting chemo-resistance and aggressive phenotype (Bizzarro et al., 2015, 2017). Other studies have reported a TGFβ- and IL-6-driven complex interplay between CAFs and prostate tumor cells, which results in EMT induction, acquisition of stemness features and cancer progression (Barcellos-de-Souza et al., 2016; Doldi et al., 2015; Du et al., 2015). Additionally, CAFs could exert an important effect on the PCSC niche trough androgen-dependent signaling. Androgens are known to regulate a variety of genes in CAFs, including *CCL2* and *CXCR8*, which increase migration of PCa cells (Cioni et al., 2018). Furthermore, IFN-γ and M-CSF1 expression is also AR regulated in CAFs, and results in increased expression of stem cell markers in PCa epithelial cells (Liao et al., 2017).

6. Therapeutic targeting of PCSCs

Current treatments for clinically localized PCa are based on surgical excision of the prostate (radical prostatectomy) and radiation therapy (Shen & Abate-Shen, 2010). However, since most prostate cancers are androgen-independent at the time of diagnosis, patients with advanced PCa are primarily treated with ADT, which reduces tumor burden and circulating PSA to low or undetectable levels (Chen, Rycaj, Liu, & Tang, 2013; Mayer, Klotz, & Venkateswaran, 2015; Yu, Yao, Jiang, & Keller, 2012). Since Huggins and Hodges described PCa hormone-dependence and reported that castration led to tumor regression in PCa patients (Huggins & Hodges, 1972), ADT has been the standard of care for PCa patients and has proven effective in limiting the disease progression in most cases (Harris, Mostaghel, Nelson, & Montgomery, 2009; Nouri et al., 2014). Nevertheless, the number of aggressive PCa with low or undetectable AR expression has increased in the clinic during the most recent years, and subsets of AR-independent tumors have been identified (Bluemn et al., 2017). These variants of cancers are completely unresponsive to ADT treatment and may emerge either from a differentiation of AR^+ adenocarcinomas into AR^{low} and AR^- tumors or from a clonal selection of rare pre-existing AR^{low} and AR^- clones (Hu, Choo, & Huang, 2015; Zou et al., 2017). Thus, while many PCa patients will initially respond to ADT, a significant number of them will become ADT-resistant and will progress to lethal metastatic castration-resistant PCa (mCRPC), for which treatment options are currently limited (Doctor, Tsao, Godbold, Galsky, & Oh, 2014; Nouri et al., 2014). It appears that current therapeutic approaches for PCa remain insufficient because they have incomplete and/or temporary effects on the tumor bulk and no effect on the PCSC niche. As we mentioned before, PCSCs exhibit self-renewal capacity, capability to proliferate, ability to be clonogenic and tumorigenic, and resistance to cytotoxic agents, including androgen ablation (Signoretti & Loda, 2007; Sun & Gao, 2011). The plasticity and genetic instability of PCSCs most likely provide a selective advantage that allows these cells to persist after standard anti-proliferative therapies (Fig. 3A) (Leão et al., 2017; Sun & Gao, 2011). For this reason, major efforts are currently focused on the development of therapeutic approaches that selectively eliminate PCSC populations, in order to reduce the likelihood of tumor relapse (Fig. 3B) (Adamowicz et al., 2017). Specifically, signaling pathways that are up-regulated in PCSCs and that are specific to their functionality represent hypothetical therapeutic targets and targeting them may represent a very promising therapeutic strategy (Leão et al., 2017).

Targeting the IL-6/JAK/STAT3 pathway is seen as a promising strategy, as this pathway is aberrantly activated in PCa and controls the PCSC niche. Notably, the inhibition of this pathway can be achieved through blockade at various levels, with inhibitors directed against IL-6 and IL-6r, against the

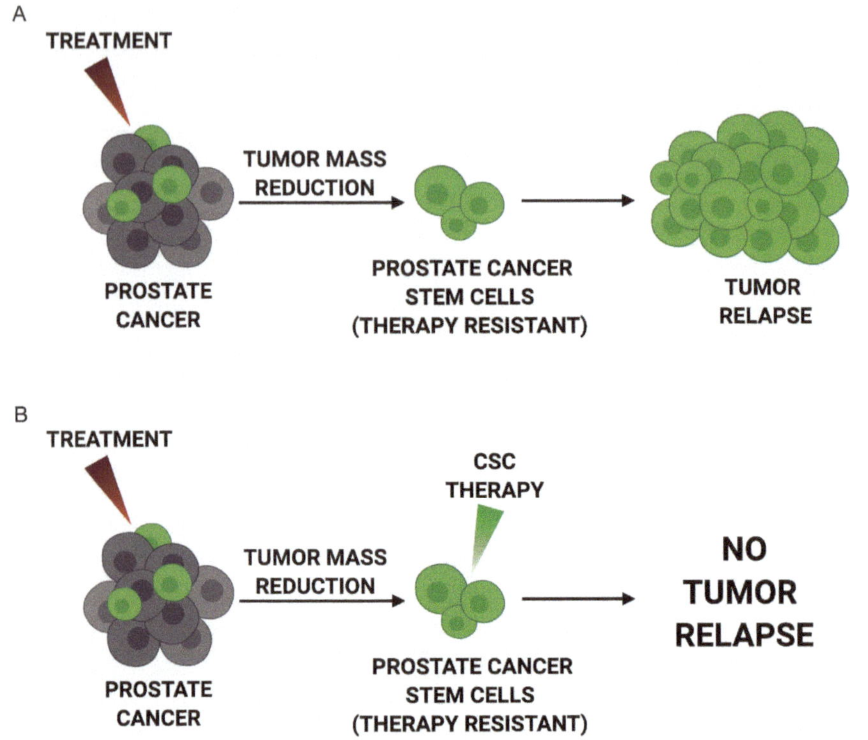

Fig. 3 Therapeutic approaches in prostate cancer. (A) Standard treatment eradicates the bulk of the tumor, resulting in a reduction of the tumor mass. However, PCSCs are intrinsically resistant to chemotherapy and other therapeutic modalities, causing the reconstitution of the original tumor cell population. (B) Targeting PCSCs after or in combination with standard anti-proliferative therapy could impair tumor re-growth and prevent tumor relapse.

tyrosine kinase or against STAT3 itself. A monoclonal antibody called Siltuximab (CNTO328) was tested as a possible therapeutic approach for PCa, due to its ability to inhibit the binding of IL-6 to its receptor (Dorff et al., 2010). The preclinical studies showed a reduction of tumor volume in CNTO328-treated mice and no drug-related adverse effects were reported (Dorff et al., 2010). However, the subsequent clinical trials showed no reduced progression-free survival after treatment and raised important concerns about the safety of IL-6 inhibitors (www.clinicaltrials.gov) (Table 1) (Fizazi et al., 2012). A different monoclonal antibody (Tocilizumab) against IL-6r was able to reverse enzalutamide-induced drug resistance in an orthotopic PCa mouse model, indicating that targeting IL-6r

Table 1 Most relevant clinical trials targeting PCSC-associated pathways.

Clinical trial	Drug	Condition	Status	Associated pathway
NCT00385827	Siltuximab (CNTO328)	Metastatic hormone-refractory prostate cancer	Phase 2 (terminated)	IL6—JAK/STAT
NCT04375228	Tocilizumab	Advanced solid tumor	Phase 2	IL6—JAK/STAT
NCT00638378	Ruxolitinib	Metastatic androgen-independent prostate cancer	Phase 2	JAK/STAT
NCT00955812	OPB-31121	Advanced solid tumors	Phase 1	STAT3
NCT00657176	OPB-31121	Solid tumors	Phase 1	STAT3
NCT01184807	OPB-51602	Malignant solid tumors	Phase 1	STAT3
NCT01423903	OPB-51602	Advanced cancers	Phase 1	STAT3
NCT02250170	OPB-111077	Advanced solid tumors	Phase 1	STAT3
NCT01711034	OPB-111077	Solid tumors	Phase 1	STAT3
NCT02462707	PF-03084014	Advanced solid tumors	Withdrawn	NOTCH
NCT01200810	RO4929097 + Bicalutamide	Patients with previously treated prostate cancer	Terminated (drug no longer available)	NOTCH
NCT02020291	Foxy-5	Metastatic prostate cancer	Phase 1	WNT
NCT02655952	Foxy-5	Metastatic prostate cancer	Phase 1	WNT

Continued

Table 1 Most relevant clinical trials targeting PCSC-associated pathways.—cont'd

Clinical trial	Drug	Condition	Status	Associated pathway
NCT01608867	OMP-54F28	Solid tumors	Phase 1	WNT
NCT02115828	GDC-0449 (Vismodegib)	Metastatic castration-resistant prostate cancer (mCRPC) with accessible metastatic lesions for tumor biopsy	Early Phase 1	Hedgehog
NCT01163084	GDC-0449 (Vismodegib) +goserelin acetate OR GDC-0449 (Vismodegib) + Leuprolide acetate	Prostate cancer that has spread to nearby tissues or lymph nodes	Phase1/Phase2	Hedgehog
NCT02465060	GDC-0449 (Vismodegib)	Recurrent prostate carcinoma	Phase 2	Hedgehog
NCT01787331	Itraconazole	Recurrent prostate carcinoma	Phase 2	Hedgehog

could work in patients with advanced or metastatic PCa, but no clinical data are currently available (Wang et al., 2018). However, because Tocilizumab has been shown effective in treating other diseases that involve immune system activation, a clinical trial has recently opened to test how effective this antibody would be in treating immune-related side effects in patients receiving immunotherapy treatment for solid tumors (www.clinicaltrials.gov) (Table 1). Similarly, a JAK inhibitor (Ruxolitinib) was tested in patients with metastatic PCa, but the clinical trial was discontinued due to lack of efficacy (www.clinicaltrials.gov) (Table 1). Thus, upstream blockers of the JAK/STAT3 pathway have not shown positive results so far. A higher specificity with reduced clinical side effects could be achieved through direct targeting of STAT3 protein instead of their upstream activators, because of the redundant mechanisms of IL-6 activation. Various direct STAT3 inhibitors have been developed and tested in PCa models. For example, OPB 31121, OPB-51602 and OPB-111077 are small molecules that directly bind to the STAT3 SH2 domain and inhibit both Tyr705 and Ser727 phosphorylation, impairing functioning of both nuclear and mitochondrial STAT3 and effectively blocking STAT3 global downstream signaling (Brambilla et al., 2015; Genini et al., 2017). Using these small molecules in PCa tumor xenografts significantly inhibited tumor growth, blocked tumor cell re-population and depleted the fraction of stem-like tumor cells as assessed by *ex vivo* flow cytometry and tumor-sphere assays (Brambilla et al., 2015; Genini et al., 2017). Remarkably, these STAT3 inhibitors have entered phase I/II clinical trials and showed some limited efficacy as single agents in advanced patients with solid tumors (www.clinicaltrials.gov) (Table 1) (Bendell et al., 2014; Oh et al., 2015; Tolcher et al., 2018; Wong et al., 2015). Another example is the small molecule galiellalactone (GL), a direct STAT3 inhibitor that binds directly to STAT3 and prevents the transcription of target genes without interfering with its upstream activation (Hellsten et al., 2008). Studies have shown that GL induces the apoptosis of pSTAT3-expressing and stem-like PCa cells *in vitro* (Hellsten, Johansson, Dahlman, Sterner, & Bjartell, 2011) and reduces the growth of primary tumors and the spread of spontaneous metastases to the lymph nodes *in vivo* (Canesin et al., 2016). Very recent data indicate that GL can decrease the viability of docetaxel-resistant and patient-derived spheres retaining stemness characteristics, emphasizing that this small molecule can target both CSCs and non-CSCs populations (Canesin et al., 2020). These results also support the hypothesis that targeting the downstream activity of STAT3 could be a relevant approach for a more effective treatment, reducing the risks of developing therapy resistance to upstream inhibitors, decreasing side effects, and increasing efficacy of the therapy.

Several inhibitors of the Notch signaling pathway are being tested for their anti-PCSCs effects, including γ-secretase inhibitors (GSI) and mAb against Notch receptors and/or ligands (Yuan et al., 2015). GSI were able to suppress tumor growth in a number of cancers including PCa (Carvalho, Simons, Eberhart, & Berman, 2014), and studies have shown that this approach can positively influence the response to other treatments (Takebe et al., 2015; Yun et al., 2016). For example, the GSI PF-03084014 in combination with docetaxel showed high anti-tumor activity, associated with decreased EMT and reduced PCSCs in the tumor (Cui et al., 2015; Revandkar et al., 2016). A clinical trial was proposed to determine the toxicity of PF-03084014 as a single-agent in Japanese patients with advanced solid tumors, but the study was withdrawn due to lack of patients (www.clinicaltrials.gov) (Table 1). However, more recent data show that this inhibitor significantly enhances the efficacy of ADT both *in vitro* and in preclinical models, indicating that Notch pathway inhibition may still represent a promising adjuvant therapy in advanced PCa (Cui et al., 2018; Mohamed et al., 2017). Of note, a clinical trial was conducted in PCa patients using the GSI RO4929097 in combination with the anti-androgen bicalutamide, but the protocol was terminated early due to lack of study drug and no conclusive results are available (www.clinicaltrials.gov) (Table 1).

A number of therapeutic agents targeting the Wnt pathway are also being tested for their efficacy in preclinical models: while they showed promising anti-cancer activity in several solid tumors, a direct anti-CSCs effect is not fully described in PCa models (Grandy et al., 2009; Gurney et al., 2012). For example, the monoclonal antibody vantictumab (OMP-18R5), which blocks canonical Wnt/β-catenin signaling by binding to several Wnt receptors, is currently being tested in solid tumors in combination with chemotherapeutic drugs such as docetaxel and paclitaxel (Gurney et al., 2012). Similarly, the hexapeptide Foxy-5 has been tested in metastatic PCa patients for whom standard therapies were not effective (www.clinicaltrials.gov) (Table 1). Foxy-5 has been characterized as a WNT5A-mimicking peptide that impairs the migration and invasion of PCa cells without affecting β-catenin activation (Canesin et al., 2017; Syed Khaja et al., 2011), but its effects on the PCSC niche are not known and the clinical trial results have not been made public yet. Furthermore, a study showed that the recombinant fusion protein OMP-54F28 (Ipafricept) antagonizes Wnt signaling by competing with the native Fzd8 receptor for its ligands, resulting in reduced CSC frequency and decreased tumor growth (Le, McDermott, & Jimeno, 2015). Interestingly, a first clinical trial was recently conducted with

OMP-54F28 in patients with solid tumors for which there is no remaining standard curative therapy: while definitive results are not yet available, previous analysis showed that Ipafricept can be safely administered with minor toxicities, indicating this is a promising compound for further testing (www.clinicaltrials.gov) (Table 1) (Jimeno et al., 2017).

Inhibition of the Hedgehog pathway is also being currently tested as a therapeutic approach in PCa. The inhibitor NVP-LDE-225 (Erismodegib) was shown to specifically suppress the expression of pluripotency factors *NANOG*, *OCT4*, *MYC* and *SOX2*, hence suppressing EMT, reducing PCSCs viability, sphere formation and tumor growth *in vivo* (Nanta et al., 2013). Analogously, the small molecule Vismodegib (GDC-0449, Erivedge®) antagonizes the Hedgehog pathway and induces apoptosis of PCSCs, indicating that this approach may enhance the efficacy of PCa treatment (Meiss, Andrlová, & Zeiser, 2018; Tong et al., 2018). Notably, Vismodegib is the first Hedgehog inhibitor approved by the Food and Drug Administration for clinical use, and it was recently tested in patients with mCRPC with accessible metastatic lesions for tumor biopsy (www.clinicaltrials.gov) (Table 1). The results of this trial showed that the Hedgehog signal was actually suppressed by Vismodegib in the majority of patients, but there was no apparent effect of clinical activity, as none of the patients achieved a PSA reduction or a measurable tumor response (Maughan et al., 2016). Vismodegib was also tested in a randomized phase I/II trial in combination with anti-hormone therapy (leuprolide acetate or goserelin acetate) in patients with PCa that has spread to nearby tissue or lymph nodes (www.clinicaltrials.gov) (Table 1). This study was very small, but the results indicate that the combination of anti-hormone therapy with this Hedgehog inhibitor may be an effective therapeutic approach for PCa patients (Karlou, Tzelepi, & Efstathiou, 2010). Remarkably, a clinical trial is currently open and recruiting patients to test Vismodegib and other compounds in patients with advanced refractory solid tumors, including PCa (www.clinicaltrials.gov) (Table 1). This phase II trial (The MATCH Screening Trial) will study Vismodegib as a targeted therapy based on genetic testing in patients that have progressed following at least one line of standard treatment, or for which no other treatment approach exists. Patients will be screened for specific genetic abnormalities (such as mutations, amplifications, or translocations), so it will be interesting to see their response to this Hedgehog inhibitor in relation to the genetic abnormalities that they carry (www.clinicaltrials.gov) (Table 1). Finally, another Hedhehog inhibitor, Itraconazole, was tested in a clinical trial involving patients with biochemically recurrent PCa: the results showed that Itraconazole can modulate serum PSA levels, but the effects were modest and associated with risk of toxicity and adverse events, such as edema, fatigue or hypertension (Lee et al., 2019) (www.clinicaltrials.gov) (Table 1).

7. Conclusions

A growing body of evidence in the last few years has demonstrated the pivotal role that PCSCs play in the progression of PCa. These cells contribute to castration resistance and disease recurrence by influencing the heterogeneity that lies beneath treatment failure. Therefore, the development of therapies targeting PCSCs is highly desirable, as these approaches may lead to more effective treatment for PCa patients. Most likely, a combination of current anti-proliferative therapies with targeted anti-PCSCs therapies would be the best approach to obtain tumor remission and avoid recurrence. A deeper knowledge of the molecular mechanisms that regulate and control the stem cell niche will be highly valuable for the design of targeted therapies aimed at eradicating PCSCs. We all look forward to more preclinical studies and clinical trials that investigate the feasibility of these therapeutic approaches that could highly benefit PCa patients.

Acknowledgments

I would like to acknowledge Dr. Agnieszka Krzyzanowska (Lund University, Sweden) for critically reviewing the text and for her suggestions. I would like to thank Prof. Anders Bjartell (Lund University, Sweden) for his continuous mentorship and support. All figures included in this chapter were created with BioRender.com.

References

Adamowicz, J., Pakravan, K., Bakhshinejad, B., Drewa, T., & Babashah, S. (2017). Prostate cancer stem cells: From theory to practice. *Scandinavian Journal of Urology*, *51*(2), 95–106. https://doi.org/10.1080/21681805.2017.1283360.

Ammirante, M., Luo, J.-L., Grivennikov, S., Nedospasov, S., & Karin, M. (2010). B-cell-derived lymphotoxin promotes castration-resistant prostate cancer. *Nature*, *464*(7286), 302–305. https://doi.org/10.1038/nature08782.

Antony, J., Thiery, J. P., & Huang, R. Y.-J. (2019). Epithelial-to-mesenchymal transition: Lessons from development, insights into cancer and the potential of EMT-subtype based therapeutic intervention. *Physical Biology*, *16*(4), 041004. https://doi.org/10.1088/1478-3975/ab157a.

Bae, K.-M., Su, Z., Frye, C., McClellan, S., Allan, R. W., Andrejewski, J. T., et al. (2010). Expression of pluripotent stem cell reprogramming factors by prostate tumor initiating cells. *Journal of Urology*, *183*(5), 2045–2053. https://doi.org/10.1016/j.juro.2009.12.092.

Balkwill, F. (2009). Tumour necrosis factor and cancer. *Nature Reviews Cancer, 9*(5), 361–371. https://doi.org/10.1038/nrc2628.

Barcellos-de-Souza, P., Comito, G., Pons-Segura, C., Taddei, M. L., Gori, V., Becherucci, V., et al. (2016). Mesenchymal stem cells are recruited and activated into carcinoma-associated fibroblasts by prostate cancer microenvironment-derived TGF-β 1. *Stem Cells, 34*(10), 2536–2547. https://doi.org/10.1002/stem.2412.

Begley, L. A., Kasina, S., Mehra, R., Adsule, S., Admon, A. J., Lonigro, R. J., et al. (2008). CXCL5 promotes prostate cancer progression. *Neoplasia (New York, N.Y.), 10*(3), 244–254. https://doi.org/10.1593/neo.07976.

Bendell, J. C., Hong, D. S., Burris, H. A., Naing, A., Jones, S. F., Falchook, G., et al. (2014). Phase 1, open-label, dose-escalation, and pharmacokinetic study of STAT3 inhibitor OPB-31121 in subjects with advanced solid tumors. *Cancer Chemotherapy and Pharmacology, 74*(1), 125–130. https://doi.org/10.1007/s00280-014-2480-2.

Berry, P. A., Maitland, N. J., & Collins, A. T. (2008). Androgen receptor signalling in prostate: Effects of stromal factors on normal and cancer stem cells. *Molecular and Cellular Endocrinology, 288*(1–2), 30–37. https://doi.org/10.1016/j.mce.2008.02.024.

Bingle, L., Brown, N. J., & Lewis, C. E. (2002). The role of tumour-associated macrophages in tumour progression: Implications for new anticancer therapies. *The Journal of Pathology, 196*(3), 254–265. https://doi.org/10.1002/path.1027.

Birnie, R., Bryce, S. D., Roome, C., Dussupt, V., Droop, A., Lang, S. H., et al. (2008). Gene expression profiling of human prostate cancer stem cells reveals a pro-inflammatory phenotype and the importance of extracellular matrix interactions. *Genome Biology, 9*(5), R83. https://doi.org/10.1186/gb-2008-9-5-r83.

Bisson, I., & Prowse, D. M. (2009). WNT signaling regulates self-renewal and differentiation of prostate cancer cells with stem cell characteristics. *Cell Research, 19*(6), 683–697. https://doi.org/10.1038/cr.2009.43.

Bizzarro, V., Belvedere, R., Migliaro, V., Romano, E., Parente, L., & Petrella, A. (2017). Hypoxia regulates ANXA1 expression to support prostate cancer cell invasion and aggressiveness. *Cell Adhesion & Migration, 11*(3), 247–260. https://doi.org/10.1080/19336918.2016.1259056.

Bizzarro, V., Belvedere, R., Milone, M. R., Pucci, B., Lombardi, R., Bruzzese, F., et al. (2015). Annexin A1 is involved in the acquisition and maintenance of a stem cell-like/aggressive phenotype in prostate cancer cells with acquired resistance to zoledronic acid. *Oncotarget, 6*(28), 25076–25092. https://doi.org/10.18632/oncotarget.4725.

Bluemn, E. G., Coleman, I. M., Lucas, J. M., Coleman, R. T., Hernandez-Lopez, S., Tharakan, R., et al. (2017). Androgen receptor pathway-independent prostate cancer is sustained through FGF signaling. *Cancer Cell, 32*(4), 474–489.e6. https://doi.org/10.1016/j.ccell.2017.09.003.

Bonkhoff, H., & Remberger, K. (1996). Differentiation pathways and histogenetic aspects of normal and abnormal prostatic growth: A stem cell model. *The Prostate, 28*(2), 98–106. https://doi.org/10.1002/(SICI)1097-0045(199602)28:2<98::AID-PROS4>3.0.CO;2-J.

Borovski, T., De Sousa E Melo, F., Vermeulen, L., & Medema, J. P. (2011). Cancer stem cell niche: The place to Be. *Cancer Research, 71*(3), 634–639. https://doi.org/10.1158/0008-5472.CAN-10-3220.

Brambilla, L., Genini, D., Laurini, E., Merulla, J., Perez, L., Fermeglia, M., et al. (2015). Hitting the right spot: Mechanism of action of OPB-31121, a novel and potent inhibitor of the signal transducer and activator of transcription 3 (STAT3). *Molecular Oncology, 9*(6), 1194–1206. https://doi.org/10.1016/j.molonc.2015.02.012.

Brusa, D., Simone, M., Gontero, P., Spadi, R., Racca, P., Micari, J., et al. (2013). Circulating immunosuppressive cells of prostate cancer patients before and after radical prostatectomy: Profile comparison. *International Journal of Urology, 20*(10). n/a-n/a https://doi.org/10.1111/iju.12086.

Camps, J. L., Chang, S. M., Hsu, T. C., Freeman, M. R., Hong, S. J., Zhau, H. E., et al. (1990). Fibroblast-mediated acceleration of human epithelial tumor growth in vivo. *Proceedings of the National Academy of Sciences of the United States of America, 87*(1), 75–79. https://doi.org/10.1073/pnas.87.1.75.

Canesin, G., Evans-Axelsson, S., Hellsten, R., Krzyzanowska, A., Prasad, C. P., Bjartell, A., et al. (2017). Treatment with the WNT5A-mimicking peptide Foxy-5 effectively reduces the metastatic spread of WNT5A-low prostate cancer cells in an orthotopic mouse model. *PLoS One, 12*(9), e0184418. https://doi.org/10.1371/journal.pone.0184418.

Canesin, G., Evans-Axelsson, S., Hellsten, R., Sterner, O., Krzyzanowska, A., Andersson, T., et al. (2016). The STAT3 inhibitor galiellalactone effectively reduces tumor growth and metastatic spread in an orthotopic xenograft mouse model of prostate cancer. *European Urology, 69*(3), 400–404. https://doi.org/10.1016/j.eururo.2015.06.016.

Canesin, G., Krzyzanowska, A., Hellsten, R., & Bjartell, A. (2020). Cytokines and Janus kinase/signal transducer and activator of transcription signaling in prostate cancer: Overview and therapeutic opportunities. *Current Opinion in Endocrine and Metabolic Research, 10*, 36–42. https://doi.org/10.1016/j.coemr.2020.02.004.

Canesin, G., Maggio, V., Palominos, M., Stiehm, A., Contreras, H., Castellón, E., et al. (2020). STAT3 inhibition with galiellalactone effectively targets the prostate cancer stem-like cell population. *Scientific Reports, 10*(1), 13958. https://doi.org/10.1038/s41598-020-70948-5.

Cano, P., Godoy, A., Escamilla, R., Dhir, R., & Onate, S. A. (2007). Stromal-epithelial cell interactions and androgen receptor-coregulator recruitment is altered in the tissue microenvironment of prostate cancer. *Cancer Research, 67*(2), 511–519. https://doi.org/10.1158/0008-5472.CAN-06-1478.

Carvalho, F. L. F., Simons, B. W., Eberhart, C. G., & Berman, D. M. (2014). Notch signaling in prostate cancer: A moving target. *The Prostate, 74*(9), 933–945. https://doi.org/10.1002/pros.22811.

Castillo, V., Valenzuela, R., Huidobro, C., Contreras, H. R., & Castellon, E. A. (2014). Functional characteristics of cancer stem cells and their role in drug resistance of prostate cancer. *International Journal of Oncology, 45*(3), 985–994. https://doi.org/10.3892/ijo.2014.2529.

Chang, H.-H., Chen, B.-Y., Wu, C.-Y., Tsao, Z.-J., Chen, Y.-Y., Chang, C.-P., et al. (2011). Hedgehog overexpression leads to the formation of prostate cancer stem cells with metastatic property irrespective of androgen receptor expression in the mouse model. *Journal of Biomedical Science, 18*(1), 6. https://doi.org/10.1186/1423-0127-18-6.

Chen, F., Chen, X., Ren, Y., Weng, G., Keng, P. C., Chen, Y., et al. (2019). Radiation-induced glucocorticoid receptor promotes CD44+ prostate cancer stem cell growth through activation of SGK1-Wnt/β-catenin signaling. *Journal of Molecular Medicine, 97*(8), 1169–1182. https://doi.org/10.1007/s00109-019-01807-8.

Chen, X., Rycaj, K., Liu, X., & Tang, D. G. (2013). New insights into prostate cancer stem cells. *Cell Cycle, 12*(4), 579–586. https://doi.org/10.4161/cc.23721.

Choi, N., Zhang, B., Zhang, L., Ittmann, M., & Xin, L. (2012). Adult murine prostate basal and luminal cells are self-sustained lineages that can both serve as targets for prostate cancer initiation. *Cancer Cell, 21*(2), 253–265. https://doi.org/10.1016/j.ccr.2012.01.005.

Cioni, B., Nevedomskaya, E., Melis, M. H. M., van Burgsteden, J., Stelloo, S., Hodel, E., et al. (2018). Loss of androgen receptor signaling in prostate cancer-associated fibroblasts (CAFs) promotes CCL2- and CXCL8-mediated cancer cell migration. *Molecular Oncology, 12*(8), 1308–1323. https://doi.org/10.1002/1878-0261.12327.

Civenni, G., Malek, A., Albino, D., Garcia-Escudero, R., Napoli, S., Di Marco, S., et al. (2013). RNAi-mediated silencing of myc transcription inhibits stem-like cell maintenance and tumorigenicity in prostate cancer. *Cancer Research, 73*(22), 6816–6827. https://doi.org/10.1158/0008-5472.CAN-13-0615.

Clark, E. L., Coulson, A., Dalgliesh, C., Rajan, P., Nicol, S. M., Fleming, S., et al. (2008). The RNA helicase p68 is a novel androgen receptor coactivator involved in splicing and is overexpressed in prostate cancer. *Cancer Research*, *68*(19), 7938–7946. https://doi.org/10.1158/0008-5472.CAN-08-0932.

Clayton, S., & Mousa, S. A. (2011). Therapeutics formulated to target cancer stem cells: Is it in our future? *Cancer Cell International*, *11*(1), 7. https://doi.org/10.1186/1475-2867-11-7.

Cocchiola, R., Rubini, E., Altieri, F., Chichiarelli, S., Paglia, G., Romaniello, D., et al. (2019). STAT3 post-translational modifications drive cellular signaling pathways in prostate cancer cells. *International Journal of Molecular Sciences*, *20*(8), 1815. https://doi.org/10.3390/ijms20081815.

Collins, A. T., Berry, P.a., Hyde, C., Stower, M. J., & Maitland, N. J. (2005). Prospective identification of tumorigenic prostate cancer stem cells cancer stem cells. *Cancer Research*, *65*, 10946–10951. https://doi.org/10.1158/0008-5472.CAN-05-2018.

Collins, A. T., Habib, F. K., Maitland, N. J., & Neal, D. E. (2001). Identification and isolation of human prostate epithelial stem cells based on alpha(2)beta(1)-integrin expression. *Journal of Cell Science*, *114*, 3865–3872. Pt 21 http://www.ncbi.nlm.nih.gov/pubmed/11719553.

Collins, A. T., & Maitland, N. J. (2006). Prostate cancer stem cells. *European Journal of Cancer*, *42*(9), 1213–1218. https://doi.org/10.1016/j.ejca.2006.01.037.

Condamine, T., Ramachandran, I., Youn, J.-I., & Gabrilovich, D. I. (2015). Regulation of tumor metastasis by myeloid-derived suppressor cells. *Annual Review of Medicine*, *66*(1), 97–110. https://doi.org/10.1146/annurev-med-051013-052304.

Cui, D., Dai, J., Keller, J. M., Mizokami, A., Xia, S., & Keller, E. T. (2015). Notch pathway inhibition using PF-03084014, a γ-Secretase inhibitor (GSI), enhances the antitumor effect of docetaxel in prostate cancer. *Clinical Cancer Research: An Official Journal of the American Association for Cancer Research*, *21*(20), 4619–4629. https://doi.org/10.1158/1078-0432.CCR-15-0242.

Cui, J., Wang, Y., Dong, B., Qin, L., Wang, C., Zhou, P., et al. (2018). Pharmacological inhibition of the Notch pathway enhances the efficacy of androgen deprivation therapy for prostate cancer. *International Journal of Cancer*, *143*(3), 645–656. https://doi.org/10.1002/ijc.31346.

DeMarzo, A. M., Nelson, W. G., Isaacs, W. B., & Epstein, J. I. (2003). Pathological and molecular aspects of prostate cancer. *Lancet (London, England)*, *361*(9361), 955–964. https://doi.org/10.1016/S0140-6736(03)12779-1.

Di Meo, S., Airoldi, I., Sorrentino, C., Zorzoli, A., Esposito, S., & Di Carlo, E. (2014). Interleukin-30 expression in prostate cancer and its draining lymph nodes correlates with advanced grade and stage. *Clinical Cancer Research*, *20*(3), 585–594. https://doi.org/10.1158/1078-0432.CCR-13-2240.

Di Mitri, D., Toso, A., Chen, J. J., Sarti, M., Pinton, S., Jost, T. R., et al. (2014). Tumour-infiltrating Gr-1+ myeloid cells antagonize senescence in cancer. *Nature*, *515*(7525), 134–137. https://doi.org/10.1038/nature13638.

Di Zazzo, E., Galasso, G., Giovannelli, P., Di Donato, M., Di Santi, A., Cernera, G., et al. (2016). Prostate cancer stem cells: The role of androgen and estrogen receptors. *Oncotarget*, 7(1), 193–208. https://doi.org/10.18632/oncotarget.6220.

Doctor, S. M., Tsao, C.-K., Godbold, J. H., Galsky, M. D., & Oh, W. K. (2014). Is prostate cancer changing?: Evolving patterns of metastatic castration-resistant prostate cancer. *Cancer*, *120*(6), 833–839. https://doi.org/10.1002/cncr.28494.

Doldi, V., Callari, M., Giannoni, E., D'Aiuto, F., Maffezzini, M., Valdagni, R., et al. (2015). Integrated gene and miRNA expression analysis of prostate cancer associated fibroblasts supports a prominent role for interleukin-6 in fibroblast activation. *Oncotarget*, *6*(31), 31441–31460. https://doi.org/10.18632/oncotarget.5056.

Domingo-Domenech, J., Vidal, S. J., Rodriguez-Bravo, V., Castillo-Martin, M., Quinn, S. A., Rodriguez-Barrueco, R., et al. (2012). Suppression of acquired docetaxel resistance in prostate cancer through depletion of notch- and hedgehog-dependent tumor-initiating cells. *Cancer Cell, 22*(3), 373–388. https://doi.org/10.1016/j.ccr.2012.07.016.

Dorff, T. B., Goldman, B., Pinski, J. K., Mack, P. C., Lara, P. N., Van Veldhuizen, P. J., et al. (2010). Clinical and correlative results of SWOG S0354: A phase II trial of CNTO328 (siltuximab), a monoclonal antibody against interleukin-6, in chemotherapy-pretreated patients with castration-resistant prostate cancer. *Clinical Cancer Research: An Official Journal of the American Association for Cancer Research, 16*(11), 3028–3034. https://doi.org/10.1158/1078-0432.CCR-09-3122.

Du, Y., Long, Q., Zhang, L., Shi, Y., Liu, X., Li, X., et al. (2015). Curcumin inhibits cancer-associated fibroblast-driven prostate cancer invasion through MAOA/mTOR/HIF-1α signaling. *International Journal of Oncology, 47*(6), 2064–2072. https://doi.org/10.3892/ijo.2015.3202.

English, H. F., Santen, R. J., & Isaacs, J. T. (1987). Response of glandular versus basal rat ventral prostatic epithelial cells to androgen withdrawal and replacement. *The Prostate, 11*(3), 229–242. https://doi.org/10.1002/pros.2990110304.

Escamilla, J., Schokrpur, S., Liu, C., Priceman, S. J., Moughon, D., Jiang, Z., et al. (2015). CSF1 receptor targeting in prostate cancer reverses macrophage-mediated resistance to androgen blockade therapy. *Cancer Research, 75*(6), 950–962. https://doi.org/10.1158/0008-5472.CAN-14-0992.

Feitelson, M. A., Arzumanyan, A., Kulathinal, R. J., Blain, S. W., Holcombe, R. F., Mahajna, J., et al. (2015). Sustained proliferation in cancer: Mechanisms and novel therapeutic targets. *Seminars in Cancer Biology, 35*, S25–S54. https://doi.org/10.1016/j.semcancer.2015.02.006.

Fizazi, K., De Bono, J. S., Flechon, A., Heidenreich, A., Voog, E., Davis, N. B., et al. (2012). Randomised phase II study of siltuximab (CNTO 328), an anti-IL-6 monoclonal antibody, in combination with mitoxantrone/prednisone versus mitoxantrone/prednisone alone in metastatic castration-resistant prostate cancer. *European Journal of Cancer (Oxford, England: 1990), 48*(1), 85–93. https://doi.org/10.1016/j.ejca.2011.10.014.

Fujii, T., Shimada, K., Asai, O., Tanaka, N., Fujimoto, K., Hirao, K., et al. (2013). Immunohistochemical analysis of inflammatory cells in benign and precancerous lesions and carcinoma of the prostate. *Pathobiology: Journal of Immunopathology, Molecular and Cellular Biology, 80*(3), 119–126. https://doi.org/10.1159/000342396.

Gabrilovich, D. I., & Nagaraj, S. (2009). Myeloid-derived suppressor cells as regulators of the immune system. *Nature Reviews. Immunology, 9*(3), 162–174. https://doi.org/10.1038/nri2506.

Galoczova, M., Coates, P., & Vojtesek, B. (2018). STAT3, stem cells, cancer stem cells and p63. *Cellular & Molecular Biology Letters, 23*(1), 12. https://doi.org/10.1186/s11658-018-0078-0.

Gannon, P. O., Poisson, A. O., Delvoye, N., Lapointe, R., Mes-Masson, A.-M., & Saad, F. (2009). Characterization of the intra-prostatic immune cell infiltration in androgen-deprived prostate cancer patients. *Journal of Immunological Methods, 348*(1–2), 9–17. https://doi.org/10.1016/j.jim.2009.06.004.

Garcia, A. J., Ruscetti, M., Arenzana, T. L., Tran, L. M., Bianci-Frias, D., Sybert, E., et al. (2014). Pten null prostate epithelium promotes localized myeloid-derived suppressor cell expansion and immune suppression during tumor initiation and progression. *Molecular and Cellular Biology, 34*(11), 2017–2028. https://doi.org/10.1128/MCB.00090-14.

Geary, L. A., Nash, K. A., Adisetiyo, H., Liang, M., Liao, C.-P., Jeong, J. H., et al. (2014). CAF-secreted annexin A1 induces prostate cancer cells to gain stem cell–like features. *Molecular Cancer Research, 12*(4), 607–621. https://doi.org/10.1158/1541-7786.MCR-13-0469.

Genini, D., Brambilla, L., Laurini, E., Merulla, J., Civenni, G., Pandit, S., et al. (2017). Mitochondrial dysfunction induced by a SH2 domain-targeting STAT3 inhibitor leads to metabolic synthetic lethality in cancer cells. *Proceedings of the National Academy of Sciences of the United States of America, 114*(25), E4924–E4933. https://doi.org/10.1073/pnas.1615730114.

Gentry, T., Foster, S., Winstead, L., Deibert, E., Fiordalisi, M., & Balber, A. (2007). Simultaneous isolation of human BM hematopoietic, endothelial and mesenchymal progenitor cells by flow sorting based on aldehyde dehydrogenase activity: Implications for cell therapy. *Cytotherapy, 9*(3), 259–274. https://doi.org/10.1080/14653240701218516.

Ginestier, C., Hur, M. H., Charafe-Jauffret, E., Monville, F., Dutcher, J., Brown, M., et al. (2007). ALDH1 is a marker of normal and malignant human mammary stem cells and a predictor of poor clinical outcome. *Cell Stem Cell, 1*(5), 555–567. https://doi.org/10.1016/j.stem.2007.08.014.

Goldstein, A. S., Huang, J., Guo, C., Garraway, I. P., & Witte, O. N. (2010). Identification of a cell of origin for human prostate cancer. *Science, 329*(5991), 568–571. https://doi.org/10.1126/science.1189992.

Gollapudi, K., Galet, C., Grogan, T., Zhang, H., Said, J. W., Huang, J., et al. (2013). Association between tumor-associated macrophage infiltration, high grade prostate cancer, and biochemical recurrence after radical prostatectomy. *American Journal of Cancer Research, 3*(5), 523–529. http://www.ncbi.nlm.nih.gov/pubmed/24224130.

Gonnissen, A., Isebaert, S., & Haustermans, K. (2013). Hedgehog signaling in prostate cancer and its therapeutic implication. *International Journal of Molecular Sciences, 14*(7), 13979–14007. https://doi.org/10.3390/ijms140713979.

Grandy, D., Shan, J., Zhang, X., Rao, S., Akunuru, S., Li, H., et al. (2009). Discovery and characterization of a small molecule inhibitor of the PDZ domain of dishevelled. *The Journal of Biological Chemistry, 284*(24), 16256–16263. https://doi.org/10.1074/jbc.M109.009647.

Grosse-Gehling, P., Fargeas, C. A., Dittfeld, C., Garbe, Y., Alison, M. R., Corbeil, D., et al. (2013). CD133 as a biomarker for putative cancer stem cells in solid tumours: Limitations, problems and challenges. *The Journal of Pathology, 229*(3), 355–378. https://doi.org/10.1002/path.4086.

Gu, G., Yuan, J., Wills, M., & Kasper, S. (2007). Prostate cancer cells with stem cell characteristics reconstitute the original human tumor in vivo. *Cancer Research, 67*(10), 4807–4815. https://doi.org/10.1158/0008-5472.CAN-06-4608.

Gurney, A., Axelrod, F., Bond, C. J., Cain, J., Chartier, C., Donigan, L., et al. (2012). Wnt pathway inhibition via the targeting of Frizzled receptors results in decreased growth and tumorigenicity of human tumors. *Proceedings of the National Academy of Sciences of the United States of America, 109*(29), 11717–11722. https://doi.org/10.1073/pnas.1120068109.

Han, Z., Wang, X., Ma, L., Chen, L., Xiao, M., Huang, L., et al. (2014). Inhibition of STAT3 signaling targets both tumor-initiating and differentiated cell populations in prostate cancer. *Oncotarget, 5*(18), 8416–8428. https://doi.org/10.18632/oncotarget.2314.

Handle, F., Erb, H. H. H., Luef, B., Hoefer, J., Dietrich, D., Parson, W., et al. (2016). SOCS3 modulates the response to enzalutamide and is regulated by androgen receptor signaling and CpG methylation in prostate cancer cells. *Molecular Cancer Research: MCR, 14*(6), 574–585. https://doi.org/10.1158/1541-7786.MCR-15-0495.

Harris, W. P., Mostaghel, E. A., Nelson, P. S., & Montgomery, B. (2009). Androgen deprivation therapy: Progress in understanding mechanisms of resistance and optimizing androgen depletion. *Nature Clinical Practice. Urology, 6*(2), 76–85. https://doi.org/10.1038/ncpuro1296.

Heinlein, C. A., & Chang, C. (2004). Androgen receptor in prostate cancer. *Endocrine Reviews*, *25*(2), 276–308. https://doi.org/10.1210/er.2002-0032.

Heinrich, P. C., Behrmann, I., Haan, S., Hermanns, H. M., Müller-Newen, G., & Schaper, F. (2003). Principles of interleukin (IL)-6-type cytokine signalling and its regulation. *The Biochemical Journal*, *374*(Pt 1), 1–20. https://doi.org/10.1042/BJ20030407.

Hellsten, R., Johansson, M., Dahlman, A., Dizeyi, N., Sterner, O., & Bjartell, A. (2008). Galiellalactone is a novel therapeutic candidate against hormone-refractory prostate cancer expressing activated Stat3. *The Prostate*, *68*(3), 269–280. https://doi.org/10.1002/pros.20699.

Hellsten, R., Johansson, M., Dahlman, A., Sterner, O., & Bjartell, A. (2011). Galiellalactone inhibits stem cell-like ALDH-positive prostate cancer cells. *PLoS One*, *6*(7), e22118. https://doi.org/10.1371/journal.pone.0022118.

Holzbeierlein, J., Lal, P., LaTulippe, E., Smith, A., Satagopan, J., Zhang, L., et al. (2004). Gene expression analysis of human prostate carcinoma during hormonal therapy identifies androgen-responsive genes and mechanisms of therapy resistance. *The American Journal of Pathology*, *164*(1), 217–227. https://doi.org/10.1016/S0002-9440(10)63112-4.

Hsieh, I.-S., Chang, K.-C., Tsai, Y.-T., Ke, J.-Y., Lu, P.-J., Lee, K.-H., et al. (2013). MicroRNA-320 suppresses the stem cell-like characteristics of prostate cancer cells by downregulating the Wnt/beta-catenin signaling pathway. *Carcinogenesis*, *34*(3), 530–538. https://doi.org/10.1093/carcin/bgs371.

Hu, C.-D., Choo, R., & Huang, J. (2015). Neuroendocrine differentiation in prostate cancer: A mechanism of radioresistance and treatment failure. *Frontiers in Oncology*, *5*, 90. https://doi.org/10.3389/fonc.2015.00090.

Huang, E. H., Hynes, M. J., Zhang, T., Ginestier, C., Dontu, G., Appelman, H., et al. (2009). Aldehyde dehydrogenase 1 is a marker for normal and malignant human colonic stem cells (SC) and tracks SC overpopulation during colon tumorigenesis. *Cancer Research*, *69*(8), 3382–3389. https://doi.org/10.1158/0008-5472.CAN-08-4418.

Huang, H., Wang, C., Liu, F., Li, H.-Z., Peng, G., Gao, X., et al. (2018). Reciprocal network between cancer stem-like cells and macrophages facilitates the progression and androgen deprivation therapy resistance of prostate cancer. *Clinical Cancer Research: An Official Journal of the American Association for Cancer Research*, *24*(18), 4612–4626. https://doi.org/10.1158/1078-0432.CCR-18-0461.

Hudson, D. L. (2004). Epithelial stem cells in human prostate growth and disease. *Prostate Cancer and Prostatic Diseases*, 7(3), 188–194. https://doi.org/10.1038/sj.pcan.4500745.

Hudson, D. L., O'Hare, M., Watt, F. M., & Masters, J. R. (2000). Proliferative heterogeneity in the human prostate: Evidence for epithelial stem cells. *Laboratory Investigation; a Journal of Technical Methods and Pathology*, *80*(8), 1243–1250. https://doi.org/10.1038/labinvest.3780132.

Huggins, C., & Hodges, C. V. (1972). Studies on prostatic cancer: I. The effect of castration, of estrogen and of androgen injection on serum phosphatases in metastatic carcinoma of the prostate. *CA: A Cancer Journal for Clinicians*, *22*(4), 232–240. https://doi.org/10.3322/canjclin.22.4.232.

Idorn, M., Køllgaard, T., Kongsted, P., Sengeløv, L., & Thor Straten, P. (2014). Correlation between frequencies of blood monocytic myeloid-derived suppressor cells, regulatory T cells and negative prognostic markers in patients with castration-resistant metastatic prostate cancer. *Cancer Immunology, Immunotherapy: CII*, *63*(11), 1177–1187. https://doi.org/10.1007/s00262-014-1591-2.

Jeter, C. R., Liu, B., Liu, X., Chen, X., Liu, C., Calhoun-Davis, T., et al. (2011). NANOG promotes cancer stem cell characteristics and prostate cancer resistance to androgen deprivation. *Oncogene*, *30*(36), 3833–3845. https://doi.org/10.1038/onc.2011.114.

Jiang, Y.-G., Luo, Y., He, D., Li, X., Zhang, L., Peng, T., et al. (2007). Role of Wnt/β-catenin signaling pathway in epithelial-mesenchymal transition of human prostate cancer induced by hypoxia-inducible factor-1α. *International Journal of Urology*, *14*(11), 1034–1039. https://doi.org/10.1111/j.1442-2042.2007.01866.x.

Jimeno, A., Gordon, M., Chugh, R., Messersmith, W., Mendelson, D., Dupont, J., et al. (2017). A first-in-human phase I study of the anticancer stem cell agent ipafricept (OMP-54F28), a decoy receptor for wnt ligands, in patients with advanced solid tumors. *Clinical Cancer Research*, *23*(24), 7490–7497. https://doi.org/10.1158/1078-0432.CCR-17-2157.

Jung, Y., Cackowski, F. C., Yumoto, K., Decker, A. M., Wang, J., Kim, J. K., et al. (2018). CXCL12γ promotes metastatic castration-resistant prostate cancer by inducing cancer stem cell and neuroendocrine phenotypes. *Cancer Research*, *78*(8), 2026–2039. https://doi.org/10.1158/0008-5472.CAN-17-2332.

Kalantari, E., Saadi, F. H., Asgari, M., Shariftabrizi, A., Roudi, R., & Madjd, Z. (2017). Increased expression of ALDH1A1 in prostate cancer is correlated with tumor aggressiveness. *Applied Immunohistochemistry & Molecular Morphology*, *25*(8), 592–598. https://doi.org/10.1097/PAI.0000000000000343.

Kalluri, R., & Weinberg, R. A. (2009). The basics of epithelial-mesenchymal transition. *The Journal of Clinical Investigation*, *119*(6), 1420–1428. https://doi.org/10.1172/JCI39104.

Karhadkar, S. S., Steven Bova, G., Abdallah, N., Dhara, S., Gardner, D., Maitra, A., et al. (2004). Hedgehog signalling in prostate regeneration, neoplasia and metastasis. *Nature*, *431*(7009), 707–712. https://doi.org/10.1038/nature02962.

Karlou, M., Tzelepi, V., & Efstathiou, E. (2010). Therapeutic targeting of the prostate cancer microenvironment. *Nature Reviews. Urology*, 7(9), 494–509. https://doi.org/10.1038/NRUROL.2010.134.

Kelly, S. E., Di Benedetto, A., Greco, A., Howard, C. M., Sollars, V. E., Primerano, D. A., et al. (2010). Rapid selection and proliferation of CD133+ cells from cancer cell lines: Chemotherapeutic implications. *PLoS One*, *5*(4), e10035. https://doi.org/10.1371/journal.pone.0010035.

Kiskowski, M. A., Jackson, R. S., Banerjee, J., Li, X., Kang, M., Iturregui, J. M., et al. (2011). Role for stromal heterogeneity in prostate tumorigenesis. *Cancer Research*, *71*(10), 3459–3470. https://doi.org/10.1158/0008-5472.CAN-10-2999.

Kregel, S., Szmulewitz, R. Z., & Vander Griend, D. J. (2014). The pluripotency factor Nanog is directly upregulated by the androgen receptor in prostate cancer cells. *The Prostate*, *74*(15), 1530–1543. https://doi.org/10.1002/pros.22870.

Kroon, P., Berry, P. A., Stower, M. J., Rodrigues, G., Mann, V. M., Simms, M., et al. (2013). JAK-STAT blockade inhibits tumor initiation and clonogenic recovery of prostate cancer stem-like cells. *Cancer Research*, *73*(16), 5288–5298. https://doi.org/10.1158/0008-5472.CAN-13-0874.

Kumar, V., Patel, S., Tcyganov, E., & Gabrilovich, D. I. (2016). The nature of myeloid-derived suppressor cells in the tumor microenvironment. *Trends in Immunology*, *37*(3), 208–220. https://doi.org/10.1016/j.it.2016.01.004.

Kwon, O.-J., Valdez, J. M., Zhang, L., Zhang, B., Wei, X., Su, Q., et al. (2014). Increased notch signalling inhibits anoikis and stimulates proliferation of prostate luminal epithelial cells. *Nature Communications*, *5*(1), 4416. https://doi.org/10.1038/ncomms5416.

Lanciotti, M., Masieri, L., Raspollini, M. R., Minervini, A., Mari, A., Comito, G., et al. (2014). The role of M1 and M2 macrophages in prostate cancer in relation to extracapsular tumor extension and biochemical recurrence after radical prostatectomy. *BioMed Research International*, *2014*, 486798. https://doi.org/10.1155/2014/486798.

Lawson, D. A., Xin, L., Lukacs, R. U., Cheng, D., & Witte, O. N. (2007). Isolation and functional characterization of murine prostate stem cells. *Proceedings of the National Academy of Sciences of the United States of America*, *104*(1), 181–186. https://doi.org/10.1073/pnas.0609684104.

Lawson, D. A., Zong, Y., Memarzadeh, S., Xin, L., Huang, J., & Witte, O. N. (2010). Basal epithelial stem cells are efficient targets for prostate cancer initiation. *Proceedings of the National Academy of Sciences of the United States of America, 107*(6), 2610–2615. https://doi.org/10.1073/pnas.0913873107.

Le, P. N., McDermott, J. D., & Jimeno, A. (2015). Targeting the Wnt pathway in human cancers: Therapeutic targeting with a focus on OMP-54F28. *Pharmacology & Therapeutics, 146*, 1–11. https://doi.org/10.1016/j.pharmthera.2014.08.005.

Leão, R., Domingos, C., Figueiredo, A., Hamilton, R., Tabori, U., & Castelo-Branco, P. (2017). Cancer stem cells in prostate cancer: Implications for targeted therapy. *Urologia Internationalis, 99*(2), 125–136. https://doi.org/10.1159/000455160.

Lee, M., Hong, H., Kim, W., Zhang, L., Friedlander, T. W., Fong, L., et al. (2019). Itraconazole as a noncastrating treatment for biochemically recurrent prostate cancer: A Phase 2 study. *Clinical Genitourinary Cancer, 17*(1), e92–e96. https://doi.org/10.1016/j.clgc.2018.09.013.

Li, H., Chen, X., Calhoun-Davis, T., Claypool, K., & Tang, D. G. (2008). PC3 human prostate carcinoma cell holoclones contain self-renewing tumor-initiating cells. *Cancer Research, 68*(6), 1820–1825. https://doi.org/10.1158/0008-5472.CAN-07-5878.

Li, J. J., & Shen, M. M. (2019). Prostate stem cells and cancer stem cells. *Cold Spring Harbor Perspectives in Medicine, 9*(6), a030395. https://doi.org/10.1101/cshperspect.a030395.

Li, T., Su, Y., Mei, Y., Leng, Q., Leng, B., Liu, Z., et al. (2010). ALDH1A1 is a marker for malignant prostate stem cells and predictor of prostate cancer patients' outcome. *Laboratory Investigation, 90*(2), 234–244. https://doi.org/10.1038/labinvest.2009.127.

Li, L., Tang, P., Li, S., Qin, X., Yang, H., Wu, C., et al. (2017). Notch signaling pathway networks in cancer metastasis: A new target for cancer therapy. *Medical Oncology, 34*(10), 180. https://doi.org/10.1007/s12032-017-1039-6.

Liao, C.-P., Adisetiyo, H., Liang, M., & Roy-Burman, P. (2010). Cancer-associated fibroblasts enhance the gland-forming capability of prostate cancer stem cells. *Cancer Research, 70*(18), 7294–7303. https://doi.org/10.1158/0008-5472.CAN-09-3982.

Liao, C.-P., Chen, L.-Y., Luethy, A., Kim, Y., Kani, K., MacLeod, A. R., et al. (2017). Androgen receptor in cancer-associated fibroblasts influences stemness in cancer cells. *Endocrine-Related Cancer, 24*(4), 157–170. https://doi.org/10.1530/ERC-16-0138.

Lin, X., Farooqi, A. A., Qureshi, M. Z., Romero, M. A., Tabassum, S., & Ismail, M. (2016). Prostate cancer stem cells: Viewing signaling cascades at a finer resolution. *Archivum Immunologiae et Therapiae Experimentalis, 64*(3), 217–223. https://doi.org/10.1007/s00005-016-0383-0.

Liu, C., Li, Z., Bi, L., Li, K., Zhou, B., Xu, C., et al. (2014). NOTCH1 signaling promotes chemoresistance via regulating ABCC1 expression in prostate cancer stem cells. *Molecular and Cellular Biochemistry, 393*(1–2), 265–270. https://doi.org/10.1007/s11010-014-2069-4.

Lu, H., Clauser, K. R., Tam, W. L., Fröse, J., Ye, X., Eaton, E. N., et al. (2014). A breast cancer stem cell niche supported by juxtacrine signalling from monocytes and macrophages. *Nature Cell Biology, 16*(11), 1105–1117. https://doi.org/10.1038/ncb3041.

Luo, W., Rodriguez, M., Valdez, J. M., Zhu, X., Tan, K., Li, D., et al. (2013). Lgr4 is a key regulator of prostate development and prostate stem cell differentiation. *Stem Cells (Dayton, Ohio), 31*(11), 2492–2505. https://doi.org/10.1002/stem.1484.

Lv, L., Yang, Z., Ma, T., & Xuan, Y. (2018). Gli1, a potential cancer stem cell marker, is strongly associated with prognosis in prostate cancer. *International Journal of Clinical and Experimental Pathology, 11*(10), 4957–4966. http://www.ncbi.nlm.nih.gov/pubmed/31949572.

Ma, F., Ye, H., He, H. H., Gerrin, S. J., Chen, S., Tanenbaum, B. A., et al. (2016). SOX9 drives WNT pathway activation in prostate cancer. *The Journal of Clinical Investigation, 126*(5), 1745–1758. https://doi.org/10.1172/JCI78815.

Maitland, N. J., Frame, F. M., Polson, E. S., Lewis, J. L., & Collins, A. T. (2011). Prostate cancer stem cells: Do they have a basal or luminal phenotype? *Hormones and Cancer*, *2*(1), 47–61. https://doi.org/10.1007/s12672-010-0058-y.

Mani, S. A., Guo, W., Liao, M.-J., Eaton, E. N., Ayyanan, A., Zhou, A. Y., et al. (2008). The epithelial-mesenchymal transition generates cells with properties of stem cells. *Cell*, *133*(4), 704–715. https://doi.org/10.1016/j.cell.2008.03.027.

Maolake, A., Izumi, K., Shigehara, K., Natsagdorj, A., Iwamoto, H., Kadomoto, S., et al. (2017). Tumor-associated macrophages promote prostate cancer migration through activation of the CCL22-CCR4 axis. *Oncotarget*, *8*(6), 9739–9751. https://doi.org/10.18632/oncotarget.14185.

Massagué, J., & Obenauf, A. C. (2016). Metastatic colonization by circulating tumour cells. *Nature*, *529*(7586), 298–306. https://doi.org/10.1038/nature17038.

Matsika, A., Srinivasan, B., Day, C., Mader, S. A., Margaret Kiernan, D., Broomfield, A., et al. (2015). Cancer stem cell markers in prostate cancer: An immunohistochemical study of ALDH1, SOX2 and EZH2. *Pathology*, *47*(7), 622–628. https://doi.org/10.1097/PAT.0000000000000325.

Maughan, B. L., Suzman, D. L., Luber, B., Wang, H., Glavaris, S., Hughes, R., et al. (2016). Pharmacodynamic study of the oral hedgehog pathway inhibitor, vismodegib, in patients with metastatic castration-resistant prostate cancer. *Cancer Chemotherapy and Pharmacology*, *78*(6), 1297–1304. https://doi.org/10.1007/s00280-016-3191-7.

Mayer, M. J., Klotz, L. H., & Venkateswaran, V. (2015). Metformin and prostate cancer stem cells: A novel therapeutic target. *Prostate Cancer and Prostatic Diseases*, *18*(4), 303–309. https://doi.org/10.1038/pcan.2015.35.

Meijer, J., Zeelenberg, I. S., Sipos, B., & Roos, E. (2006). The CXCR5 chemokine receptor is expressed by carcinoma cells and promotes growth of colon carcinoma in the liver. *Cancer Research*, *66*(19), 9576–9582. https://doi.org/10.1158/0008-5472.CAN-06-1507.

Meiss, F., Andrlová, H., & Zeiser, R. (2018). Vismodegib. In *Recent results in cancer research. Fortschritte der Krebsforschung. Progres dans les recherches sur le cancer* (pp. 125–139). Vol. 211 https://doi.org/10.1007/978-3-319-91442-8_9.

Mitsui, K., Tokuzawa, Y., Itoh, H., Segawa, K., Murakami, M., Takahashi, K., et al. (2003). The homeoprotein nanog is required for maintenance of pluripotency in mouse epiblast and ES cells. *Cell*, *113*(5), 631–642. https://doi.org/10.1016/S0092-8674(03)00393-3.

Mohamed, A. A., Tan, S.-H., Xavier, C. P., Katta, S., Huang, W., Ravindranath, L., et al. (2017). Synergistic activity with NOTCH inhibition and androgen ablation in ERG-positive prostate cancer cells. *Molecular Cancer Research: MCR*, *15*(10), 1308–1317. https://doi.org/10.1158/1541-7786.MCR-17-0058.

Mongan, N., Scotland, K. B., Boorjian, S. A., Nilsson, E. M., Persson, J. L., Abrahamsson, P. A., et al. (2011). The androgen receptor and stem cell pathways in prostate and bladder cancers (Review). *International Journal of Oncology*, *40*(1), 5–12. https://doi.org/10.3892/ijo.2011.1212.

Montanari, M., Rossetti, S., Cavaliere, C., D'Aniello, C., Malzone, M. G., Vanacore, D., et al. (2017). Epithelial-mesenchymal transition in prostate cancer: An overview. *Oncotarget*, *8*(21), 35376–35389. https://doi.org/10.18632/oncotarget.15686.

Murillo-Garzón, V., & Kypta, R. (2017). WNT signalling in prostate cancer. *Nature Reviews. Urology*, *14*(11), 683–696. https://doi.org/10.1038/nrurol.2017.144.

Murray, P. J. (2017). Macrophage polarization. *Annual Review of Physiology*, *79*(1), 541–566. https://doi.org/10.1146/annurev-physiol-022516-034339.

Nanta, R., Kumar, D., Meeker, D., Rodova, M., Van Veldhuizen, P. J., Shankar, S., et al. (2013). NVP-LDE-225 (Erismodegib) inhibits epithelial–mesenchymal transition and human prostate cancer stem cell growth in NOD/SCID IL2Rγ null mice by regulating Bmi-1 and microRNA-128. *Oncogenesis*, *2*(4), e42. https://doi.org/10.1038/oncsis.2013.5.

Nelson, P. S., Clegg, N., Arnold, H., Ferguson, C., Bonham, M., White, J., et al. (2002). The program of androgen-responsive genes in neoplastic prostate epithelium. *Proceedings of the National Academy of Sciences of the United States of America*, *99*(18), 11890–11895. https://doi.org/10.1073/pnas.182376299.

Nguyen, D. P., Li, J., & Tewari, A. K. (2014). Inflammation and prostate cancer: The role of interleukin 6 (IL-6). *BJU International*, *113*(6), 986–992. https://doi.org/10.1111/bju.12452.

Ni, J., Cozzi, P. J., Hao, J. L., Beretov, J., Chang, L., Duan, W., et al. (2014). CD44 variant 6 is associated with prostate cancer metastasis and chemo-/radioresistance. *The Prostate*, *74*(6), 602–617. https://doi.org/10.1002/pros.22775.

Nonomura, N., Takayama, H., Nakayama, M., Nakai, Y., Kawashima, A., Mukai, M., et al. (2011). Infiltration of tumour-associated macrophages in prostate biopsy specimens is predictive of disease progression after hormonal therapy for prostate cancer. *BJU International*, *107*(12), 1918–1922. https://doi.org/10.1111/j.1464-410X.2010.09804.x.

Nouri, M., Ratther, E., Stylianou, N., Nelson, C. C., Hollier, B. G., & Williams, E. D. (2014). Androgen-targeted therapy-induced epithelial mesenchymal plasticity and neuroendocrine transdifferentiation in prostate cancer: An opportunity for intervention. *Frontiers in Oncology*, *4*, 370. https://doi.org/10.3389/fonc.2014.00370.

Noy, R., & Pollard, J. W. (2014). Tumor-associated macrophages: From mechanisms to therapy. *Immunity*, *41*(1), 49–61. https://doi.org/10.1016/j.immuni.2014.06.010.

O'Brien, R., & Marignol, L. (2017). The notch-1 receptor in prostate tumorigenesis. *Cancer Treatment Reviews*, *56*, 36–46. https://doi.org/10.1016/j.ctrv.2017.04.003.

Oh, D.-Y., Lee, S.-H., Han, S.-W., Kim, M.-J., Kim, T.-M., Kim, T.-Y., et al. (2015). Phase I study of OPB-31121, an Oral STAT3 inhibitor, in patients with advanced solid tumors. *Cancer Research and Treatment: Official Journal of Korean Cancer Association*, *47*(4), 607–615. https://doi.org/10.4143/crt.2014.249.

Okita, K., Ichisaka, T., & Yamanaka, S. (2007). Generation of germline-competent induced pluripotent stem cells. *Nature*, *448*(7151), 313–317. https://doi.org/10.1038/nature05934.

Olumi, A., Grossfeld, G., Hayward, S., Carroll, P., Cunha, G., Hein, P., et al. (2000). Carcinoma-associated fibroblasts stimulate tumor progression of initiated human epithelium. *Breast Cancer Research*, *2*(S1), S.19. https://doi.org/10.1186/bcr138.

Orr, B., Riddick, A. C. P., Stewart, G. D., Anderson, R. A., Franco, O. E., Hayward, S. W., et al. (2012). Identification of stromally expressed molecules in the prostate by tag-profiling of cancer-associated fibroblasts, normal fibroblasts and fetal prostate. *Oncogene*, *31*(9), 1130–1142. https://doi.org/10.1038/onc.2011.312.

Pan, Q., Li, Q., Liu, S., Ning, N., Zhang, X., Xu, Y., et al. (2015). Concise review: Targeting cancer stem cells using immunologic approaches. *Stem Cells (Dayton, Ohio)*, *33*(7), 2085–2092. https://doi.org/10.1002/stem.2039.

Park, S.-M., Gaur, A. B., Lengyel, E., & Peter, M. E. (2008). The miR-200 family determines the epithelial phenotype of cancer cells by targeting the E-cadherin repressors ZEB1 and ZEB2. *Genes & Development*, *22*(7), 894–907. https://doi.org/10.1101/gad.1640608.

Patrawala, L., Calhoun, T., Schneider-Broussard, R., Li, H., Bhatia, B., Tang, S., et al. (2006). Highly purified CD44+ prostate cancer cells from xenograft human tumors are enriched in tumorigenic and metastatic progenitor cells. *Oncogene*, *25*(12), 1696–1708. https://doi.org/10.1038/sj.onc.1209327.

Patrawala, L., Calhoun-Davis, T., Schneider-Broussard, R., & Tang, D. G. (2007). Hierarchical organization of prostate cancer cells in xenograft tumors: The CD44+ alpha2beta1+ cell population is enriched in tumor-initiating cells. *Cancer Research*, *67*(14), 6796–6805. https://doi.org/10.1158/0008-5472.CAN-07-0490.

Peitzsch, C., Cojoc, M., Hein, L., Kurth, I., Mäbert, K., Trautmann, F., et al. (2016). An epigenetic reprogramming strategy to resensitize radioresistant prostate cancer cells. *Cancer Research*, *76*(9), 2637–2651. https://doi.org/10.1158/0008-5472.CAN-15-2116.

Pellacani, D., Oldridge, E. E., Collins, A. T., & Maitland, N. J. (2013). Prominin-1 (CD133) expression in the prostate and prostate cancer: A marker for quiescent stem cells. In *Vol. 777*. *Advances in experimental medicine and biology* (pp. 167–184). https://doi.org/10.1007/978-1-4614-5894-4_11.

Qian, B.-Z., & Pollard, J. W. (2010). Macrophage diversity enhances tumor progression and metastasis. *Cell*, *141*(1), 39–51. https://doi.org/10.1016/j.cell.2010.03.014.

Raggi, C., Correnti, M., Sica, A., Andersen, J. B., Cardinale, V., Alvaro, D., et al. (2017). Cholangiocarcinoma stem-like subset shapes tumor-initiating niche by educating associated macrophages. *Journal of Hepatology*, *66*(1), 102–115. https://doi.org/10.1016/j.jhep.2016.08.012.

Rajan, P., Gaughan, L., Dalgliesh, C., El-Sherif, A., Robson, C. N., Leung, H. Y., et al. (2008). The RNA-binding and adaptor protein Sam68 modulates signal-dependent splicing and transcriptional activity of the androgen receptor. *The Journal of Pathology*, *215*(1), 67–77. https://doi.org/10.1002/path.2324.

Rajasekhar, V. K., Studer, L., Gerald, W., Socci, N. D., & Scher, H. I. (2011). Tumour-initiating stem-like cells in human prostate cancer exhibit increased NF-κB signalling. *Nature Communications*, *2*(1), 162. https://doi.org/10.1038/ncomms1159.

Revandkar, A., Perciato, M. L., Toso, A., Alajati, A., Chen, J., Gerber, H., et al. (2016). Inhibition of notch pathway arrests PTEN-deficient advanced prostate cancer by triggering p27-driven cellular senescence. *Nature Communications*, 7(1), 13719. https://doi.org/10.1038/ncomms13719.

Rice, M. A., Malhotra, S. V., & Stoyanova, T. (2019). Second-generation antiandrogens: From discovery to standard of care in castration resistant prostate cancer. *Frontiers in Oncology*, *9*, 801. https://doi.org/10.3389/fonc.2019.00801.

Richardson, G. D., Robson, C. N., Lang, S. H., Neal, D. E., Maitland, N. J., & Collins, A. T. (2004). CD133, a novel marker for human prostatic epithelial stem cells. *Journal of Cell Science*, *117*(16), 3539–3545. https://doi.org/10.1242/jcs.01222.

Ries, C. H., Cannarile, M. A., Hoves, S., Benz, J., Wartha, K., Runza, V., et al. (2014). Targeting tumor-associated macrophages with anti-CSF-1R antibody reveals a strategy for cancer therapy. *Cancer Cell*, *25*(6), 846–859. https://doi.org/10.1016/j.ccr.2014.05.016.

Rizzo, S., Attard, G., & Hudson, D. L. (2005). Prostate epithelial stem cells. *Cell Proliferation*, *38*(6), 363–374. https://doi.org/10.1111/j.1365-2184.2005.00356.x.

Ruscetti, M., Quach, B., Dadashian, E. L., Mulholland, D. J., & Wu, H. (2015). Tracking and functional characterization of epithelial-mesenchymal transition and mesenchymal tumor cells during prostate cancer metastasis. *Cancer Research*, *75*(13), 2749–2759. https://doi.org/10.1158/0008-5472.CAN-14-3476.

Sawant, A., & Ponnazhagan, S. (2013). Myeloid-derived suppressor cells as osteoclast progenitors: A novel target for controlling osteolytic bone metastasis. *Cancer Research*, *73*(15), 4606–4610. https://doi.org/10.1158/0008-5472.CAN-13-0305.

Sawicki, J. A., & Rothman, C. J. (2002). Evidence for stem cells in cultures of mouse prostate epithelial cells. *The Prostate*, *50*(1), 46–53. https://doi.org/10.1002/pros.10031.

Schmelz, M., Moll, R., Hesse, U., Prasad, A. R., Gandolfi, J. A., Hasan, S. R., et al. (2005). Identification of a stem cell candidate in the normal human prostate gland. *European Journal of Cell Biology*, *84*(2–3), 341–354. https://doi.org/10.1016/j.ejcb.2004.12.019.

Schroeder, A., Herrmann, A., Cherryholmes, G., Kowolik, C., Buettner, R., Pal, S., et al. (2014). Loss of androgen receptor expression promotes a stem-like cell phenotype in prostate cancer through STAT3 signaling. *Cancer Research*, *74*(4), 1227–1237. https://doi.org/10.1158/0008-5472.CAN-13-0594.

Setlur, S. R., & Rubin, M. A. (2005). Current thoughts on the role of the androgen receptor and prostate cancer progression. *Advances in Anatomic Pathology*, *12*(5), 265–270. https://doi.org/10.1097/01.pap.0000184179.50672.f7.

Shang, Z., Cai, Q., Zhang, M., Zhu, S., Ma, Y., Sun, L., et al. (2015). A switch from $CD44^+$ cell to EMT cell drives the metastasis of prostate cancer. *Oncotarget*, *6*(2), 1202–1216. https://doi.org/10.18632/oncotarget.2841.

Shang, Y., Myers, M., & Brown, M. (2002). Formation of the androgen receptor transcription complex. *Molecular Cell*, *9*(3), 601–610. https://doi.org/10.1016/s1097-2765(02)00471-9.

Shen, M. M., & Abate-Shen, C. (2010). Molecular genetics of prostate cancer: New prospects for old challenges. *Genes & Development*, *24*(18), 1967–2000. https://doi.org/10.1101/gad.1965810.

Sica, A., Schioppa, T., Mantovani, A., & Allavena, P. (2006). Tumour-associated macrophages are a distinct M2 polarised population promoting tumour progression: Potential targets of anti-cancer therapy. *European Journal of Cancer (Oxford, England: 1990)*, *42*(6), 717–727. https://doi.org/10.1016/j.ejca.2006.01.003.

Siegel, R. L., Miller, K. D., & Jemal, A. (2019). Cancer statistics, 2019. *CA: A Cancer Journal for Clinicians*, *69*(1), 7–34. https://doi.org/10.3322/caac.21551.

Signoretti, S., & Loda, M. (2007). Prostate stem cells: From development to cancer. *Seminars in Cancer Biology*, *17*(3), 219–224. https://doi.org/10.1016/j.semcancer.2006.04.004.

Song, X.-L., Huang, B., Zhou, B.-W., Wang, C., Liao, Z.-W., Yu, Y., et al. (2018). miR-1301-3p promotes prostate cancer stem cell expansion by targeting SFRP1 and GSK3β. *Biomedicine & Pharmacotherapy = Biomedecine & Pharmacotherapie*, *99*, 369–374. https://doi.org/10.1016/j.biopha.2018.01.086.

Sorrentino, C., Ciummo, S. L., Cipollone, G., Caputo, S., Bellone, M., & Di Carlo, E. (2018). Interleukin-30/IL27p28 shapes prostate cancer stem-like cell behavior and is critical for tumor onset and metastasization. *Cancer Research*, *78*(10), 2654–2668. https://doi.org/10.1158/0008-5472.CAN-17-3117.

Sotomayor, P., Godoy, A., Smith, G. J., & Huss, W. J. (2009). Oct4A is expressed by a subpopulation of prostate neuroendocrine cells. *The Prostate*, *69*(4), 401–410. https://doi.org/10.1002/pros.20895.

Sun, Y., & Gao, W.-Q. (2011). Human prostate cancer stem cells: New features unveiled. *Asian Journal of Andrology*, *13*(3), 355–356. https://doi.org/10.1038/aja.2011.17.

Syed Khaja, A. S., Helczynski, L., Edsjö, A., Ehrnström, R., Lindgren, A., Ulmert, D., et al. (2011). Elevated level of Wnt5a protein in localized prostate cancer tissue is associated with better outcome. *PLoS One*, *6*(10), e26539. https://doi.org/10.1371/journal.pone.0026539.

Takebe, N., Miele, L., Harris, P. J., Jeong, W., Bando, H., Kahn, M., et al. (2015). Targeting notch, hedgehog, and Wnt pathways in cancer stem cells: Clinical update. *Nature Reviews. Clinical Oncology*, *12*(8), 445–464. https://doi.org/10.1038/nrclinonc.2015.61.

Talmadge, J. E., & Gabrilovich, D. I. (2013). History of myeloid-derived suppressor cells. *Nature Reviews Cancer*, *13*(10), 739–752. https://doi.org/10.1038/nrc3581.

Talukdar, S., Das, S. K., Pradhan, A. K., Emdad, L., Windle, J. J., Sarkar, D., et al. (2019). MDA-9/syntenin (SDCBP) is a critical regulator of chemoresistance, survival and stemness in prostate cancer stem cells. *Cancers*, *12*(1), 53. https://doi.org/10.3390/cancers12010053.

Thiery, J. P. (2002). Epithelial-mesenchymal transitions in tumour progression. *Nature Reviews. Cancer*, *2*(6), 442–454. https://doi.org/10.1038/nrc822.

Tolcher, A., Flaherty, K., Shapiro, G. I., Berlin, J., Witzig, T., Habermann, T., et al. (2018). A first-in-human phase I study of OPB-111077, a small-molecule STAT3 and oxidative phosphorylation inhibitor, in patients with advanced cancers. *The Oncologist*, *23*(6), 658–e72. https://doi.org/10.1634/theoncologist.2017-0325.

Tomita, H., Tanaka, K., Tanaka, T., & Hara, A. (2016). Aldehyde dehydrogenase 1A1 in stem cells and cancer. *Oncotarget*, 7(10), 11018–11032. https://doi.org/10.18632/oncotarget.6920.

Tong, W., Qiu, L., Qi, M., Liu, J., Hu, K., Lin, W., et al. (2018). GANT-61 and GDC-0449 induce apoptosis of prostate cancer stem cells through a GLI-dependent mechanism. *Journal of Cellular Biochemistry*, *119*(4), 3641–3652. https://doi.org/10.1002/jcb.26572.

Twillie, D. A., Eisenberger, M. A., Carducci, M. A., Hseih, W.-S., Kim, W. Y., & Simons, J. W. (1995). Interleukin-6: A candidate mediator of human prostate cancer morbidity. *Urology*, *45*(3), 542–549. https://doi.org/10.1016/S0090-4295(99)80034-X.

van den Hoogen, C., van der Horst, G., Cheung, H., Buijs, J. T., Lippitt, J. M., Guzmán-Ramírez, N., et al. (2010). High aldehyde dehydrogenase activity identifies tumor-initiating and metastasis-initiating cells in human prostate cancer. *Cancer Research*, *70*(12), 5163–5173. https://doi.org/10.1158/0008-5472.CAN-09-3806.

van Leenders, G. J. L. H., & Schalken, J. A. (2003). Epithelial cell differentiation in the human prostate epithelium: Implications for the pathogenesis and therapy of prostate cancer. *Critical Reviews in Oncology/Hematology*, *46*(Suppl), S3–10. https://doi.org/10.1016/s1040-8428(03)00059-3.

Varambally, S., Yu, J., Laxman, B., Rhodes, D. R., Mehra, R., Tomlins, S. A., et al. (2005). Integrative genomic and proteomic analysis of prostate cancer reveals signatures of metastatic progression. *Cancer Cell*, *8*(5), 393–406. https://doi.org/10.1016/j.ccr.2005.10.001.

Vassalli, G. (2019). Aldehyde dehydrogenases: Not just markers, but functional regulators of stem cells. *Stem Cells International*, *2019*, 1–15. https://doi.org/10.1155/2019/3904645.

Vlachostergios, P. J., Puca, L., & Beltran, H. (2017). Emerging variants of castration-resistant prostate cancer. *Current Oncology Reports*, *19*(5), 32. https://doi.org/10.1007/s11912-017-0593-6.

Vuk-Pavlović, S., Bulur, P. A., Lin, Y., Qin, R., Szumlanski, C. L., Zhao, X., et al. (2010). Immunosuppressive CD14+HLA-DRlow/-monocytes in prostate cancer. *The Prostate*, *70*(4), 443–455. https://doi.org/10.1002/pros.21078.

Wang, H., Huang, B., Li, B. M., Cao, K. Y., Mo, C. Q., Jiang, S. J., et al. (2018). ZEB1-mediated vasculogenic mimicry formation associates with epithelial-mesenchymal transition and cancer stem cell phenotypes in prostate cancer. *Journal of Cellular and Molecular Medicine*, *22*(8), 3768–3781. https://doi.org/10.1111/jcmm.13637.

Wang, X., Julio, M. K., Economides, K. D., Walker, D., Yu, H., Halili, M. V., et al. (2009). A luminal epithelial stem cell that is a cell of origin for prostate cancer. *Nature*, *461*(7263), 495–500. https://doi.org/10.1038/nature08361.

Wang, G., Lu, X., Dey, P., Deng, P., Wu, C. C., Jiang, S., et al. (2016). Targeting YAP-dependent MDSC infiltration impairs tumor progression. *Cancer Discovery*, *6*(1), 80–95. https://doi.org/10.1158/2159-8290.CD-15-0224.

Wang, Z. A., Mitrofanova, A., Bergren, S. K., Abate-Shen, C., Cardiff, R. D., Califano, A., et al. (2013). Lineage analysis of basal epithelial cells reveals their unexpected plasticity and supports a cell-of-origin model for prostate cancer heterogeneity. *Nature Cell Biology*, *15*(3), 274–283. https://doi.org/10.1038/ncb2697.

Wang, C., Peng, G., Huang, H., Liu, F., Kong, D.-P., Dong, K.-Q., et al. (2018). Blocking the feedback loop between neuroendocrine differentiation and macrophages improves the therapeutic effects of enzalutamide (MDV3100) on prostate cancer. *Clinical Cancer Research: An Official Journal of the American Association for Cancer Research*, *24*(3), 708–723. https://doi.org/10.1158/1078-0432.CCR-17-2446.

Wang, X.-D., Shou, J., Wong, P., French, D. M., & Gao, W.-Q. (2004). Notch1-expressing cells are indispensable for prostatic branching morphogenesis during development and re-growth following castration and androgen replacement. *The Journal of Biological Chemistry*, *279*(23), 24733–24744. https://doi.org/10.1074/jbc.M401602200.

Wang, G., Zhao, D., Spring, D. J., & DePinho, R. A. (2018). Genetics and biology of prostate cancer. *Genes & Development*, *32*(17–18), 1105–1140. https://doi.org/10.1101/gad.315739.118.

Watson, P. A., Arora, V. K., & Sawyers, C. L. (2015). Emerging mechanisms of resistance to androgen receptor inhibitors in prostate cancer. *Nature Reviews. Cancer*, *15*(12), 701–711. https://doi.org/10.1038/nrc4016.

Wen, J., Huang, G., Liu, S., Wan, J., Wang, X., Zhu, Y., et al. (2020). Polymorphonuclear MDSCs are enriched in the stroma and expanded in metastases of prostate cancer. *The Journal of Pathology: Clinical Research*. cjp2.160 https://doi.org/10.1002/cjp2.160.

Williams, K., Motiani, K., Giridhar, P. V., & Kasper, S. (2013). CD44 integrates signaling in normal stem cell, cancer stem cell and (pre)metastatic niches. *Experimental Biology and Medicine (Maywood, N.J.)*, *238*(3), 324–338. https://doi.org/10.1177/1535370213480714.

Williamson, S. C., Hepburn, A. C., Wilson, L., Coffey, K., Ryan-Munden, C. A., Pal, D., et al. (2012). Human α(2)β(1)(HI) CD133(+VE) epithelial prostate stem cells express low levels of active androgen receptor. *PLoS One*, 7(11), e48944. https://doi.org/10.1371/journal.pone.0048944.

Wong, A. L., Soo, R. A., Tan, D. S., Lee, S. C., Lim, J. S., Marban, P. C., et al. (2015). Phase I and biomarker study of OPB-51602, a novel signal transducer and activator of transcription (STAT) 3 inhibitor, in patients with refractory solid malignancies. *Annals of Oncology: Official Journal of the European Society for Medical Oncology*, *26*(5), 998–1005. https://doi.org/10.1093/annonc/mdv026.

Wu, C.-T., Huang, Y.-C., Chen, W.-C., & Chen, M.-F. (2019). Effect of tumor burden on tumor aggressiveness and immune modulation in prostate cancer: Association with IL-6 signaling. *Cancers*, *11*(7), 992. https://doi.org/10.3390/cancers11070992.

Wu, X., Schulte, B. C., Zhou, Y., Haribhai, D., Mackinnon, A. C., Plaza, J. A., et al. (2014). Depletion of M2-like tumor-associated macrophages delays cutaneous T-cell lymphoma development in vivo. *The Journal of Investigative Dermatology*, *134*(11), 2814–2822. https://doi.org/10.1038/jid.2014.206.

Wu, K., Xu, H., Tian, Y., Yuan, X., Wu, H., Liu, Q., et al. (2015). The role of CD44 in epithelial–mesenchymal transition and cancer development. *Oncotargets and Therapy*, *8*, 3783. https://doi.org/10.2147/OTT.S95470.

Yadav, A. K., & Desai, N. S. (2019). Cancer stem cells: Acquisition, characteristics, therapeutic implications, targeting strategies and future prospects. *Stem Cell Reviews and Reports*, *15*(3), 331–355. https://doi.org/10.1007/s12015-019-09887-2.

Yardy, G. W., & Brewster, S. F. (2005). Wnt signalling and prostate cancer. *Prostate Cancer and Prostatic Diseases*, *8*(2), 119–126. https://doi.org/10.1038/sj.pcan.4500794.

Yasumizu, Y., Rajabi, H., Jin, C., Hata, T., Pitroda, S., Long, M. D., et al. (2020). MUC1-C regulates lineage plasticity driving progression to neuroendocrine prostate cancer. *Nature Communications*, *11*(1), 338. https://doi.org/10.1038/s41467-019-14219-6.

Ye, X., & Weinberg, R. A. (2015). Epithelial–mesenchymal plasticity: A central regulator of cancer progression. *Trends in Cell Biology*, *25*(11), 675–686. https://doi.org/10.1016/j.tcb.2015.07.012.

Yoo, Y. A., Roh, M., Naseem, A. F., Lysy, B., Desouki, M. M., Unno, K., et al. (2016). Bmi1 marks distinct castration-resistant luminal progenitor cells competent for prostate regeneration and tumour initiation. *Nature Communications*, 7(1), 12943. https://doi.org/10.1038/ncomms12943.

Yoshida, A., Hsu, L. C., & Davé, V. (1992). Retinal oxidation activity and biological role of human cytosolic aldehyde dehydrogenase. *Enzyme*, *46*(4–5), 239–244. https://doi.org/10.1159/000468794.

Yu, C., Yao, Z., Jiang, Y., & Keller, E. T. (2012). Prostate cancer stem cell biology. *Minerva Urologica e Nefrologica = The Italian Journal of Urology and Nephrology*, *64*(1), 19–33. http://www.ncbi.nlm.nih.gov/pubmed/22402315.

Yuan, X., Wu, H., Xu, H., Xiong, H., Chu, Q., Yu, S., et al. (2015). Notch signaling: An emerging therapeutic target for cancer treatment. *Cancer Letters*, *369*(1), 20–27. https://doi.org/10.1016/j.canlet.2015.07.048.

Yun, E.-J., Zhou, J., Lin, C.-J., Hernandez, E., Fazli, L., Gleave, M., et al. (2016). Targeting cancer stem cells in castration-resistant prostate cancer. *Clinical Cancer Research*, *22*(3), 670–679. https://doi.org/10.1158/1078-0432.CCR-15-0190.

Zhang, J., Chen, M., Zhu, Y., Dai, X., Dang, F., Ren, J., et al. (2019). SPOP promotes nanog destruction to suppress stem cell traits and prostate cancer progression. *Developmental Cell*, *48*(3), 329–344.e5. https://doi.org/10.1016/j.devcel.2018.11.035.

Zhang, K., Guo, Y., Wang, X., Zhao, H., Ji, Z., Cheng, C., et al. (2017). WNT/β-catenin directs self-renewal symmetric cell division of hTERThigh prostate cancer stem cells. *Cancer Research*, 77(9), 2534–2547. https://doi.org/10.1158/0008-5472.CAN-16-1887.

Zhang, Q., Huang, H., Liu, A., Li, J., Liu, C., Sun, B., et al. (2019). Cell division cycle 20 (CDC20) drives prostate cancer progression via stabilization of β-catenin in cancer stem-like cells. *eBioMedicine*, *42*, 397–407. https://doi.org/10.1016/j.ebiom.2019.03.032.

Zhang, L., Sha, J., Yang, G., Huang, X., Bo, J., & Huang, Y. (2017). Activation of notch pathway is linked with epithelial-mesenchymal transition in prostate cancer cells. *Cell Cycle*, *16*(10), 999–1007. https://doi.org/10.1080/15384101.2017.1312237.

Zhang, D., Zhao, S., Li, X., Kirk, J. S., & Tang, D. G. (2018). Prostate luminal progenitor cells in development and cancer. *Trends in Cancer*, *4*(11), 769–783. https://doi.org/10.1016/j.trecan.2018.09.003.

Zhou, W., Ke, S. Q., Huang, Z., Flavahan, W., Fang, X., Paul, J., et al. (2015). Periostin secreted by glioblastoma stem cells recruits M2 tumour-associated macrophages and promotes malignant growth. *Nature Cell Biology*, *17*(2), 170–182. https://doi.org/10.1038/ncb3090.

Zhou, J., Tang, Z., Gao, S., Li, C., Feng, Y., & Zhou, X. (2020). Tumor-associated macrophages: Recent insights and therapies. *Frontiers in Oncology*, *10*, 188. https://doi.org/10.3389/fonc.2020.00188.

Zou, M., Toivanen, R., Mitrofanova, A., Floch, N., Hayati, S., Sun, Y., et al. (2017). Transdifferentiation as a mechanism of treatment resistance in a mouse model of castration-resistant prostate cancer. *Cancer Discovery*, 7(7), 736–749. https://doi.org/10.1158/2159-8290.CD-16-1174.

CHAPTER SIX

Impact of prostate cancer stem cell niches on prostate cancer tumorigenesis and progression

Taosha Li[a], Xiaozeng Lin[b,c,d], Anil Kapoor[b,c,d], Jingyi Peng[b,c,d], Yan Gu[b,c,d], and Damu Tang[b,c,d,*]
[a]Life-Tech Industry Alliance, Shenzhen, China
[b]Department of Medicine, McMaster University, Hamilton, ON, Canada
[c]The Research Institute of St Joe's Hamilton, Hamilton, ON, Canada
[d]Urological Cancer Center for Research and Innovation (HCCRI), St. Joseph's Hospital, Hamilton, ON, Canada
*Corresponding author: e-mail address: damut@mcmaster.ca

Contents

1. Introduction

Prostate cancer (PC) is the most frequently diagnosed male malignancy and the second leading cause of cancer death in men in the developed world (Siegel, Ward, Brawley, & Jemal, 2011). Abnormalities in prostate epithelial cells lead to high grade prostatic intra-epithelial neoplasia (HGPIN), which progresses to invasive carcinoma and then metastatic

Advances in Stem Cells and their Niches, Volume 5
ISSN 2468-5097
https://doi.org/10.1016/bs.asn.2020.10.001

cancer (Moon, Park, Chae, Yun, & Kim, 2008; Ross, 2007). Bone is the most common site of PC metastasis (Bagnall, 2014). The prostate gland consists of two types of epithelial cells: basal and luminal epithelial cells (Shen & Abate-Shen, 2010). While the latter cells are androgen receptor (AR)-positive and require AR signaling for survival, basal epithelial cells are AR-negative and castration-resistant (Isaacs & Coffey, 1989). As first observed by Charles Huggins (Lytton, 2001; Wong, Ferraldeschi, Attard, & de Bono, 2014), the majority of PC cells express AR and rely on AR signaling for survival. This property resembles that of luminal epithelial cells, leading to the concept of PC being originated from prostate luminal epithelial cells. This view is supported by the observation that mouse luminal epithelial cells harbor prostate stem cells (PSCs), as single mouse luminal epithelial cell can generate prostate organoids (Chua et al., 2014).

However, classic evidence supports the existence of PSCs within basal epithelial cells. These cells are AR-negative and can survive castration; castration-induced prostate gland involution can be fully regenerated upon androgen supplementation. This involution-regeneration cycle can undergo multiple rounds, supporting the presence of PSCs in basal epithelial cells (Isaacs & Coffey, 1989). These PSCs bear a variety of surface antigens, including Sca-l (stem cell antigen 1), CD49f, and Trop2, and are capable of regenerating prostate glands (Burger et al., 2005; Goldstein et al., 2008; Lawson, Xin, Lukacs, Cheng, & Witte, 2007). Furthermore, a single basal epithelial cell defined by a panel of surface antigens $Lin^{-}Sca\text{-}1^{+}$ $CD133^{+}CD44^{+}CD117^{+}$ can generate prostate upon implantation into the renal capsule space of nude mice (Leong, Wang, Johnson, & Gao, 2008).

Collectively, both luminal and basal epithelial cells of mouse prostate contain PSCs which is in accordance with PC being initiated in mice from luminal and basal epithelial cells (Skvortsov, Skvortsova, Tang, & Dubrovska, 2018). In this chapter, we will discuss the influence of niches on PSCs-derived PC tumorigenesis.

Metastasis causes more than 90% of cancer deaths (Gupta & Massague, 2006; Weigelt, Peterse, & van't Veer, 2005), including prostate cancer (Gundem et al., 2015). PC commonly metastasizes to the bone with 90% of metastatic cases involving the bone (Bubendorf et al., 2000). The standard of care for metastatic PC (mPC) remains androgen deprivation therapy (ADT), which was based on the landmark discovery by Charles Huggins in the 1940s (see reviews by Lytton and Want et al. for details) (Lytton, 2001; Wong et al., 2014). While ADT results in remarkable responses in more than 80% of patients, resistance in the form of castration-resistant PCs (mCRPCs) commonly occurs (Heinlein & Chang, 2004); mCRPCs can be treated with multiple options, including those involving docetaxel (Berthold et al., 2008; Tannock et al., 2004) and the second generation anti-androgens approved by FDA in 2011 and 2012 (de Bono et al., 2011; Scher et al., 2012). However, these treatments only modestly extend median overall survival. Metastasis to the bone thus remains the overarching challenge in the management of PC patients.

Metastasis is one of the most complex processes of tumorigenesis. Primary cancer cells are disseminated from primary sites, enter the blood stream through intravasation, cross the vessel walls into the target organ (extravasation), home in foreign environment, and initiate metastatic growth (colonization) (Chambers, Groom, & MacDonald, 2002; Chiang & Massague, 2008). For epithelial cancerous cells, including those of PC, the relocation process from the primary site to the target organ, with bone being the most frequent site for PC, requires epithelial-mesenchymal transition (EMT) (De Craene & Berx, 2013; Tsai & Yang, 2013). To home and form overt metastasis in bone, PC and other epithelial cancer cells undergo a reverse transition, i.e., mesenchymal-epithelial transition (MET), to resume their epithelial cell properties (Gunasinghe, Wells, Thompson, & Hugo, 2012; Yao, Dai, & Peng, 2011). Additionally, PC cells need to adapt to the bone marrow environment first (dormancy) and exit from dormancy to overt metastasis. Accumulative evidence supports that the plasticity of prostate cancer stem cells (PCSCs) is at the heart for PC to undergo these transitions (Mei et al., 2019; Nassar & Blanpain, 2016; Oskarsson, Batlle, & Massague, 2014). This knowledge is in agreement with the notion that the stem cell properties of PCSCs likely underlie their ability to utilize the niches of hematopoietic stem cells (HSCs) in bone marrow for PC metastasis (Shiozawa et al., 2011). In this chapter, we will discuss the contributions of HSC niches and their interactions with PCSCs during PC metastasis to the bone.

To provide a comprehensive evaluation of the topic of PCSCs-niches during PC tumorigenesis and metastasis, we have systemically organized the materials based on the PRISMA Guidelines (Moher et al., 2015; Shamseer et al., 2015). We performed a literature search through the PubMed database for "Prostate Cancer Cell Niche." A total of 105 articles were retrieved; among these, 20 were deemed not-directly relevant and 18 review articles were redundant, which left n = 67 articles being evaluated.

2. Prostate cancer stem cells

Following the identification of cancer stem cells (CSCs) in acute myeloid leukemia (AML) as those bearing the surface antigen profile of $CD34^+CD38^-$ in 1994 (Lapidot et al., 1994), their counterparts in solid tumors were first isolated from primary breast cancer based on the surface antigens of $CD44^+CD24^{-/low}$ in 2013 (Al-Hajj, Wicha, Benito-Hernandez, Morrison, & Clarke, 2003). This led to the identification of CSCs in a variety of carcinomas (Baccelli & Trumpp, 2012; Kreso & Dick, 2014), including cancers of the bladder (She, Zhang, Wang, Gan, & Che, 2008), brain (Singh et al., 2004), colon (Ovalle et al., 2007; Ricci-Vitiani et al., 2007), head and neck (Prince et al., 2007), liver (Ma et al., 2007), lung (Eramo et al., 2008), skin (Fang et al., 2005), and pancreas (Hermann et al., 2007; Li et al., 2007). The common feature of these identifications was based on the potential shared surface antigens of CSCs with their tissue stem cells (Visvader & Lindeman, 2008).

Human prostate stem cells (PSCs) display the surface antigen profile of $\alpha_2\beta_1{}^{hi}CD133^+$ (Richardson et al., 2004). This knowledge underlie the identification of PCSCs by Collins et al. in 2005, which were $CD44^+\alpha_2\beta_1{}^{hi}CD133^+$ (Collins, Berry, Hyde, Stower, & Maitland, 2005). PCSCs from primary human PCs were also defined as expressing the breast cancer resistance protein/BCPR (ABCG2) (Huss, Gray, Greenberg, Mohler, & Smith, 2005). $CD44^+\alpha_2\beta_1{}^{hi}CD133^+$ and $BCPR^+$ PCSCs can generate other PC cell types including those of AR^+ (Collins et al., 2005; Huss et al., 2005). Subsequently, $Trop2^{hi}$, CD166/ALCAM, $PSA^{-/low}$ (prostate specific antigen), and ALDH1A1 were also reported to mark the PCSC populations in human primary PC tissues (Chen et al., 2016; Goldstein et al., 2008; Jiao et al., 2012; Qin et al., 2012). Importantly, a large body of evidence reveals critical roles of PCSCs in PC tumorigenesis, metastasis and CRPC progression (Mei et al., 2019); this would immediately suggest the requirement of niches to maintain and regulate PCSCs for their actions in PC initiation and progression.

3. Niche involvement in PCSC-mediated tumor initiation

3.1 PC initiation through PCSCs

One of key issues in PC research is the cell origin of the disease; this knowledge is critical to rationalize a better management of PC. While this issue needs substantial further investigations, accumulative evidence supports PCSCs being the origin of PC. Mouse PSCs are defined by the surface antigens of Sca-l, CD49f, and Trop2 (Burger et al., 2005; Goldstein et al., 2008; Lawson et al., 2007). Prostate-specific knockout of PTEN using the PB-Cre4;PTEN$^{loxp/loxp}$ system induces PC in mice (Wang et al., 2003); the Lin^{-}Sca-1^{+}CD49f^{hi} cell population is elevated following PC initiation and progression in these animals (Mulholland et al., 2009). Furthermore, co-expression of an active form of Kras (G12D) in PB-Cre4;PTEN$^{loxp/loxp}$ mice enhances PC tumorigenesis compared to PB-Cre4;PTEN$^{loxp/loxp}$ mice and Lin^{-}Sca-1^{+}CD49f^{hi} cells from the Kras mice form large sphere than the cells from PB-Cre4;PTEN$^{loxp/loxp}$ mice (Mulholland et al., 2012). Sphere formation is an established assay to analyze PCSCs (Rybak, He, Kapoor, Cutz, & Tang, 2011). Collectively, these correlational observations support the concept of PCSCs as an origin of PC.

Direct evidence also exists in both human and mouse-based PC research. In human prostate, $\alpha_2\beta_1{}^{hi}$CD133^{+} PSCs reside in the basal layer (Richardson et al., 2004); PSCs has been isolated based on the antigen profile of CD49f^{hi}Trop2^{hi}; these human basal, but not luminal, epithelial cells generated prostate glandular structure in the subcutaneous (s.c.) space of NOD/SCID/IL-2Rγ^{null} (NSG) mice, confirming their stem cell properties (Goldstein, Huang, Guo, Garraway, & Witte, 2010). Furthermore, concurrent expression of activated AKT (myristoylated AKT), ERG, and AR only in the basal epithelial cells marked with CD49f^{hi}Trop2^{hi} (indicative of these cells being PSCs) produced PINs and PCs in NSG mice (Goldstein et al., 2010) (Fig. 1A). Considering the basal layer of human prostate is also the home for PSCs defined by the surface markers CD44$^{+}\alpha_2\beta_1{}^{hi}$CD133^{+} or ABCG2 (Collins et al., 2005; Huss et al., 2005), it will be interesting to examine their potential in PC initiation. It remains unknown whether CD49f^{hi}Trop2^{hi}, CD44$^{+}\alpha_2\beta_1{}^{hi}$CD133^{+} and ABCG2 are the same PCSC population or result of PCSC heterogeneity. This will be a critical issue to address in the future. Nonetheless, these observations are consistent with PSCSs being an origin of PC.

Similar observations were also derived from murine PSCs. Mouse PSCs within the basal layer express the surface antigens of Sca-l, CD49f, or Trop2 (Burger et al., 2005; Goldstein et al., 2008; Lawson et al., 2007). Basal PSCs were genetically confirmed using the lineage-tracing approach (Kwon & Xin, 2014; Ousset et al., 2012; Wang et al., 2013). Single Lin^{-}Sca-1^{+}CD133^{+}CD44^{+}CD117^{+} basal cell produces prostate gland structure when implanted in renal capsule of nude mice (Leong et al., 2008). Upon ectopic

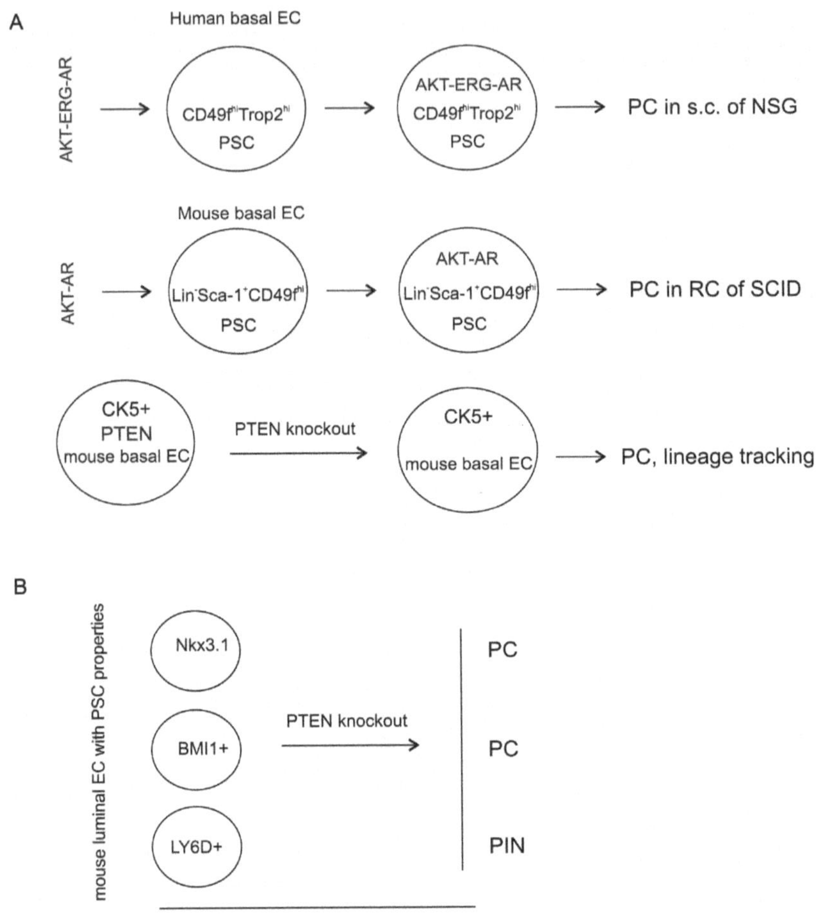

Fig. 1 Oncogenic actions in PSC initiate PC tumorigenesis. (A) Human and mouse basal epithelial cell (EC) PSC are ectopically expressed with AKT-ERG-AR or AKT-AR, followed by implantation into the subcutaneous (s.c.) or renal capsule (RC) space in the indicated immunocompromised mice. PCs were subsequently resulted. PTEN was specific knockout in mouse $CK5^+$ basal EC, which leads to PC development. (B) The indicated luminal cell populations possess PSC properties; specific knockout of PTEN in these cell populations using lineage tracking approach led to the formation of PC or PIN (prostatic intra-epithelial neoplasia).

expression of AKT1 in $Sca\text{-}1^+$ PSCs, these cells form PIN lesions following engraftment in renal capsule of SCID mice (Xin, Lawson, & Witte, 2005). Similarly, co-expression of AKT and AR in $Lin^-Sca\text{-}1^+CD49f^{hi}$ basal PSCs resulted in poorly differentiated PC in SCID mice (Fig. 1A) (Lawson et al., 2010). Furthermore, specific deletion of PTEN in the $CK5^+$ (cytokeratin 5) basal epithelial cells leads to PC pathogenesis in transgenic mice (Fig. 1A) (Lu et al., 2013; Wang et al., 2013). Collectively, evidence from both human and murine research supports that abnormalities in basal PSCs are capable of PC generation.

Murine luminal epithelial cells also contain PSCs, based on their ability to generate prostate organoids (Chua et al., 2014). By lineage tracing, mouse prostate luminal epithelial cells expressing the Nkx3.1 tumor suppressor possess capacity of self-renewal and initiate PC following PTEN knockout (Fig. 1B) (Wang, Kruithof-de Julio, et al., 2009; Wang, Lathia, et al., 2009). Similarly, the $BMI1^+$ luminal epithelial cells can self-renew, are resistant to castration (Yoo et al., 2016), and initiate PC following PTEN deletion (Fig. 1B) (Yoo et al., 2019). The mouse $LY6D^+$ luminal epithelial cells display resistance to castration, PSC properties, and an ability of inducing PIN lesions when PTEN is knocked-out in these cells (Fig. 1B) (Barros-Silva et al., 2018). Taken together, different luminal PSCs ($Nkx3.1^+$, $BMI1^+$, and $LY6D^+$) are capable of PC initiation when an oncogenic event of PTEN deletion occurs (Barros Silva et al., 2018; Yoo et al., 2016).

3.2 PSC niche

While the current knowledge on PSC niche remain rudimentary, evidence supports the niches to reside within the proximal ductal region connecting to urethra in mouse prostate. Basal epithelial cells in these regions are quiescent, possess high proliferation potential in vitro, and produce branched glandular ductal structures in vitro (Tsujimura et al., 2002). Proximal epithelial cells regenerate prostate and the proximal duct region following implantation into the renal capsule space of recipient mice (Goto et al., 2006). These epithelial cells are $Sca\text{-}1^+$ with concurrent expression of α6 integrin and BCL-2 (Burger et al., 2005; Xin et al., 2005). CD49f (α6 integrin) is a well-demonstrated surface markers of PSCs in both human and mice (Burger et al., 2005; Goldstein et al., 2010); it is also a marker of epidermal stem cells (Li, Simmons, & Kaur, 1998; Tani, Morris, & Kaur, 2000). Importantly, adhesion of stem cells to stromal cells or extracellular matrix is a major mechanism of communication between stem cells and their niches; α6 integrin likely contributes to this event (Jones & Wagers, 2008). This possibility is supported by the broad range expression of α6 integrin in spermatogonial stem cells (Shinohara, Avarbock, & Brinster, 1999), pancreatic islet progenitor cells (Sugiyama, Rodriguez, McLean, & Kim, 2007), corneal epithelial stem cells (Hayashi et al., 2008), hepatic progenitor cells (Suzuki et al., 2000), and mammary epithelial stem cells (Stingl et al., 2006). Additionally, an enrichment in Trop2$^+$ (a well-demonstrated PSC marker) was also demonstrated in the proximal region (Hofner et al., 2016). Collectively, accumulating evidence supports the location of PSC niche in proximal prostate ducts in mice.

Niche makes an essential contribution to stem cell physiology in terms of their self-renewal, tissue development, homeostasis, and repair. For instance, in an elegant recombination experiments using the pharyngeal epithelium isolated from 5-day old chicken embryos and mesenchymal-derived from the first mandibular molars of 16- to 18-day old mouse embryos, complete teeth development was achieved upon engrafting the recombinant materials into nude mice (Kollar & Fisher, 1980). This clearly supports the astonishing capacity of stroma in supporting stem cell-mediated tissue development, particularly considering the loss of odontogenic potential in birds approximately 100 million years ago during the evolution process (Strand & Goldstein, 2015). In a recombination grafting experiment using the mesenchymal of mouse embryonic urogenital sinus (UGS) and the epithelium derived from mouse bladder, it was demonstrated that androgen-sensitive mesenchymal of UGS but not the androgen-insensitive counterpart was essential to instruct the epithelium to form prostate structure (Cunha & Lung, 1978), revealing a critical role of stroma or niche-derived androgen signaling in directing PSCs to develop prostate. This early study fits well with a recent report for an essential role of androgen receptor (AR) signaling from Gil expressing stromal cells (Le et al., 2020) (also see next section on stroma-derived hedgehog signaling). In addition to androgen signaling, interactions with stromal cells was deemed to have critical roles in prostate development (Hayward, Del Buono, Deshpande, & Hall, 1992), which is in line with the view that interactions with niche components are essential in stem cell regulation (Jones & Wagers, 2008).

The common expression of α6 integrin in a variety of niches of epithelial stem cells fits the theme of conserveness in stem cell niches; this would be expected owing to the functional conserveness in maintaining tissue homeostasis among individual types of stem cell. In this regard, the Wnt signaling is one of the typical pathways that regulate the communications between stem cells and their niches (Jones & Wagers, 2008). Wnt is well-established for its critical roles in development, stem cells, and tissue regeneration or renewal (Clevers, Loh, & Nusse, 2014; Korinek et al., 1998; Nusse & Clevers, 2017). In prostate, the Wnt/β-catenin signaling specifies the prostate epithelial linage during development (Francis, Thomsen, Taketo, & Swain, 2013; Lee et al., 2015; Simons et al., 2012) and is required for basal stem/progenitor cells to produce luminal cells (Lu & Chen, 2015). Mouse PSCs resides in proximal ducts and are maintained at quiescent stage (Tsujimura et al., 2002), suggesting that niche signals inhibit PSC proliferation. In support of this notion, stromal cells in this location express a high level of Wnt/β-catenin signaling which inhibits the proliferation of Sca-1$^+$ CD49f^{hi} basal PSCs and Sca-1$^+$ luminal PSCs through two mechanisms (Fig. 2); activation of the non-canonical Wnt receptor ROR2 through Wnt5a ligand and TGFβ suppress PSC proliferation (Wei et al., 2019). The stromal cells concurrently express active Wnt via a canonical Wnt pathway, which maintains a repressive niche environment to reserve PSCs quiescence (Fig. 2) (Wei et al., 2019). These observations further support the functionality of PSC niches in the proximal prostate ducts and are also in line with a general role of TGFβ in maintaining stem cells quiescence (Mendez-Ferrer et al., 2020).

In collaboration with the above studies, multipotent basal stem cells in human prostate reside in niches located in the proximal regions within the juxta-urethral ducts (Moad et al., 2017). These stem cells are positive for delta homolog 1 (DLK1) and generate bipotent base progenitor cells along the entire length of epithelium from the proximal ducts to the branching gland regions (Moad et al., 2017). DLK1 is a non-canonical Notch ligand (delta like non-canonical Notch ligand 1) and plays a role in stem cell regulations (Traustadottir et al., 2019). DLK1 was reported as

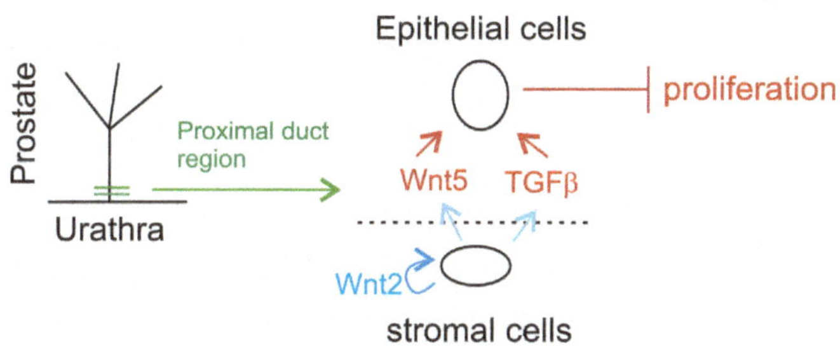

Fig. 2 Inhibition of epithelial cell proliferation through Wnt signaling produced from stroma cell in the proximal duct region. The mouse urethra and prostate are simplified on the left with the proximal duct region adjacent the urethra indicated; the distal prostate ducts are also represented. Wnt signaling from stromal cells within the proximal ducts produce Wnt5 and TGFβ; both suppress epithelial cell proliferation through non-canonical Wnt signaling and TGFβ, respectively. Stromal cells also produce Wnt2 to activate the Wnt pathway in these cells.

a marker of human PSCs (Ceder, Jansson, Helczynski, & Abrahamsson, 2008), and found to be co-expressed with α6 integrin (CD49f) in basal epithelial cells from the proximal juxta-urethra ducts (Moad et al., 2017), supporting the association of DLK1 with PSCs in the proximal region of human prostate. However, detailed mechanisms underlying how niches in the proximal juxta-urethra ducts regulate human PSCs remain to be investigated. For instance, is DLK1 signaling functionally important? Does α6 integrin provide an important role in adhesion? Most importantly, will α6 integrin-mediated adhesion regulate DLK1 signaling in PSCs?

3.3 The effects of niches on PC tumorigenesis via regulating PSCs

Current knowledge on PSCs niche remains limited, but its impacts on PC tumorigenesis is now gaining more attention. Evidence supports the importance of niche abnormalities in PC tumorigenesis through affecting PSCs. Aging is the most significant risk factor for PC development, and is generally regarded to facilitate malignancy in part via influencing stem cell niches (Naik, Larsen, Cowley, & Fuchs, 2018). For instance, the contributions of niches to HSC-initiated hematological cancers are being increasingly recognized (Mendez-Ferrer et al., 2020). In PC, it was reported that carcinoma-associated fibroblasts (CAFs) isolated from human PC tissues supports a non-tumorigenic human prostate epithelial cell line BPH-1 to form tumors in nude mice (Hayward et al., 2001), providing evidence for oncogenic effects of abnormalities in niche. In a reverse manner, urogenital sinus mesenchyme and normal prostate fibroblast of mice enhanced $Lin^{-}Sca\text{-}1^{+}$ PCSCs isolated from *PB-Cre4;PTEN*$^{lox1/loxp}$ mice to form spheres (Liao, Adisetiyo, Liang, & Roy-Burman, 2010), suggesting that PCSCs modulate niche and that altered niche activities in turn support the "stemness" of PCSCs.

Stromal impact on PSC-initiated PC has been more directly demonstrated. In a transgenic mouse model for PC, specific co-knockout of PTEN and p53 in androgen-responsive (luminal) prostate epithelial cells using PB-Cre4 resulted in PC only in the proximal ducts, in which PSCs reside, but not in the distal region of prostate, despite the system deleting both tumor suppressors in all AR-positive epithelial cells (Zhou, Flesken-Nikitin, & Nikitin, 2007). In the distal regions, only PIN lesions were observed, which did not progress to invasive carcinoma at least in the duration of experiments. Intriguingly, upon reimplantation of epithelial cells isolated from either the proximal or distal regions, comparable xenograft tumors were formed (Zhou et al., 2007). These results suggest that the niches in the proximal prostate ducts support PC formation, while the environment in distal prostate region does not (see the coming paragraph for details). This concept is further corroborated by the observation that tumors generated from the proximal regions are Sca-1^{+} (Zhou et al., 2007), a well-regarded PSC marker in mouse PSCs. Since Sca-1^{+} luminal PSCs confined in the niches within the proximal prostate ducts are maintained in quiescence through stroma cells-emanated Wnt signaling (Wei et al., 2019), it will be interesting to see whether PSCs-initiated oncogenic signaling (knockout of PTEN and p53 in this case) employ the stroma-derived inhibitory signaling (Wnt5a and TGFβ) for oncogenesis, as both Wnt and TGFβ can be oncogenic. Alternatively, as stroma-derived AR signaling is critical in prostate development (Le et al., 2020), PB-Cre4 may also knockout both p53 and PTEN in the niche compartment, which may contribute to PC development.

PSC niche can also suppress PC through the hedgehog (HH) signaling pathway. Mouse prostate stromal cells express the well-established HH target Gli1; sonic hedgehog (SHH) signals from basal epithelial cells induce stromal Gli1 expression (Peng, Levine, Zahid, Wilson, & Joyner, 2013). SHH signaling contributes to androgen-stimulated stroma regeneration, including the smooth muscle regeneration, after castration-induced involution (Peng et al., 2013). HH signaling resulted from the epithelial cell to stromal cell communication is important in maintaining integrity of the smooth muscle cell layer which decreases PC progression to invasive disease in PB-MYC-initiated PC (Yang et al., 2017). Similar niche-derived tumor inhibitory roles were also observed in epithelial stem cells of the bladder, mammary gland, pancreas, and colon (Roberts, Kershner, & Beachy, 2017). The general concept suggests that activation of Gli in stroma or niche leads to (1) Wnt signaling-mediated epithelial proliferation and (2) Bmp-derived differentiation which inhibits stem cell proliferation. The balance of both signals enables epithelial stem cells to maintain tissue homeostasis while suppressing tumorigenesis (Roberts et al., 2017).

4. Utilization of HSC niches by PCSC for bone metastasis

4.1 Contributions of PCSCs to PC bone metastasis

PC commonly metastasizes to the bone. Analogous to PCSCs as the origin of PC, a large body of evidence supports a major role of PCSCs in bone metastasis. There is a general relationship between PCSCs and PC metastasis

(Harris & Kerr, 2017; Mei et al., 2019). The stromal-derived factor 1 (SDF-1)/CXCL12 and its chemokine receptor CXCR4 are a major pathway for hematopoietic stem cells (HSCs) to home in bone marrow (Juarez & Bendall, 2004; Karpova & Bonig, 2015). This system is utilized for bone metastasis by non-small cell lung cancer (NSCLC) (Su et al., 2005) and breast cancer (Smith et al., 2004). Similarly, the SDF-1/CXCR4 pathway promotes PC bone metastasis (Taichman et al., 2002). The expression of CXCR4 is associated with PCSCs. In human PC cells immortalized by hTERT an elevation of CXCR4 expression occurs in $CD133^{+}$ cells (Miki et al., 2007); NANOG confers CSC propensities in PC cells along with upregulation of CD133 and CXCR4 (Jeter et al., 2011). CXCL12γ, a ligand of CXCR4, promotes PC metastasis in part via induction of PCSCs (Jung et al., 2018). Furthermore, ectopic expression of CD133 in PC3 cells upregulates ALDHA1, OCT4, and NANOG (D. Kim et al., 2019), a set of well-established CSC factors (Mei et al., 2019), suggesting the acquisition of PCSC properties. Importantly, these PCSC-like cells cause osteoblastic bone lesion (Kim et al., 2019), supporting a critical role of PCSCs in PC bone metastasis. This concept is well in line with the utilization of HSC niche by PCSCs for bone metastasis.

4.2 Cancer stem cell niche

Cancer stem cells (CSCs) are generally regarded for their essential roles in cancer progression, metastasis, and resistance to therapy. At the center of CSCs is their plasticity which directs cancer evolution; this property of CSCs is shared with their tissue counterparts (Mei et al., 2019; Rodriguez et al., 2019). In a similar relationship between stem cells and niches, CSCs drive cancer progression in response to cues derived from stroma. CSC niches consist of non-tumor cells including cancer-associated fibroblasts (CAFs), endothelial cells, immune cells as well as non-CSC cancer cells (Kise, Kinugasa-Katayama, & Takakura, 2016). For instance, non-CSC PC cells secrete the protein SPARC which enhances the metastatic potential of PC3-derived PCSCs (Mateo et al., 2014). Nonetheless the compositions of CSC niches are complex and dynamic, which fits the tasks of CSCs in promoting cancer progression. There are generally four types of niches: hypoxic, immune, perivascular, and infiltrating (Prager, Xie, Bao, & Rich, 2019); clearly these niches cover the typical features and process of tumorigenesis. The details of these niches have been thoroughly reviewed recently (Prager et al., 2019); we will provide a brief summary to facilitate the discussions to follow. The hypoxic niche plays an essential role in programming and maintaining CSCs. For instance, the hypoxia-inducible factor 2 (HIF-2) contributes to these functions in part by induction of key stem cell factor expression, KLF4, SOX2, and OCT4 (Mathieu et al., 2011). CSCs possess properties to escape immune response, support tumor migration and invasion in the leading edge, and enter the circulation; the last two processes are essential for cancer metastasis; avoiding immune surveillance is an essential aspect of tumorigenesis, progression, metastasis, and resistance to immune therapy (Naik et al., 2018). While these individual properties of CSCs are regulated by individual niches, the intrinsic connections among these niches are apparent; for example, CSCs resulted from the hypoxic niches will equip these cells with the properties to communicate with other niches (Prager et al., 2019). In this regard, expression of HIF2 was observed in neuroendocrine PC (NEPC) (Monsef, Helczynski, Lundwall, Pahlman, & Anders, 2007), an androgen-independent and aggressive PC that is trans-differentiated from prostate adenocarcinoma through PCSCs (Mei et al., 2019). Generation of NEPC in response to ADR treatment of metastatic PC is a mechanism of CRPC development (Ojo, Lin, Wong, Gu, & Tang, 2015). Collectively, evidence supports a critical role of niche in supporting PCSCs-derived PC bone metastasis.

4.3 Hijack of HSC niches by PCSCs for the development of bone metastasis

4.3.1 Homing PCSCs to HSC niche in bone marrow

Conceptually, PC metastasis to the bone is supported by the hypoxic, immune, infiltrating, and perivascular niches. However, knowledge on these niches in PCSCs remains limited. The most thoroughly investigated niches for PC metastasis are those of HSC (Shiozawa et al., 2011). Bone marrow niches provide the essential environment to retain HSCs, maintain their quiescence, and activate HSCs for hematopoiesis (Mendez-Ferrer et al., 2020). Mice have two spatially located HSC niches: endosteal and central bone marrow (BM) niche (Mendez-Ferrer et al., 2020). These are perivascular niches evident by the dependence of HSC retention in bone marrow on stem cell factor (KITL) expressed by perivascular cells (Ding, Saunders, Enikolopov, & Morrison, 2012) and CXCL12 expressed in perivascular stromal cells (Ding & Morrison, 2013; Greenbaum et al., 2013). The endosteal niches preserve HSCs by promoting their quiescence, while the central BM niches contain both quiescent and active HSCs for blood production (see individual references cited in the review by Mendez-Ferrer et al., 2020).

Disseminated tumor cells of prostate cancer (DTCs) in bone marrow possess PCSC properties (Shiozawa et al., 2016). Furthermore, using a zebrafish tumor model, PCSC-like PC3 cells defined by $ALDH^{hi}$ metastasize to caudal hematopoietic tissues more dynamically than $ALDH^{low}$ (non-cancer stem) PC3 cells and all metastases including those resulted from $ALDH^{low}$ PC3 cells exhibited PCSC properties as evident by the expression of NANOG and OCT4 (Chen, Boleslaw Olszewski, Kruithof-de Julio, & Snaar-Jagalska, 2020). SCF (Stem cell factor) and G-CSF (granulocyte colony-stimulating factor) of bone marrow induce PC3 and DU145 cells to acquire PCSC properties in vitro (Ma et al., 2012). Collectively, evidence supports a direct role of PCSCs in bone metastasis. This concept is further supported by the utilization of HSC niches by PCSCs. In a pioneer study by Shiozawa et al., human PC cells metastasized to mouse bone marrow by taking over HSC niches; and increases or decreases of HSC niche size enhanced or inhibited bone metastasis respectively (Shiozawa et al., 2011). Consistent with the important role of CXCL12-CXCR4 axis in homing HSCs to bone marrow (Juarez & Bendall, 2004; Karpova & Bonig, 2015), DTCs upregulate CXCR4 and blocking of the interaction of CXCR4 to BM CXCL12 with AMD3100 mobilizes DTCs out of niches, leading to the return of DTCs into circulation (Shiozawa et al., 2011). A sub-population of PC3 cells was observed to express an elevated level of CXCR4; these cells homed to bone more efficiently compared to PC3 cells without CXCR4 upregulation (Wang et al., 2015). In addition to CXCR4, PC cells also use annexin 2 (ANXA2) to home DTCs to bone marrow via binding to CXCL12 (Jung et al., 2015; Shiozawa et al., 2008), consistent with the critical role of CXCL12 in homing HSCs and PCSCs to bone marrow (Table 1). Limited evidence suggests an association of DTCs of PC with perivascular bone marrow stromal cells and the association facilitates PC bone metastasis (Funari et al., 2019). Taken together, collective evidence suggests a role for PCSCs in exploring the HSC niche for bone metastasis.

4.3.2 Regulation of PCSCs (DTCs) metastatic process by bone marrow niche components

Mesenchymal stem cells (MSCs) are an essential component of HSC niche and play a critical role in maintaining HSCs (Mendez-Ferrer et al., 2010; Sacchetti et al., 2007). Analogous to this function, MSCs forms bone in a 3-D in vitro culture system which supports PC cells to undergo mesenchymal to epithelial transition, an essential transition for PC metastasis to the bone, and form tumoroids containing hypoxic core regions and features

Table 1 Mechanisms underlie PC metastasis to the bone.

Factors	Functions	References
CXCR4	PC-derived; homing PC cells to bone marrow via binding to CXCL12	Shiozawa et al. (2011)
Annexin 2	PC-derived; seeding PC cells to bone marrow via binding to CXCL12	Shiozawa et al. (2011), Jung et al. (2015)
GAS6	Bone marrow derived; inhibition of PC cell proliferation and induction of PCSCs	Shiozawa et al. (2010), Shiozawa et al. (2016)
TBK1	Induction by bone marrow; suppression of PC cell proliferation via inhibition of mTOR	Kim et al. (2013)
TGFβ	Osteoblast-produced; inhibition of PC cell proliferation	Yumoto et al. (2016)
RANKL	PC-derived; promotion of PC cell metastatic outgrowth in bone marrow via interactions with RANK and induction of PCSC	Chu et al. (2014)
IL-1β	PC-derived; interactions with its receptor on MSC and enhancing metastatic growth of PC in bone marrow	Shahriari et al. (2017)

of osteogenesis (Molla, Katti, & Katti, 2018). However, a sub-population of MSCs expressing both endothelial and pericytic cell surface markers reduces PC cells homing to bone marrow (Rossnagl et al., 2018), indicating a protective function of normal HSC niche to suppress PC bone metastasis.

PCSCs or DTCs need to survive in HSC niche in order to form metastatic tumors. Dormancy in bone marrow is viewed as a survival mechanism (Cackowski & Taichman, 2019). A variety of signals emanated from HSC niche inhibit DTC proliferation (Table 1). GAS6 (growth arrest specific 6) from bone marrow represses DTC proliferation (Shiozawa et al., 2010), and promotes the reprograming of PCs to PCSCs (Shiozawa et al., 2016), providing additional evidence for PCSCs as an origin of bone metastasis. HSC niche osteoblasts upregulate TBK1 (TANK binding kinase 1) expression in PC cells which arrests PC cell cycle progression via inhibiting mTOR, implicating its roles in inducing DTC dormancy (Kim et al., 2013) (Table 1). TBK1 is activated by Toll-like receptors (TLRs) and contributes to antiviral innate immune reactions (Kawai & Akira, 2007), supporting an important role of HSC niche-derived immune modulators in the regulation of PCSCs. The transforming growth factor-β (TGFβ), a pleiotropic cytokine, is a major contributor to HSCs quiescence in both the endosteal and central BM niches (Mendez-Ferrer et al., 2020). Evidence suggests a role of TGFβ produced by osteoblasts in suppressing DTCs proliferation (Yumoto et al., 2016) (Table 1).

While mechanisms underlying DTCs outgrowth from their dormancy are not clear, evidence supports IL6 and the RANKL-RANK pathway in promoting this process. The receptor activator of nuclear factor-κB ligand (RNAKL) and its receptor RANK play an essential role in osteoclast differentiation and bone resorption (Danks & Takayanagi, 2013). Upregulation of RANKL occurs in human PCs, which is associated with PC progression; the upregulation facilitates homing of PC cells to bone and the outgrowth of bone metastasis (Chu et al., 2014) (Table 1). RANKL upregulation confers LNCaP cells with stem cell properties, evident by increases in SOX2, MYC, OCT3, and NANOG expression (Chu et al., 2014), consistent with PCSCs being critical in PC bone metastasis. IL6 plays a critical role in maintaining CSCs, including glioma (Wang, Kruithof-de Julio, et al., 2009; Wang, Lathia, et al., 2009), lung cancer (Yi et al., 2012), colon cancer (Lin et al., 2011), and breast cancer (Marotta et al., 2011). In PC, elevations of circulating IL-6 occur in patients with metastatic PC (Adler et al., 1999; Akimoto, Okumura, & Fuse, 1998); reprograming of premalignant human prostate EPT2-D5 cells to PCSCs requires IL-6/STAT3 signaling (Qu et al., 2013). In line with these observations, evidence suggests that IL6 signaling from either PC or bone marrow environment facilitates PC bone outgrowth (Decker, Jung, Cackowski, & Taichman, 2016).

Upregulations of IL6 were observed in human PC bone metastasis (Morrissey et al., 2010), suggesting that PC-derived IL6 facilitates bone metastasis in part through regulation of bone marrow activities. PC3-derived bone metastases are osteolytic in which IL6 plays a role (Joseph et al., 2012; Morrissey et al., 2010). Analogously, the interplay between PC-derived IL-1β with its receptor expressed in bone marrow MSCs supports the outgrowth of bone metastasis (Table 1) (Shahriari et al., 2017).

5. Conclusions and perspectives

With the emerging role of CSCs in tumorigenesis, progression and resistance to therapy comes with increasing recognition of the importance of CSC niche in these processes. Nonetheless, in comparison of our understanding of cancer cells and CSCs, the current knowledge on microenvironment and niche remains far behind, despite the long-standing knowledge that the large proportion of solid tumor mass consists of non-tumor cells. This is in part attributable to the complex and dynamic nature of CSC niche. Clearly, the field of PCSC niche faces the same challenge and actually remains less advanced compared to other CSC niches, including those of hematologic malignancies and glioma (Mendez-Ferrer et al., 2020; Prager et al., 2019).

Nonetheless, available evidence supports the notion of PCSC niche being critical in PC tumorigenesis. This concept is particularly strengthened by the demonstrations of PSCs and their niches in the proximal region connecting to urethra in both humans (Moad et al., 2017) and mice (Tsujimura et al., 2002), the maintenance of PSCs quiescence through the stroma Wnt singling (Wei et al., 2019), and inhibition of tumorigenesis by stromal HH signaling (Yang et al., 2017). Of note, the niche-derived HH signaling likely functions as a general tumor suppression mechanisms in epithelium-originated malignancies (Roberts et al., 2017), which further supports the importance of PSC niche in regulating PC tumorigenesis. Even with these developments, a substantial advance will certainly be required for a better appreciation of the impacts of PCS niche on tumorigenesis and the utilization of this knowledge in PC management. With our understanding of the disease course, it can be certain that intervention of PC initiation is more effective in controlling the disease than targeting PC progression.

Metastasis to the bone is the most impactful progression of PC because of its frequency and lethality. In comparison to PC initiation, we know more on the utilization of HSC niches by PCSCs for PC metastasis to the bone, including seeding PC cells to bone marrow, entering of PC into dormancy, and exiting dormancy for metastatic outgrowth (Cackowski & Taichman, 2019; Decker et al., 2016). Clearly much more remains to be discovered. Bone metastasis remains lethal and therapies targeting HSC niches to prevent metastatic outgrowth are not effective. The current therapies can reduce bone events but not extend patient's overall survival (Decker et al., 2016). For instance, while the standard of care for PC metastasis is androgen deprivation therapy, castration induces bone loss and triggers disseminated PC cell growth in the bone in animal models for bone metastasis (Ottewell et al., 2014; Schneider et al., 2005). While the underlying mechanisms remain unknown, the plasticity of PCSCs is likely a contributing factor. It has been observed that RANKL upregulation converted LNCaP cells to PCSC-like cells with enhanced bone metastasis capacity and a concurrent downregulation of AR (Chu et al., 2014).

Stem cells of epithelia are heterogeneous (Morrissey et al., 2010). CSCs evolve following cancer progression and are likely heterogeneous, which might be attributable to the association of multiple markers with individual CSC types. For instance, PCSC markers include $CD44^{+}\alpha_2\beta_1{}^{hi}CD133^{+}$ (Collins et al., 2005), $BCPR^{+}$ (Huss et al., 2005) as well as $Trop2^{hi}$, CD166/ALCAM, $PSA^{-/low}$ (prostate specific antigen), and ALDH1A1 (Chen et al., 2016; Goldstein et al., 2008; Jiao et al., 2012; Qin et al., 2012). Recently, rBC2LN was reported to associate with PC3-derived PCSCs (Mawaribuchi et al., 2019). There is thus a good possibility that PCSCs are heterogeneous and it will be important to consider this heterogeneity in addressing the impact of their niche on PC progression, including metastasis and therapy resistance. The same issue also applies for PSCs-originated PC; a different set of markers also exist in PSCs (Mei et al., 2019).

Author contributions

Conceptualization, T.L., A.K. and D.T.; material preparation, T.L., X.L., J.P. and Y.G.; writing—original draft preparation, T.L. and D.T.; writing—review, all authors; writing editing—Y.G. and D.T.; supervision, D.T.

Funding

D.T. is supported by grants from Cancer Research Society, Canadian Cancer Society (grant #: 319412) and CIHR and by funds from Urological Cancer Center for Research and Innovation (UCCRI). Y.G. is supported by Studentship provided by Ontario Graduate Scholarships and Research Institute of St Joe's Hamilton.

Conflicts of interest

The authors declare no conflict of interest.

References

Adler, H. L., McCurdy, M. A., Kattan, M. W., Timme, T. L., Scardino, P. T., & Thompson, T. C. (1999). Elevated levels of circulating interleukin-6 and transforming growth factor-beta1 in patients with metastatic prostatic carcinoma. *The Journal of Urology*, *161*(1), 182–187.

Akimoto, S., Okumura, A., & Fuse, H. (1998). Relationship between serum levels of interleukin-6, tumor necrosis factor-alpha and bone turnover markers in prostate cancer patients. *Endocrine Journal*, *45*(2), 183–189.

Al-Hajj, M., Wicha, M. S., Benito-Hernandez, A., Morrison, S. J., & Clarke, M. F. (2003). Prospective identification of tumorigenic breast cancer cells. *Proceedings of the National Academy of Sciences of the United States of America*, *100*(7), 3983–3988. https://doi.org/10.1073/pnas.0530291100.

Baccelli, I., & Trumpp, A. (2012). The evolving concept of cancer and metastasis stem cells. *The Journal of Cell Biology*, *198*(3), 281–293. https://doi.org/10.1083/jcb.201202014.

Bagnall, P. (2014). Diagnosis and treatment of prostate cancer. *Nursing Times*, *110*(9), 12–15.

Barros-Silva, J. D., Linn, D. E., Steiner, I., Guo, G., Ali, A., Pakula, H., et al. (2018). Single-cell analysis identifies LY6D as a marker linking castration-resistant prostate luminal cells to prostate progenitors and cancer. *Cell Reports*, *25*(12), 3504–3518. e3506. https://doi.org/10.1016/j.celrep.2018.11.069.

Berthold, D. R., Pond, G. R., Soban, F., de Wit, R., Eisenberger, M., & Tannock, I. F. (2008). Docetaxel plus prednisone or mitoxantrone plus prednisone for advanced prostate cancer: Updated survival in the TAX 327 study. *Journal of Clinical Oncology*, *26*(2), 242–245. https://doi.org/10.1200/JCO.2007.12.4008.

Bubendorf, L., Schopfer, A., Wagner, U., Sauter, G., Moch, H., Willi, N., et al. (2000). Metastatic patterns of prostate cancer: An autopsy study of 1,589 patients. *Human Pathology*, *31*(5), 578–583. https://doi.org/10.1053/hp.2000.6698.

Burger, P. E., Xiong, X., Coetzee, S., Salm, S. N., Moscatelli, D., Goto, K., et al. (2005). Sca-1 expression identifies stem cells in the proximal region of prostatic ducts with high capacity to reconstitute prostatic tissue. *Proceedings of the National Academy of Sciences of the United States of America*, *102*(20), 7180–7185. https://doi.org/10.1073/pnas.0502761102.

Cackowski, F. C., & Taichman, R. S. (2019). Parallels between hematopoietic stem cell and prostate cancer disseminated tumor cell regulation. *Bone*, *119*, 82–86. https://doi.org/10.1016/j.bone.2018.02.025.

Ceder, J. A., Jansson, L., Helczynski, L., & Abrahamsson, P. A. (2008). Delta-like 1 (Dlk-1), a novel marker of prostate basal and candidate epithelial stem cells, is downregulated by notch signalling in intermediate/transit amplifying cells of the human prostate. *European Urology*, *54*(6), 1344–1353. https://doi.org/10.1016/j.eururo.2008.03.006.

Chambers, A. F., Groom, A. C., & MacDonald, I. C. (2002). Dissemination and growth of cancer cells in metastatic sites. *Nature Reviews. Cancer*, *2*(8), 563–572. https://doi.org/10.1038/nrc865.

Chen, L., Boleslaw Olszewski, M., Kruithof-de Julio, M., & Snaar-Jagalska, B. E. (2020). Zebrafish microenvironment elevates EMT and CSC-like phenotype of engrafted prostate cancer cells. *Cell*, *9*(4). https://doi.org/10.3390/cells9040797.

Chen, X., Li, Q., Liu, X., Liu, C., Liu, R., Rycaj, K., et al. (2016). Defining a population of stem-like human prostate cancer cells that can generate and propagate castration-resistant prostate cancer. *Clinical Cancer Research*, *22*(17), 4505–4516. https://doi.org/10.1158/1078-0432.CCR-15-2956.

Chiang, A. C., & Massague, J. (2008). Molecular basis of metastasis. *The New England Journal of Medicine*, *359*(26), 2814–2823. https://doi.org/10.1056/NEJMra0805239.

Chu, G. C., Zhau, H. E., Wang, R., Rogatko, A., Feng, X., Zayzafoon, M., et al. (2014). RANK- and c-Met-mediated signal network promotes prostate cancer metastatic colonization. *Endocrine-Related Cancer*, *21*(2), 311–326. https://doi.org/10.1530/ERC-13-0548.

Chua, C. W., Shibata, M., Lei, M., Toivanen, R., Barlow, L. J., Bergren, S. K., et al. (2014). Single luminal epithelial progenitors can generate prostate organoids in culture. *Nature Cell Biology*, *16*(10), 951–961. 951–954 https://doi.org/10.1038/ncb3047.

Clevers, H., Loh, K. M., & Nusse, R. (2014). Stem cell signaling. An integral program for tissue renewal and regeneration: Wnt signaling and stem cell control. *Science*, *346*(6205), 1248012. https://doi.org/10.1126/science.1248012.

Collins, A. T., Berry, P. A., Hyde, C., Stower, M. J., & Maitland, N. J. (2005). Prospective identification of tumorigenic prostate cancer stem cells. *Cancer Research*, *65*(23), 10946–10951. https://doi.org/10.1158/0008-5472.CAN-05-2018.

Cunha, G. R., & Lung, B. (1978). The possible influence of temporal factors in androgenic responsiveness of urogenital tissue recombinants from wild-type and androgen-insensitive (Tfm) mice. *The Journal of Experimental Zoology*, *205*(2), 181–193. https://doi.org/10.1002/jez.1402050203.

Danks, L., & Takayanagi, H. (2013). Immunology and bone. *Journal of Biochemistry*, *154*(1), 29–39. https://doi.org/10.1093/jb/mvt049.

de Bono, J. S., Logothetis, C. J., Molina, A., Fizazi, K., North, S., Chu, L., et al. (2011). Abiraterone and increased survival in metastatic prostate cancer. *The New England Journal of Medicine*, *364*(21), 1995–2005. https://doi.org/10.1056/NEJMoa1014618.

De Craene, B., & Berx, G. (2013). Regulatory networks defining EMT during cancer initiation and progression. *Nature Reviews. Cancer*, *13*(2), 97–110. https://doi.org/10.1038/nrc3447.

Decker, A. M., Jung, Y., Cackowski, F., & Taichman, R. S. (2016). The role of hematopoietic stem cell niche in prostate cancer bone metastasis. *The Journal of Bone Oncology*, *5*(3), 117–120. https://doi.org/10.1016/j.jbo.2016.02.005.

Ding, L., & Morrison, S. J. (2013). Haematopoietic stem cells and early lymphoid progenitors occupy distinct bone marrow niches. *Nature*, *495*(7440), 231–235. https://doi.org/10.1038/nature11885.

Ding, L., Saunders, T. L., Enikolopov, G., & Morrison, S. J. (2012). Endothelial and perivascular cells maintain haematopoietic stem cells. *Nature*, *481*(7382), 457–462. https://doi.org/10.1038/nature10783.

Eramo, A., Lotti, F., Sette, G., Pilozzi, E., Biffoni, M., Di Virgilio, A., et al. (2008). Identification and expansion of the tumorigenic lung cancer stem cell population. *Cell Death and Differentiation*, *15*(3), 504–514. https://doi.org/10.1038/sj.cdd.4402283.

Fang, D., Nguyen, T. K., Leishear, K., Finko, R., Kulp, A. N., Hotz, S., et al. (2005). A tumorigenic subpopulation with stem cell properties in melanomas. *Cancer Research*, *65*(20), 9328–9337. https://doi.org/10.1158/0008-5472.CAN-05-1343.

Francis, J. C., Thomsen, M. K., Taketo, M. M., & Swain, A. (2013). Beta-catenin is required for prostate development and cooperates with Pten loss to drive invasive carcinoma. *PLoS Genetics*, *9*(1), e1003180. https://doi.org/10.1371/journal.pgen.1003180.

Funari, A., Alimandi, M., Pierelli, L., Pino, V., Gentileschi, S., & Sacchetti, B. (2019). Human sinusoidal subendothelial cells regulate homing and invasion of circulating metastatic prostate cancer cells to bone marrow. *Cancers (Basel)*, *11*(6), 763. https://doi.org/10.3390/cancers11060763.

Goldstein, A. S., Huang, J., Guo, C., Garraway, I. P., & Witte, O. N. (2010). Identification of a cell of origin for human prostate cancer. *Science*, *329*(5991), 568–571. https://doi.org/10.1126/science.1189992.

Goldstein, A. S., Lawson, D. A., Cheng, D., Sun, W., Garraway, I. P., & Witte, O. N. (2008). Trop2 identifies a subpopulation of murine and human prostate basal cells with stem cell characteristics. *Proceedings of the National Academy of Sciences of the United States of America*, *105*(52), 20882–20887. https://doi.org/10.1073/pnas.0811411106.

Goto, K., Salm, S. N., Coetzee, S., Xiong, X., Burger, P. E., Shapiro, E., et al. (2006). Proximal prostatic stem cells are programmed to regenerate a proximal-distal ductal axis. *Stem Cells*, *24*(8), 1859–1868. https://doi.org/10.1634/stemcells.2005-0585.

Greenbaum, A., Hsu, Y. M., Day, R. B., Schuettpelz, L. G., Christopher, M. J., Borgerding, J. N., et al. (2013). CXCL12 in early mesenchymal progenitors is required for haematopoietic stem-cell maintenance. *Nature*, *495*(7440), 227–230. https://doi.org/10.1038/nature11926.

Gunasinghe, N. P., Wells, A., Thompson, E. W., & Hugo, H. J. (2012). Mesenchymal-epithelial transition (MET) as a mechanism for metastatic colonisation in breast cancer. *Cancer Metastasis Reviews*, *31*(3–4), 469–478. https://doi.org/10.1007/s10555-012-9377-5.

Gundem, G., Van Loo, P., Kremeyer, B., Alexandrov, L. B., Tubio, J. M. C., Papaemmanuil, E., et al. (2015). The evolutionary history of lethal metastatic prostate cancer. *Nature*, *520*(7547), 353–357. https://doi.org/10.1038/nature14347.

Gupta, G. P., & Massague, J. (2006). Cancer metastasis: Building a framework. *Cell, 127*(4), 679–695. https://doi.org/10.1016/j.cell.2006.11.001.

Harris, K. S., & Kerr, B. A. (2017). Prostate cancer stem cell markers drive progression, therapeutic resistance, and bone metastasis. *Stem Cells International, 2017*. https://doi.org/10.1155/2017/8629234, 8629234.

Hayashi, R., Yamato, M., Saito, T., Oshima, T., Okano, T., Tano, Y., et al. (2008). Enrichment of corneal epithelial stem/progenitor cells using cell surface markers, integrin alpha6 and CD71. *Biochemical and Biophysical Research Communications, 367*(2), 256–263. https://doi.org/10.1016/j.bbrc.2007.12.077.

Hayward, S. W., Del Buono, R., Deshpande, N., & Hall, P. A. (1992). A functional model of adult human prostate epithelium. The role of androgens and stroma in architectural organisation and the maintenance of differentiated secretory function. *Journal of Cell Science, 102*(Pt. 2), 361–372.

Hayward, S. W., Wang, Y., Cao, M., Hom, Y. K., Zhang, B., Grossfeld, G. D., et al. (2001). Malignant transformation in a nontumorigenic human prostatic epithelial cell line. *Cancer Research, 61*(22), 8135–8142.

Heinlein, C. A., & Chang, C. (2004). Androgen receptor in prostate cancer. *Endocrine Reviews, 25*(2), 276–308. https://doi.org/10.1210/er.2002-0032.

Hermann, P. C., Huber, S. L., Herrler, T., Aicher, A., Ellwart, J. W., Guba, M., et al. (2007). Distinct populations of cancer stem cells determine tumor growth and metastatic activity in human pancreatic cancer. *Cell Stem Cell, 1*(3), 313–323. https://doi.org/10.1016/j.stem.2007.06.002.

Hofner, T., Klein, C., Eisen, C., Rigo-Watermeier, T., Haferkamp, A., Trumpp, A., et al. (2016). The influence of prostatic anatomy and neurotrophins on basal prostate epithelial progenitor cells. *Prostate, 76*(1), 114–121. https://doi.org/10.1002/pros.23109.

Huss, W. J., Gray, D. R., Greenberg, N. M., Mohler, J. L., & Smith, G. J. (2005). Breast cancer resistance protein-mediated efflux of androgen in putative benign and malignant prostate stem cells. *Cancer Research, 65*(15), 6640–6650. https://doi.org/10.1158/0008-5472.CAN-04-2548.

Isaacs, J. T., & Coffey, D. S. (1989). Etiology and disease process of benign prostatic hyperplasia. *The Prostate. Supplement, 2*, 33–50.

Jeter, C. R., Liu, B., Liu, X., Chen, X., Liu, C., Calhoun-Davis, T., et al. (2011). NANOG promotes cancer stem cell characteristics and prostate cancer resistance to androgen deprivation. *Oncogene, 30*(36), 3833–3845. https://doi.org/10.1038/onc.2011.114.

Jiao, J., Hindoyan, A., Wang, S., Tran, L. M., Goldstein, A. S., Lawson, D., et al. (2012). Identification of CD166 as a surface marker for enriching prostate stem/progenitor and cancer initiating cells. *PLoS One*, 7(8), e42564. https://doi.org/10.1371/journal.pone.0042564.

Jones, D. L., & Wagers, A. J. (2008). No place like home: Anatomy and function of the stem cell niche. *Nature Reviews. Molecular Cell Biology, 9*(1), 11–21. https://doi.org/10.1038/nrm2319.

Joseph, J., Shiozawa, Y., Jung, Y., Kim, J. K., Pedersen, E., Mishra, A., et al. (2012). Disseminated prostate cancer cells can instruct hematopoietic stem and progenitor cells to regulate bone phenotype. *Molecular Cancer Research, 10*(3), 282–292. https://doi.org/10.1158/1541-7786.MCR-11-0404.

Juarez, J., & Bendall, L. (2004). SDF-1 and CXCR4 in normal and malignant hematopoiesis. *Histology and Histopathology, 19*(1), 299–309. https://doi.org/10.14670/HH-19.299.

Jung, Y., Cackowski, F. C., Yumoto, K., Decker, A. M., Wang, J., Kim, J. K., et al. (2018). CXCL12gamma promotes metastatic castration-resistant prostate cancer by inducing cancer stem cell and neuroendocrine phenotypes. *Cancer Research, 78*(8), 2026–2039. https://doi.org/10.1158/0008-5472.CAN-17-2332.

Jung, Y., Wang, J., Lee, E., McGee, S., Berry, J. E., Yumoto, K., et al. (2015). Annexin 2-CXCL12 interactions regulate metastatic cell targeting and growth in the bone marrow. *Molecular Cancer Research, 13*(1), 197–207. https://doi.org/10.1158/1541-7786.MCR-14-0118.

Karpova, D., & Bonig, H. (2015). Concise Review: CXCR4/CXCL12 signaling in immature hematopoiesis—Lessons from pharmacological and genetic models. *Stem Cells, 33*(8), 2391–2399. https://doi.org/10.1002/stem.2054.

Kawai, T., & Akira, S. (2007). Signaling to NF-kappaB by Toll-like receptors. *Trends in Molecular Medicine, 13*(11), 460–469. https://doi.org/10.1016/j.molmed.2007.09.002.

Kim, J. K., Jung, Y., Wang, J., Joseph, J., Mishra, A., Hill, E. E., et al. (2013). TBK1 regulates prostate cancer dormancy through mTOR inhibition. *Neoplasia, 15*(9), 1064–1074. https://doi.org/10.1593/neo.13402.

Kim, D., Ko, Y., Park, M., Kim, B., Sohn, H., & Lim, W. (2019). Regulation of osteosclerosis by inoculated Cd133(+) PC3 cells in bone-marrow microenvironmental niches. *JBMR Plus, 3*(7). https://doi.org/10.1002/jbm4.10189, e10189.

Kise, K., Kinugasa-Katayama, Y., & Takakura, N. (2016). Tumor microenvironment for cancer stem cells. *Advanced Drug Delivery Reviews, 99*(Pt. B), 197–205. https://doi.org/10.1016/j.addr.2015.08.005.

Kollar, E. J., & Fisher, C. (1980). Tooth induction in chick epithelium: Expression of quiescent genes for enamel synthesis. *Science, 207*(4434), 993–995. https://doi.org/10.1126/science.7352302.

Korinek, V., Barker, N., Moerer, P., van Donselaar, E., Huls, G., Peters, P. J., et al. (1998). Depletion of epithelial stem-cell compartments in the small intestine of mice lacking Tcf-4. *Nature Genetics, 19*(4), 379–383. https://doi.org/10.1038/1270.

Kreso, A., & Dick, J. E. (2014). Evolution of the cancer stem cell model. *Cell Stem Cell, 14*(3), 275–291. https://doi.org/10.1016/j.stem.2014.02.006.

Kwon, O. J., & Xin, L. (2014). Prostate epithelial stem and progenitor cells. *American Journal of Clinical and Experimental Urology, 2*(3), 209–218.

Lapidot, T., Sirard, C., Vormoor, J., Murdoch, B., Hoang, T., Caceres-Cortes, J., et al. (1994). A cell initiating human acute myeloid leukaemia after transplantation into SCID mice. *Nature, 367*(6464), 645–648. https://doi.org/10.1038/367645a0.

Lawson, D. A., Xin, L., Lukacs, R. U., Cheng, D., & Witte, O. N. (2007). Isolation and functional characterization of murine prostate stem cells. *Proceedings of the National Academy of Sciences of the United States of America, 104*(1), 181–186. https://doi.org/10.1073/pnas.0609684104.

Lawson, D. A., Zong, Y., Memarzadeh, S., Xin, L., Huang, J., & Witte, O. N. (2010). Basal epithelial stem cells are efficient targets for prostate cancer initiation. *Proceedings of the National Academy of Sciences of the United States of America, 107*(6), 2610–2615. https://doi.org/10.1073/pnas.0913873107.

Le, V., He, Y., Aldahl, J., Hooker, E., Yu, E. J., Olson, A., et al. (2020). Loss of androgen signaling in mesenchymal sonic hedgehog responsive cells diminishes prostate development, growth, and regeneration. *PLoS Genetics, 16*(1), e1008588. https://doi.org/10.1371/journal.pgen.1008588.

Lee, S. H., Johnson, D. T., Luong, R., Yu, E. J., Cunha, G. R., Nusse, R., et al. (2015). Wnt/beta-catenin-responsive cells in prostatic development and regeneration. *Stem Cells, 33*(11), 3356–3367. https://doi.org/10.1002/stem.2096.

Leong, K. G., Wang, B. E., Johnson, L., & Gao, W. Q. (2008). Generation of a prostate from a single adult stem cell. *Nature, 456*(7223), 804–808. https://doi.org/10.1038/nature07427.

Li, C., Heidt, D. G., Dalerba, P., Burant, C. F., Zhang, L., Adsay, V., et al. (2007). Identification of pancreatic cancer stem cells. *Cancer Research, 67*(3), 1030–1037. https://doi.org/10.1158/0008-5472.CAN-06-2030.

Li, A., Simmons, P. J., & Kaur, P. (1998). Identification and isolation of candidate human keratinocyte stem cells based on cell surface phenotype. *Proceedings of the National Academy of Sciences of the United States of America*, *95*(7), 3902–3907. https://doi.org/10.1073/pnas.95.7.3902.

Liao, C. P., Adisetiyo, H., Liang, M., & Roy-Burman, P. (2010). Cancer-associated fibroblasts enhance the gland-forming capability of prostate cancer stem cells. *Cancer Research*, *70*(18), 7294–7303. https://doi.org/10.1158/0008-5472.CAN-09-3982.

Lin, L., Liu, A., Peng, Z., Lin, H. J., Li, P. K., Li, C., et al. (2011). STAT3 is necessary for proliferation and survival in colon cancer-initiating cells. *Cancer Research*, *71*(23), 7226–7237. https://doi.org/10.1158/0008-5472.CAN-10-4660.

Lu, T. L., & Chen, C. M. (2015). Differential requirements for beta-catenin in murine prostate cancer originating from basal versus luminal cells. *The Journal of Pathology*, *236*(3), 290–301. https://doi.org/10.1002/path.4521.

Lu, T. L., Huang, Y. F., You, L. R., Chao, N. C., Su, F. Y., Chang, J. L., et al. (2013). Conditionally ablated Pten in prostate basal cells promotes basal-to-luminal differentiation and causes invasive prostate cancer in mice. *The American Journal of Pathology*, *182*(3), 975–991. https://doi.org/10.1016/j.ajpath.2012.11.025.

Lytton, B. (2001). Prostate cancer: A brief history and the discovery of hormonal ablation treatment. *The Journal of Urology*, *165*(6 Pt. 1), 1859–1862.

Ma, S., Chan, K. W., Hu, L., Lee, T. K., Wo, J. Y., Ng, I. O., et al. (2007). Identification and characterization of tumorigenic liver cancer stem/progenitor cells. *Gastroenterology*, *132*(7), 2542–2556. https://doi.org/10.1053/j.gastro.2007.04.025.

Ma, Y., Liang, D., Liu, J., Axcrona, K., Kvalheim, G., Giercksky, K. E., et al. (2012). Synergistic effect of SCF and G-CSF on stem-like properties in prostate cancer cell lines. *Tumour Biology*, *33*(4), 967–978. https://doi.org/10.1007/s13277-012-0325-3.

Marotta, L. L., Almendro, V., Marusyk, A., Shipitsin, M., Schemme, J., Walker, S. R., et al. (2011). The JAK2/STAT3 signaling pathway is required for growth of CD44(+) CD24(-) stem cell-like breast cancer cells in human tumors. *The Journal of Clinical Investigation*, *121*(7), 2723–2735. https://doi.org/10.1172/JCI44745.

Mateo, F., Meca-Cortes, O., Celia-Terrassa, T., Fernandez, Y., Abasolo, I., Sanchez-Cid, L., et al. (2014). SPARC mediates metastatic cooperation between CSC and non-CSC prostate cancer cell subpopulations. *Molecular Cancer*, *13*, 237. https://doi.org/10.1186/1476-4598-13-237.

Mathieu, J., Zhang, Z., Zhou, W., Wang, A. J., Heddleston, J. M., Pinna, C. M., et al. (2011). HIF induces human embryonic stem cell markers in cancer cells. *Cancer Research*, *71*(13), 4640–4652. https://doi.org/10.1158/0008-5472.CAN-10-3320.

Mawaribuchi, S., Onuma, Y., Aiki, Y., Kuriyama, Y., Mutoh, M., Fujii, G., et al. (2019). The rBC2LCN-positive subpopulation of PC-3cells exhibits cancer stem-like properties. *Biochemical and Biophysical Research Communications*, *515*(1), 176–182. https://doi.org/10.1016/j.bbrc.2019.05.108.

Mei, W., Lin, X., Kapoor, A., Gu, Y., Zhao, K., & Tang, D. (2019). The contributions of prostate cancer stem cells in prostate cancer initiation and metastasis. *Cancers (Basel)*, *11*(4). https://doi.org/10.3390/cancers11040434.

Mendez-Ferrer, S., Bonnet, D., Steensma, D. P., Hasserjian, R. P., Ghobrial, I. M., Gribben, J. G., et al. (2020). Bone marrow niches in haematological malignancies. *Nature Reviews. Cancer*, *20*(5), 285–298. https://doi.org/10.1038/s41568-020-0245-2.

Mendez-Ferrer, S., Michurina, T. V., Ferraro, F., Mazloom, A. R., Macarthur, B. D., Lira, S. A., et al. (2010). Mesenchymal and haematopoietic stem cells form a unique bone marrow niche. *Nature*, *466*(7308), 829–834. https://doi.org/10.1038/nature09262.

Miki, J., Furusato, B., Li, H., Gu, Y., Takahashi, H., Egawa, S., et al. (2007). Identification of putative stem cell markers, CD133 and CXCR4, in hTERT-immortalized primary nonmalignant and malignant tumor-derived human prostate epithelial cell lines and in prostate cancer specimens. *Cancer Research*, *67*(7), 3153–3161. https://doi.org/10.1158/0008-5472.CAN-06-4429.

Moad, M., Hannezo, E., Buczacki, S. J., Wilson, L., El-Sherif, A., Sims, D., et al. (2017). Multipotent basal stem cells, maintained in localized proximal niches, support directed long-ranging epithelial flows in human prostates. *Cell Reports*, *20*(7), 1609–1622. https://doi.org/10.1016/j.celrep.2017.07.061.

Moher, D., Shamseer, L., Clarke, M., Ghersi, D., Liberati, A., Petticrew, M., et al. (2015). Preferred reporting items for systematic review and meta-analysis protocols (PRISMA-P) 2015 statement. *Systematic Reviews*, *4*, 1. https://doi.org/10.1186/2046-4053-4-1.

Molla, M. S., Katti, D. R., & Katti, K. S. (2018). In vitro design of mesenchymal to epithelial transition of prostate cancer metastasis using 3D nanoclay bone-mimetic scaffolds. *Journal of Tissue Engineering and Regenerative Medicine*, *12*(3), 727–737. https://doi.org/10.1002/term.2492.

Monsef, N., Helczynski, L., Lundwall, A., Pahlman, S., & Anders, B. (2007). Localization of immunoreactive HIF-1alpha and HIF-2alpha in neuroendocrine cells of both benign and malignant prostate glands. *Prostate*, *67*(11), 1219–1229. https://doi.org/10.1002/pros.20594.

Moon, C., Park, J. C., Chae, Y. K., Yun, J. H., & Kim, S. (2008). Current status of experimental therapeutics for prostate cancer. *Cancer Letters*, *266*(2), 116–134. https://doi.org/10.1016/j.canlet.2008.02.065.

Morrissey, C., Lai, J. S., Brown, L. G., Wang, Y. C., Roudier, M. P., Coleman, I. M., et al. (2010). The expression of osteoclastogenesis-associated factors and osteoblast response to osteolytic prostate cancer cells. *Prostate*, *70*(4), 412–424. https://doi.org/10.1002/pros.21075.

Mulholland, D. J., Kobayashi, N., Ruscetti, M., Zhi, A., Tran, L. M., Huang, J., et al. (2012). Pten loss and RAS/MAPK activation cooperate to promote EMT and metastasis initiated from prostate cancer stem/progenitor cells. *Cancer Research*, *72*(7), 1878–1889. https://doi.org/10.1158/0008-5472.CAN-11-3132.

Mulholland, D. J., Xin, L., Morim, A., Lawson, D., Witte, O., & Wu, H. (2009). Lin-Sca-1+CD49fhigh stem/progenitors are tumor-initiating cells in the Pten-null prostate cancer model. *Cancer Research*, *69*(22), 8555–8562. https://doi.org/10.1158/0008-5472.CAN-08-4673.

Naik, S., Larsen, S. B., Cowley, C. J., & Fuchs, E. (2018). Two to tango: Dialog between immunity and stem cells in health and disease. *Cell*, *175*(4), 908–920. https://doi.org/10.1016/j.cell.2018.08.071.

Nassar, D., & Blanpain, C. (2016). Cancer stem cells: Basic concepts and therapeutic implications. *Annual Review of Pathology*, *11*, 47–76. https://doi.org/10.1146/annurev-pathol-012615-044438.

Nusse, R., & Clevers, H. (2017). Wnt/beta-catenin signaling, disease, and emerging therapeutic modalities. *Cell*, *169*(6), 985–999. https://doi.org/10.1016/j.cell.2017.05.016.

Ojo, D., Lin, X., Wong, N., Gu, Y., & Tang, D. (2015). Prostate cancer stem-like cells contribute to the development of castration-resistant prostate cancer. *Cancers (Basel)*, 7(4), 2290–2308. https://doi.org/10.3390/cancers7040890.

Oskarsson, T., Batlle, E., & Massague, J. (2014). Metastatic stem cells: Sources, niches, and vital pathways. *Cell Stem Cell*, *14*(3), 306–321. https://doi.org/10.1016/j.stem.2014.02.002.

Ottewell, P. D., Wang, N., Meek, J., Fowles, C. A., Croucher, P. I., Eaton, C. L., et al. (2014). Castration-induced bone loss triggers growth of disseminated prostate cancer cells in bone. *Endocrine-Related Cancer*, *21*(5), 769–781. https://doi.org/10.1530/ERC-14-0199.

Ousset, M., Van Keymeulen, A., Bouvencourt, G., Sharma, N., Achouri, Y., Simons, B. D., et al. (2012). Multipotent and unipotent progenitors contribute to prostate postnatal development. *Nature Cell Biology, 14*(11), 1131–1138. https://doi.org/10.1038/ncb2600.

Ovalle, S., Gutierrez-Lopez, M. D., Olmo, N., Turnay, J., Lizarbe, M. A., Majano, P., et al. (2007). The tetraspanin CD9 inhibits the proliferation and tumorigenicity of human colon carcinoma cells. *International Journal of Cancer, 121*(10), 2140–2152. https://doi.org/10.1002/ijc.22902.

Peng, Y. C., Levine, C. M., Zahid, S., Wilson, E. L., & Joyner, A. L. (2013). Sonic hedgehog signals to multiple prostate stromal stem cells that replenish distinct stromal subtypes during regeneration. *Proceedings of the National Academy of Sciences of the United States of America, 110*(51), 20611–20616. https://doi.org/10.1073/pnas.1315729110.

Prager, B. C., Xie, Q., Bao, S., & Rich, J. N. (2019). Cancer stem cells: The architects of the tumor ecosystem. *Cell Stem Cell, 24*(1), 41–53. https://doi.org/10.1016/j.stem.2018.12.009.

Prince, M. E., Sivanandan, R., Kaczorowski, A., Wolf, G. T., Kaplan, M. J., Dalerba, P., et al. (2007). Identification of a subpopulation of cells with cancer stem cell properties in head and neck squamous cell carcinoma. *Proceedings of the National Academy of Sciences of the United States of America, 104*(3), 973–978. https://doi.org/10.1073/pnas.0610117104.

Qin, J., Liu, X., Laffin, B., Chen, X., Choy, G., Jeter, C. R., et al. (2012). The PSA(-/lo) prostate cancer cell population harbors self-renewing long-term tumor-propagating cells that resist castration. *Cell Stem Cell, 10*(5), 556–569. https://doi.org/10.1016/j.stem.2012.03.009.

Qu, Y., Oyan, A. M., Liu, R., Hua, Y., Zhang, J., Hovland, R., et al. (2013). Generation of prostate tumor-initiating cells is associated with elevation of reactive oxygen species and IL-6/STAT3 signaling. *Cancer Research, 73*(23), 7090–7100. https://doi.org/10.1158/0008-5472.CAN-13-1560.

Ricci-Vitiani, L., Lombardi, D. G., Pilozzi, E., Biffoni, M., Todaro, M., Peschle, C., et al. (2007). Identification and expansion of human colon-cancer-initiating cells. *Nature, 445*(7123), 111–115. https://doi.org/10.1038/nature05384.

Richardson, G. D., Robson, C. N., Lang, S. H., Neal, D. E., Maitland, N. J., & Collins, A. T. (2004). CD133, a novel marker for human prostatic epithelial stem cells. *Journal of Cell Science, 117*(Pt. 16), 3539–3545. https://doi.org/10.1242/jcs.01222.

Roberts, K. J., Kershner, A. M., & Beachy, P. A. (2017). The stromal niche for epithelial stem cells: A template for regeneration and a brake on malignancy. *Cancer Cell, 32*(4), 404–410. https://doi.org/10.1016/j.ccell.2017.08.007.

Rodriguez, D., Ramkairsingh, M., Lin, X., Kapoor, A., Major, P., & Tang, D. (2019). The central contributions of breast cancer stem cells in developing resistance to endocrine therapy in Estrogen Receptor (ER)-positive breast cancer. *Cancers (Basel), 11*(7), 1028. https://doi.org/10.3390/cancers11071028.

Ross, J. S. (2007). The androgen receptor in prostate cancer: Therapy target in search of an integrated diagnostic test. *Advances in Anatomic Pathology, 14*(5), 353–357. https://doi.org/10.1097/PAP.0b013e31814a52c4.

Rossnagl, S., Ghura, H., Groth, C., Altrock, E., Jakob, F., Schott, S., et al. (2018). A Subpopulation of stromal cells controls cancer cell homing to the bone marrow. *Cancer Research, 78*(1), 129–142. https://doi.org/10.1158/0008-5472.CAN-16-3507.

Rybak, A. P., He, L., Kapoor, A., Cutz, J. C., & Tang, D. (2011). Characterization of sphere-propagating cells with stem-like properties from DU145 prostate cancer cells. *Biochimica et Biophysica Acta, 1813*(5), 683–694. https://doi.org/10.1016/j.bbamcr.2011.01.018.

Sacchetti, B., Funari, A., Michienzi, S., Di Cesare, S., Piersanti, S., Saggio, I., et al. (2007). Self-renewing osteoprogenitors in bone marrow sinusoids can organize a hematopoietic microenvironment. *Cell*, *131*(2), 324–336. https://doi.org/10.1016/j.cell.2007.08.025.

Scher, H. I., Fizazi, K., Saad, F., Taplin, M. E., Sternberg, C. N., Miller, K., et al. (2012). Increased survival with enzalutamide in prostate cancer after chemotherapy. *The New England Journal of Medicine*, *367*(13), 1187–1197. https://doi.org/10.1056/NEJMoa1207506.

Schneider, A., Kalikin, L. M., Mattos, A. C., Keller, E. T., Allen, M. J., Pienta, K. J., et al. (2005). Bone turnover mediates preferential localization of prostate cancer in the skeleton. *Endocrinology*, *146*(4), 1727–1736. https://doi.org/10.1210/en.2004-1211.

Shahriari, K., Shen, F., Worrede-Mahdi, A., Liu, Q., Gong, Y., Garcia, F. U., et al. (2017). Cooperation among heterogeneous prostate cancer cells in the bone metastatic niche. *Oncogene*, *36*(20), 2846–2856. https://doi.org/10.1038/onc.2016.436.

Shamseer, L., Moher, D., Clarke, M., Ghersi, D., Liberati, A., Petticrew, M., et al. (2015). Preferred reporting items for systematic review and meta-analysis protocols (PRISMA-P) 2015: Elaboration and explanation. *BMJ*, *350*, g7647. https://doi.org/10.1136/bmj.g7647.

She, J. J., Zhang, P. G., Wang, Z. M., Gan, W. M., & Che, X. M. (2008). Identification of side population cells from bladder cancer cells by DyeCycle Violet staining. *Cancer Biology & Therapy*, 7(10), 1663–1668.

Shen, M. M., & Abate-Shen, C. (2010). Molecular genetics of prostate cancer: New prospects for old challenges. *Genes & Development*, *24*(18), 1967–2000. https://doi.org/10.1101/gad.1965810.

Shinohara, T., Avarbock, M. R., & Brinster, R. L. (1999). beta1- and alpha6-integrin are surface markers on mouse spermatogonial stem cells. *Proceedings of the National Academy of Sciences of the United States of America*, *96*(10), 5504–5509. https://doi.org/10.1073/pnas.96.10.5504.

Shiozawa, Y., Berry, J. E., Eber, M. R., Jung, Y., Yumoto, K., Cackowski, F. C., et al. (2016). The marrow niche controls the cancer stem cell phenotype of disseminated prostate cancer. *Oncotarget*, 7(27), 41217–41232. https://doi.org/10.18632/oncotarget.9251.

Shiozawa, Y., Havens, A. M., Jung, Y., Ziegler, A. M., Pedersen, E. A., Wang, J., et al. (2008). Annexin II/annexin II receptor axis regulates adhesion, migration, homing, and growth of prostate cancer. *Journal of Cellular Biochemistry*, *105*(2), 370–380. https://doi.org/10.1002/jcb.21835.

Shiozawa, Y., Pedersen, E. A., Havens, A. M., Jung, Y., Mishra, A., Joseph, J., et al. (2011). Human prostate cancer metastases target the hematopoietic stem cell niche to establish footholds in mouse bone marrow. *The Journal of Clinical Investigation*, *121*(4), 1298–1312. https://doi.org/10.1172/JCI43414.

Shiozawa, Y., Pedersen, E. A., Patel, L. R., Ziegler, A. M., Havens, A. M., Jung, Y., et al. (2010). GAS6/AXL axis regulates prostate cancer invasion, proliferation, and survival in the bone marrow niche. *Neoplasia*, *12*(2), 116–127. https://doi.org/10.1593/neo.91384.

Siegel, R., Ward, E., Brawley, O., & Jemal, A. (2011). Cancer statistics, 2011: The impact of eliminating socioeconomic and racial disparities on premature cancer deaths. *CA: A Cancer Journal for Clinicians*, *61*(4), 212–236. https://doi.org/10.3322/caac.20121.

Simons, B. W., Hurley, P. J., Huang, Z., Ross, A. E., Miller, R., Marchionni, L., et al. (2012). Wnt signaling though beta-catenin is required for prostate lineage specification. *Developmental Biology*, *371*(2), 246–255. https://doi.org/10.1016/j.ydbio.2012.08.016.

Singh, S. K., Hawkins, C., Clarke, I. D., Squire, J. A., Bayani, J., Hide, T., et al. (2004). Identification of human brain tumour initiating cells. *Nature*, *432*(7015), 396–401. https://doi.org/10.1038/nature03128.

Skvortsov, S., Skvortsova, I. I., Tang, D. G., & Dubrovska, A. (2018). Concise review: Prostate cancer stem cells: Current understanding. *Stem Cells*, *36*(10), 1457–1474. https://doi.org/10.1002/stem.2859.

Smith, M. C., Luker, K. E., Garbow, J. R., Prior, J. L., Jackson, E., Piwnica-Worms, D., et al. (2004). CXCR4 regulates growth of both primary and metastatic breast cancer. *Cancer Research*, *64*(23), 8604–8612. https://doi.org/10.1158/0008-5472.CAN-04-1844.

Stingl, J., Eirew, P., Ricketson, I., Shackleton, M., Vaillant, F., Choi, D., et al. (2006). Purification and unique properties of mammary epithelial stem cells. *Nature*, *439*(7079), 993–997. https://doi.org/10.1038/nature04496.

Strand, D. W., & Goldstein, A. S. (2015). The many ways to make a luminal cell and a prostate cancer cell. *Endocrine-Related Cancer*, *22*(6), T187–T197. https://doi.org/10.1530/ERC-15-0195.

Su, L., Zhang, J., Xu, H., Wang, Y., Chu, Y., Liu, R., et al. (2005). Differential expression of CXCR4 is associated with the metastatic potential of human non-small cell lung cancer cells. *Clinical Cancer Research*, *11*(23), 8273–8280. https://doi.org/10.1158/1078-0432.CCR-05-0537.

Sugiyama, T., Rodriguez, R. T., McLean, G. W., & Kim, S. K. (2007). Conserved markers of fetal pancreatic epithelium permit prospective isolation of islet progenitor cells by FACS. *Proceedings of the National Academy of Sciences of the United States of America*, *104*(1), 175–180. https://doi.org/10.1073/pnas.0609490104.

Suzuki, A., Zheng, Y., Kondo, R., Kusakabe, M., Takada, Y., Fukao, K., et al. (2000). Flow-cytometric separation and enrichment of hepatic progenitor cells in the developing mouse liver. *Hepatology*, *32*(6), 1230–1239. https://doi.org/10.1053/jhep.2000.20349.

Taichman, R. S., Cooper, C., Keller, E. T., Pienta, K. J., Taichman, N. S., & McCauley, L. K. (2002). Use of the stromal cell-derived factor-1/CXCR4 pathway in prostate cancer metastasis to bone. *Cancer Research*, *62*(6), 1832–1837.

Tani, H., Morris, R. J., & Kaur, P. (2000). Enrichment for murine keratinocyte stem cells based on cell surface phenotype. *Proceedings of the National Academy of Sciences of the United States of America*, *97*(20), 10960–10965. https://doi.org/10.1073/pnas.97.20.10960.

Tannock, I. F., de Wit, R., Berry, W. R., Horti, J., Pluzanska, A., Chi, K. N., et al. (2004). Docetaxel plus prednisone or mitoxantrone plus prednisone for advanced prostate cancer. *The New England Journal of Medicine*, *351*(15), 1502–1512. https://doi.org/10.1056/NEJMoa040720.

Traustadottir, G. A., Lagoni, L. V., Ankerstjerne, L. B. S., Bisgaard, H. C., Jensen, C. H., & Andersen, D. C. (2019). The imprinted gene Delta like non-canonical Notch ligand 1 (Dlk1) is conserved in mammals, and serves a growth modulatory role during tissue development and regeneration through Notch dependent and independent mechanisms. *Cytokine & Growth Factor Reviews*, *46*, 17–27. https://doi.org/10.1016/j.cytogfr.2019.03.006.

Tsai, J. H., & Yang, J. (2013). Epithelial-mesenchymal plasticity in carcinoma metastasis. *Genes & Development*, *27*(20), 2192–2206. https://doi.org/10.1101/gad.225334.113.

Tsujimura, A., Koikawa, Y., Salm, S., Takao, T., Coetzee, S., Moscatelli, D., et al. (2002). Proximal location of mouse prostate epithelial stem cells: A model of prostatic homeostasis. *The Journal of Cell Biology*, *157*(7), 1257–1265. https://doi.org/10.1083/jcb.200202067.

Visvader, J. E., & Lindeman, G. J. (2008). Cancer stem cells in solid tumours: Accumulating evidence and unresolved questions. *Nature Reviews. Cancer*, *8*(10), 755–768. https://doi.org/10.1038/nrc2499.

Wang, N., Docherty, F., Brown, H. K., Reeves, K., Fowles, A., Lawson, M., et al. (2015). Mitotic quiescence, but not unique "stemness," marks the phenotype of bone metastasis-initiating cells in prostate cancer. *The FASEB Journal*, *29*(8), 3141–3150. https://doi.org/10.1096/fj.14-266379.

Wang, S., Gao, J., Lei, Q., Rozengurt, N., Pritchard, C., Jiao, J., et al. (2003). Prostate-specific deletion of the murine Pten tumor suppressor gene leads to metastatic prostate cancer. *Cancer Cell*, *4*(3), 209–221.

Wang, X., Kruithof-de Julio, M., Economides, K. D., Walker, D., Yu, H., Halili, M. V., et al. (2009). A luminal epithelial stem cell that is a cell of origin for prostate cancer. *Nature, 461*(7263), 495–500. https://doi.org/10.1038/nature08361.

Wang, H., Lathia, J. D., Wu, Q., Wang, J., Li, Z., Heddleston, J. M., et al. (2009). Targeting interleukin 6 signaling suppresses glioma stem cell survival and tumor growth. *Stem Cells, 27*(10), 2393–2404. https://doi.org/10.1002/stem.188.

Wang, Z. A., Mitrofanova, A., Bergren, S. K., Abate-Shen, C., Cardiff, R. D., Califano, A., et al. (2013). Lineage analysis of basal epithelial cells reveals their unexpected plasticity and supports a cell-of-origin model for prostate cancer heterogeneity. *Nature Cell Biology, 15*(3), 274–283. https://doi.org/10.1038/ncb2697.

Wei, X., Zhang, L., Zhou, Z., Kwon, O. J., Zhang, Y., Nguyen, H., et al. (2019). Spatially restricted stromal Wnt signaling restrains prostate epithelial progenitor growth through direct and indirect mechanisms. *Cell Stem Cell, 24*(5), 753–768. e756 https://doi.org/10.1016/j.stem.2019.03.010.

Weigelt, B., Peterse, J. L., & van 't Veer, L. J. (2005). Breast cancer metastasis: Markers and models. *Nature Reviews. Cancer, 5*(8), 591–602. https://doi.org/10.1038/nrc1670.

Wong, Y. N., Ferraldeschi, R., Attard, G., & de Bono, J. (2014). Evolution of androgen receptor targeted therapy for advanced prostate cancer. *Nature Reviews. Clinical Oncology, 11*(6), 365–376. https://doi.org/10.1038/nrclinonc.2014.72.

Xin, L., Lawson, D. A., & Witte, O. N. (2005). The Sca-1 cell surface marker enriches for a prostate-regenerating cell subpopulation that can initiate prostate tumorigenesis. *Proceedings of the National Academy of Sciences of the United States of America, 102*(19), 6942–6947. https://doi.org/10.1073/pnas.0502320102.

Yang, Z., Peng, Y. C., Gopalan, A., Gao, D., Chen, Y., & Joyner, A. L. (2017). Stromal hedgehog signaling maintains smooth muscle and hampers micro-invasive prostate cancer. *Disease Models & Mechanisms, 10*(1), 39–52. https://doi.org/10.1242/dmm.027417.

Yao, D., Dai, C., & Peng, S. (2011). Mechanism of the mesenchymal-epithelial transition and its relationship with metastatic tumor formation. *Molecular Cancer Research, 9*(12), 1608–1620. https://doi.org/10.1158/1541-7786.MCR-10-0568.

Yi, H., Cho, H. J., Cho, S. M., Jo, K., Park, J. A., Kim, N. H., et al. (2012). Blockade of interleukin-6 receptor suppresses the proliferation of H460 lung cancer stem cells. *International Journal of Oncology, 41*(1), 310–316. https://doi.org/10.3892/ijo.2012.1447.

Yoo, Y. A., Roh, M., Naseem, A. F., Lysy, B., Desouki, M. M., Unno, K., et al. (2016). Bmi1 marks distinct castration-resistant luminal progenitor cells competent for prostate regeneration and tumour initiation. *Nature Communications*, 7, 12943. https://doi.org/10.1038/ncomms12943.

Yoo, Y. A., Vatapalli, R., Lysy, B., Mok, H., Desouki, M. M., & Abdulkadir, S. A. (2019). The role of castration-resistant Bmi1+Sox2+ cells in driving recurrence in prostate cancer. *Journal of the National Cancer Institute, 111*(3), 311–321. https://doi.org/10.1093/jnci/djy142.

Yumoto, K., Eber, M. R., Wang, J., Cackowski, F. C., Decker, A. M., Lee, E., et al. (2016). Axl is required for TGF-beta2-induced dormancy of prostate cancer cells in the bone marrow. *Scientific Reports, 6*, 36520. https://doi.org/10.1038/srep36520.

Zhou, Z., Flesken-Nikitin, A., & Nikitin, A. Y. (2007). Prostate cancer associated with p53 and Rb deficiency arises from the stem/progenitor cell-enriched proximal region of prostatic ducts. *Cancer Research, 67*(12), 5683–5690. https://doi.org/10.1158/0008-5472.CAN-07-0768.

CHAPTER SEVEN

The testicular cancer stem cell niche

Lucía Jimenez-Rojo[a], Alejandro Díez-Torre[b], Unai Silván[c,d,*], and Juan Aréchaga[a,b]

[a]Department of Cell Biology and Histology, Faculty of Medicine and Nursery, University of the Basque Country (UPV/EHU), Leioa, Spain
[b]Analytical & High Resolution Biomedical Microscopy Core Facility (SGIker), University of the Basque Country (UPV/EHU), Leioa, Spain
[c]BCMaterials, Basque Center for Materials, Applications and Nanostructures, UPV/EHU Science Park, Leioa, Spain
[d]IKERBASQUE, Basque Foundation for Science, Bilbao, Spain
*Corresponding author: e-mail address: unai.silvan@bcmaterials.net

Contents

Abbreviations

ALP alkaline phosphatase
BTB blood-testis barrier
d.p.c. days post coitum
EC embryonal carcinoma
ES cell embryonic stem cell
GCNIS germ cell neoplasia *in situ*
GCT germ cell tumor
MMP matrix metalloproteinase
NSGCT non-seminomatous germ cell tumor
PGC primordial germ cell
PMC peritubular myoid cell
SSC spermatogonia stem cell
TGCT testicular germ cell tumor

Advances in Stem Cells and their Niches, Volume 5
ISSN 2468-5097
https://doi.org/10.1016/bs.asn.2020.09.004

1. Introduction

According to the GLOBOCAN database, there are around 71,000 cases of testicular cancer worldwide each year, causing approximately 9500 deaths per year (Bray et al., 2018). Germ cell tumors (GCTs) account for 95% of primary testicular cancers, being the rest testicular tumors of non-germ cell origin. Nowadays GCTs represent one of the most frequent tumors among adolescents and young adults with 19% of patients displaying metastatic spread to the regional lymph nodes and 12% distant metastases at the time of diagnosis (Baroni, Arato, Mancuso, Calafiore, & Luca, 2019).

GCTs form a heterogeneous group of tumors that originate from germ cells. Although these malignancies mainly arise in the gonads, and receive the name of testicular germ cell tumors (TGCTs), GCTs can also be found in extragonadal tissues (Oosterhuis, Stoop, Honecker, & Looijenga, 2007). TGCTs are thought to be initiated from an intratubular malignant germ cell. These 'tumor initiating cells' have been classically referred to as 'carcinoma *in situ*' (CIS) or 'intratubular germ cell neoplasia unclassified' (IGCNU), however, a recent terminology revision has proposed 'germ cell neoplasia *in situ*' (GCNIS) as an alternative term that better defines the nature of these cells (Moch, Cubilla, Humphrey, Reuter, & Ulbright, 2016). GCNIS are gonocyte-like cells with large nuclei and abundant glycogen that can be histologically identified inside seminiferous tubules as well-delimited and enlarged cells with a pale cytoplasm (Dieckmann & Skakkebaek, 1999).

The latest WHO classification of tumors of the testis divides TGCTs in two main groups: those derived from GCNIS and those which are not (Table 1). Among the TGCTs derived from GCNIS, there are two main types of neoplasms: seminomas and non-seminomatous germ cell tumors (NSGCTs) (Moch et al., 2016). NSGCTs consist of pure or mixed types of teratoma, yolk sac tumor, trophoblastic tumors and embryonal carcinoma (EC). EC is composed of pluripotent cells and displays a very aggressive behavior with tendency to metastasize. Although it is the most common type of NSGCTs, it generally appears combined with other types of TGCT.

The age-related incidence of GCNIS-derived TGCTs (also referred to as Type II GCTs) reaches a maximum in men in their late 20s and 30s, in contrast with GCNIS non-derived forms of TGCTs which arise mostly at prepubertal age and in elderly men. Among GCNIS non-related TGTCs that appear at prepubertal age, prepubertal-type teratomas, prepubertal-type yolk-sac tumors and prepubertal-type mixed teratoma/yolk-sac tumors are

Table 1 Main subtypes of TGCTs.

GCNIS-derived TGCTs (Type II GCTs)
Seminomas
Non-seminomatous germ cell tumors (NSGCTs)
Teratoma (postpubertal-type)
Yolk sac tumors (postpubertal-type)
Trophoblastic tumors (postpubertal-type)
Embryonal carcinoma
Mixed germ cell tumors
GCNIS non-related TGCTs
Teratoma (prepubertal-type) (*Type I GCTs*)
Yolk sac tumors (prepubertal-type) (*Type I GCTs*)
Mixed teratoma/yolk sac tumors (prepubertal-type) (*Type I GCTs*)
Spermatocytic tumors (*Type III GCTs*)

With data from Moch, H., Cubilla, A. L., Humphrey, P. A., Reuter, V. E., & Ulbright, T. M. (2016). The 2016 WHO classification of tumours of the urinary system and male genital organs—Part a: Renal, penile, and testicular tumours. *European Urology*, *70*(1), 93–105. https://doi.org/10.1016/j.eururo.2016.02.029; Oosterhuis, J. W., & Looijenga, L. H. J. (2005). Testicular germ-cell tumours in a broader perspective. In *Nature Reviews Cancer* (Vol. 5, issue 3, pp. 210–222). https://doi.org/10.1038/nrc1568.

included (also considered Type I GCTs). In elderly men, TGCTs are rare, and when present they are usually classified as spermatocytic tumors (also considered Type III GCTs). Although GCNIS non-related TGTCs, and more specifically prepubertal-types, histologically may resemble GCNIS-derived TGCTs, they often show differences in their genome and behavior, which makes necessary the design of specific treatments after the proper classification has been made (Eble, Sauter, Epstein, & Sesterhenn, 2004; Moch et al., 2016 ; Oosterhuis & Looijenga, 2005).

2. Establishment of the Spermatogonial stem cell (SSC) niche

In mammals, as in most higher animals, germ cells are set apart from the somatic lineages during the early stages of embryonic development.

The first cells of the germ line that are detectable during embryonic development are called primordial germ cells (PGCs). In mice, PGCs are first detected at 7.5 days post coitum (d.p.c.) as a group of around 40 alkaline phosphatase (ALP)-positive cells that arise from a population of pluripotent somatic cells at the base of the allantois. At around 7.5 d.p.c. PGCs are passively driven into the embryo during invagination of the visceral endoderm to form the hindgut. Later, they proliferate and actively migrate along the hindgut wall and the dorsal mesentery until they reach and colonize the genital ridges at 10.5 d.p.c. Throughout this process of migration and proliferation, the PGC genome undergoes a demethylation process that involves the full deletion of the genomic imprinting (Saitou & Yamaji, 2012). In female embryos, PGCs proliferate until 13.5 d.p.c., when they enter into meiosis I and give rise to primary oocytes that remain arrested at the diplotene stage of prophase I until females acquire sexual maturity. Then, primary oocytes enter meiosis II giving rise to secondary oocytes that remain arrested at metaphase II and will not complete meiosis II until fertilization. Unlike those in the XX embryo, male PGCs enter mitotic arrest when arriving at the genital ridges. PGCs stay in that state until day 5 after birth, when many of them start proliferating, whereas some others give rise to the spermatogonial stem cells (SSCs) (Yoshida, 2010). In humans, PGCs arise next to the allantois evagination by the third week after conception, start migrating from the dorsal mesentery of the hindgut at week four and colonize de developing gonads at around the fifth week of development (Mamsen, Brøchner, Byskov, & Møllgard, 2012).

A number of factors are known to influence the migration process of PGCs, including the presence of specific ECM components, the proteolytic activity of matrix metalloproteinases (MMPs), and the expression of stem cell factor (SCF, also known as KIT ligand) by the somatic cells surrounding migratory and post-migratory PGCs (A Díez-Torre, Díaz-Núñez, Eguizábal, Silván, & Aréchaga, 2013; Fujimoto, Yoshinaga, & Kono, 1985; Laird, Altshuler-Keylin, Kissner, Zhou, & Anderson, 2011). Binding of SCF to the c-Kit receptor on the membrane of PGCs facilitates migration and colonization of gonadal ridges, and indeed, when stimulated *ex vivo* with that factor PGCs adopt a polarized morphology characteristic of cell migration (Laird et al., 2011). Shortly after the PGCs colonize the genital ridges, those cells that have not been able to reach their destination and remain in extragonadal sites, downregulate the expression of SCF and are eliminated by entering the intrinsic Bax apoptotic pathway (Runyan et al., 2006).

The early stages of gonad development in the embryo occur concurrently with PGC migration. In mammals, the development of both male and female gonads starts at the intermediate mesoderm with the thickening of the coelomic epithelium into two multilayer structures, named genital ridges, which eventually give rise to the bipotential gonads. In mice, this process starts at around 10 d.p.c. and it takes approximately 2 days to be completed. In humans, bipotential gonad formation occurs during weeks 4–6 of embryonic development on the dorsal wall of the body cavity (Mäkelä, Koskenniemi, Virtanen, & Toppari, 2019). Once PGCs colonize the genital ridges and get confined by somatic cells, at 12 d.p.c. in mice and week 6 of gestation in humans, bipotential gonads start to differentiate into either testes or ovaries in response to the expression of sex-determining genes. In males, the first gene to be expressed is *SRY*, followed by the up-regulation of high-mobility group box 9 (*SOX9*) and the activation of fibroblast growth factor 9 (*FGF9*), the main drivers of bipotential supporting cell differentiation into Sertoli cells (Sekido & Lovell-Badge, 2013).

In humans, the precursor Sertoli cells first appear in the gonadal primordia at around weeks 7–8 of gestation. Subsequently, fetal Sertoli cells start proliferating and aggregate surrounding the germ cells also known as gonocytes at this stage. This cell association ultimately gives rise to the formation of the testicular cords, which are distinguishable from weeks 9–10. The differentiation of Leydig cells and peritubular myoid cells (PMCs) begins approximately 1 week after that of Sertoli cells (Ostrer, Huang, Masch, & Shapiro, 2007). All of these somatic cell lineages create the necessary microenvironment to support the proper development of gonocytes. At this stage, a rapid proliferation of the gonocytes occurs and their entry into meiosis is inhibited. Before weeks 16–20 of gestation, human gonocytes express pluripotency factors, such as *OCT4*, *NANOG* and *ALP*, which are progressively downregulated as gonocytes progress into pre-spermatogonia (Mitchell et al., 2008). At this point, the so-called germ cell determination process occurs: the differentiation potential of gonocytes is restricted and they lose the capacity to give rise to pluripotent cell lines, known as embryonic germ (EG) cells, *in vitro*, and to teratomas when transplanted in extra-embryonic locations. Although the precise mechanisms that regulate this process are not fully understood, the Nodal/Activin signaling pathway has been identified as a key mechanism in the regulation of gonocyte differentiation in both mouse and humans (Jørgensen et al., 2018; Wu et al., 2009). Alterations in this differentiation process can potentially result in the prevalence of an undifferentiated gonocyte population, which could lead to the formation of germ cell neoplasia *in situ* (GCNIS), the precursor cell of some types of TGCTs (Harpelunde Poulsen et al., 2020; Rajpert-De Meyts, McGlynn, Okamoto, Jewett, & Bokemeyer, 2016).

Before birth, Sertoli cells cover the inner surface of the seminiferous tubules and are in direct contact with the basement membrane, while gonocytes are restricted to the luminal area. During the first week after birth, gonocytes migrate towards the basement membrane and establish contact with Sertoli cells. In the course of this process, gonocytes enter mitosis and some of them directly turn into differentiating spermatogonia that initiate the first round of spermatogenesis. The rest differentiate into primary SSCs, which constitute the self-renewing pool and the origin of all subsequent rounds of spermatogenesis throughout life (Yoshida et al., 2006). This process is tightly regulated by Sertoli cell-derived factors. Thus, gonocyte migration towards the basement membrane within the seminiferous tubules is induced by stem cell factor (SCF) and platelet-derived growth factor (PDGF). Anti-Müllerian hormone (AMH) stimulates the differentiation of gonocytes into SSCs, and retinoic acid promotes differentiation of both gonocytes and SSCs. Lastly, glial cell-line derived neurotrophic factor (GDNF) and connexin 43 are required for normal Sertoli cell maturation and SSC niche maintenance (Sridharan et al., 2007).

Leydig cells, also known as interstitial cells, are located in the connective tissue between seminiferous tubules. During embryonic development, fetal Leydig cells (FLCs) are responsible for the production of androgens and other hormones that are indispensable for the early development of the testis and for the establishment of the SSC niche. In spite of their capital role, our knowledge on their origin and development is scarce. FLCs arise shortly after sex determination, reach a maximum number before birth, and their number gradually descents postnatally (Griswold & Behringer, 2009). It has been speculated that FLCs and adult Leydig cells (ALCs) have different origins, albeit this assertion has not yet been fully proven. FLCs are thought to derive from either the mesonephros or the coelomic epithelium covering the primitive gonad. At 11.5 d.p.c., highly proliferative coelomic epithelial cells expressing SF1 migrate into the developing gonad in male mice embryos. At 12.5 d.p.c., this population gives rise to Sertoli cells and interstitial cells that have been suggested to be FLC precursors, since SF1 is essential for the activation of genes related with steroid production. ALCs, instead, seem to arise shortly before puberty from undifferentiated mesenchymal-like stem cells located in the interstitial compartment of the testis, or from a spindle-shaped peritubular stem cell population, and are responsible for androgen production during adult life (Chen, Wang, Ge, & Zirkin, 2017; Ge et al., 2006). A third intermediate population, the infantile Leydig cells (ILCs), has been described in humans and other primates (Griswold & Behringer, 2009). Again, little experimental evidence on the origin of these cells is available, however a direct relationship between ILC specification and Sertoli cell Desert Hedgehog (DHH) signaling has been identified (Yao, Whoriskey, & Capel, 2002).

As will be discussed later, testis vascularization is another key process in the establishment and maintenance of the SSC niche. In male embryos the male-specific vascularization pattern of the coelomic region has been shown to be essential for the correct formation of the testicular cords (Cool, DeFalco, & Capel, 2011). In this process, endothelial cells from the mesonephric vessel break down and migrate through the gonads until the coelomic region. This remodeling mechanism requires the participation of the Sertoli cell-derived PDGF-B, and vascular endothelial growth factor A (VEGF-A), expressed chiefly by endothelial cells (Brennan, Tilmann, & Capel, 2003).

An important feature in many mammals, with direct effects on the SSC niche, is the location of the testes in the scrotal sac, outside the body cavity. In humans and mice, the descent of testis from the abdominal cavity to their final extra-abdominal position occurs at later stages of fetal development through a two-stage process of transabdominal and inguino-scrotal migration (Kaleva & Toppari, 2003). This results in a 2–6 °C lower temperature of the testicles compared with other organs. The maintenance of this temperature difference is central for the proper functioning of the testis and correct differentiation of SSCs into functional spermatozoa. In fact, the anomalous disruption of normal testicular descent is clinically known as cryptorchidism, and has been directly linked with infertility and with increased risk of testicular cancer in a number of species, including human (Rodprasert, Virtanen, Mäkelä, & Toppari, 2019).

In short, the microenvironment for the SSC pool maintenance and the correct spermatozoa production is controlled by a complex and tightly regulated signaling network provided by the diverse somatic cell populations around the SSC niche. Other factors such as vascular perfusion and temperature have been as well shown to be directly related with tissue homeostasis and with the development of testicular malignancies. Nevertheless, there are still many unknowns about the precise mechanisms that regulate and coordinate the establishment of the initial SSC niche during both fetal and neonatal development.

3. Anatomical and histological organization of the SSC niche

In adult humans, testes are suspended in the scrotal sac by the spermatic cords, which run from the deep inguinal ring down to each gonad. The spermatic cord contains the deferent ducts, that are the channels through which the newly formed spermatozoa exit the testicle, and the blood and lymphatic vessels that irrigate and drain the organ, respectively (Fig. 1A and D). The testis is surrounded by the tunica albuginea, a fibromuscular connective tissue capsule that appears thickened at the posterior side to form the mediastinum testis. Unlike in murine testis, in human, fibrous septa extend from the mediastinum and divide the body of the organ into 250–350 conical lobules, each one containing between 1 and 4 seminiferous tubules.

The seminiferous tubules occupy practically the entire volume of the testes, forming a dense network of helicoid close loops with occasional bifurcations (Nakata et al., 2015) (Fig. 1A and C). While in humans the rete testis is constituted by a complex system of channels ('rete' in Latin meaning 'network'), in mice it consists of a single collection reservoir with an approximate size of 400 by 600 μm and around 10 tubule connections (Silván & Aréchaga, 2012). From this location, the newly formed spermatozoa leave the testes through the efferent ducts and reach the epididymis, where they finish their maturation and acquire their motility and fertilizing capacity.

The seminiferous epithelium and the interstitial tissue are the two main components that define the histological organization of mammalian testes. During embryonic development somatic Sertoli cells and germ cells establish a close association that leads to the formation of the seminiferous cords, the precursors of the seminiferous tubules (Jørgensen et al., 2018). During childhood, Sertoli cells constitute the most abundant cell population within the seminiferous tubules, but just after puberty, and as result of the subsequent proliferation of the germ line, Sertoli cells are limited to the 10% of the epithelium. In histological sections of adult testis, Sertoli cells can be observed as tall, columnar cells with their long axis perpendicular to the basement membrane, that contain lobulated nuclei with a noticeable central nucleolus. The cytoplasm of Sertoli cells is highly branched and forms abundant projections that surround the developing germ cells and generate numerous contacts between neighboring Sertoli cells. During the first round of spermatogenesis, some of these contacts form tight

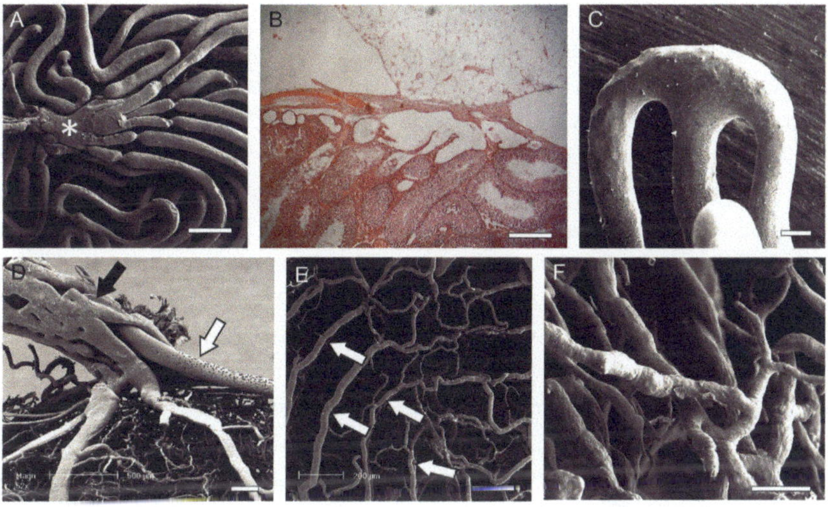

Fig. 1 Anatomical and vascular organization of the testis and TGCTs. (A) In corrosion casting preparations the overall anatomy of the testicular tubule network can be observed. In the image the connection of the seminiferous tubules to the rete testis (asterisk) can be detected. (B) In the histological preparations of the same region, the murine rete testis appears as a reservoir located beneath the tunica albuginea that drives the newly formed gametes to the efferent ducts. (C) Although rare, tubule bifurcations can be found in corrosion casting preparations of the seminiferous network. (D) Blood supply enters the testis through the testicular artery (white arrow), which is covered by the pampiniform plexus (black arrow), as seen in vascular corrosion casts. (E) The capillary architecture of the testis is characterized by the presence of vessels that run parallel to the seminiferous tubules (white arrows) and other vessels that connect these every few hundred micrometers. (F) In turn, TGCTs display a highly chaotic vascular network with characteristics of immature vessels and frequent blind endings. Scale bars represent 700 μm in (A), 200 μm in (B–E) and 100 μm in (F). *Panel (E) is reproduced with permission from Silván, U., Díez-Torre, A., Moreno, P., Arluzea, J., Andrade, R., Silió, M., et al. (2013). The spermatogonial stem cell niche in testicular germ cell tumors.* The International Journal of Developmental Biology, *57(2–4), 185–195. doi: 10.1387/ijdb.130068ja.*

junctions that give rise to the blood-testis barrier (BTB), a structure that divides the seminiferous epithelium into basal and adluminal compartments.

Spermatogonial cells are located in the basal compartment of the epithelium and are classified into A, Intermediate, and B subtypes. Eventually, some of the type A spermatogonia turn into a differentiating subtype, whereas the rest remains undifferentiated (Griswold, 2016). The self-renewal ability of Spermatogonia Stem Cell (SSC) is necessary to maintain a stem cell pool with potential to produce differentiating spermatogonia and, consequently, preserve the balance between self-renewal and differentiation, and sustain spermatogenesis at optimal levels. In most mammals, including mice, some of the undifferentiated spermatogonia constitute the SSC population, formed by individual cells (A_s). This SSC population represents up to 0.03% of the total germ cells present in the gonads, and upon division give rise to two daughter cells, called A_{pr} spermatogonia, that stay connected by intercellular bridges. Further divisions result in the formation of longer chains composed of 4, 8 or 16 cells, and are referred as A_{al} spermatogonia. As spermatogonia are responsible for SSCs self-renewal, conversely, A_{al} spermatogonia divide to produce A1 to A4 spermatogonia, which further give rise to intermediate spermatogonia, B spermatogonia, and then preleptotene spermatocytes (Takashima et al., 2015). Unlike other mammals, in humans and primates the A_s spermatogonia are classified in only two subtypes, A_{dark} (A_d) and A_{pale} (A_p) spermatogonia. The nuclei of the former have high affinity for hematoxylin and, in humans, contain in their central region a chromatin rarefaction zone, which is a 'cavity' or 'vacuolar space' visible when stained with dyes with chromatin affinity. In turn, nuclei of A_p spermatogonia are more subtly stained with hematoxylin and lack such rarefaction zone. A_d spermatogonia are considered either a stem cell pool that only divides to replace damaged A_p cells, or the actual SSCs population with a very low cell cycle, being however, the differences in the nature and behavior of these two spermatogonia subpopulations still under discussion (Hermann, Sukhwani, Hansel, & Orwig, 2010; von Kopylow & Spiess, 2017). In mammals, the SSC population is confined within the basal compartment of the seminiferous epithelium. This histological localization points to elements of the basement membrane and Sertoli cell-secreted growth factors as key for the correct self-renewal and differentiation of SSCs. In fact, Sertoli cells have been identified as the main source of two growth factors that play a critical role in the regulation of spermatogonial self-renewal, the Glial cell line-derived neurotrophic factor (GDNF) and the basic fibroblast growth factor 2 (bFGF, encoded by the *FGF2* gene) (Masaki et al., 2018). The Gonadotropin-releasing hormone (GnRH) is cyclically produced in the hypothalamus and induces the secretion of follicle-stimulating hormone (FSH) and luteinizing hormone (LH) in the pituitary gland. FSH is a major regulator of Sertoli cell function, and induces the secretion of growth factors by the Sertoli cells, including GDNF (Fig. 2) (Oduwole et al., 2018).

The basal location in the seminiferous tubules makes SSCs highly exposed to factors from interstitial tissue. Several cell populations are found

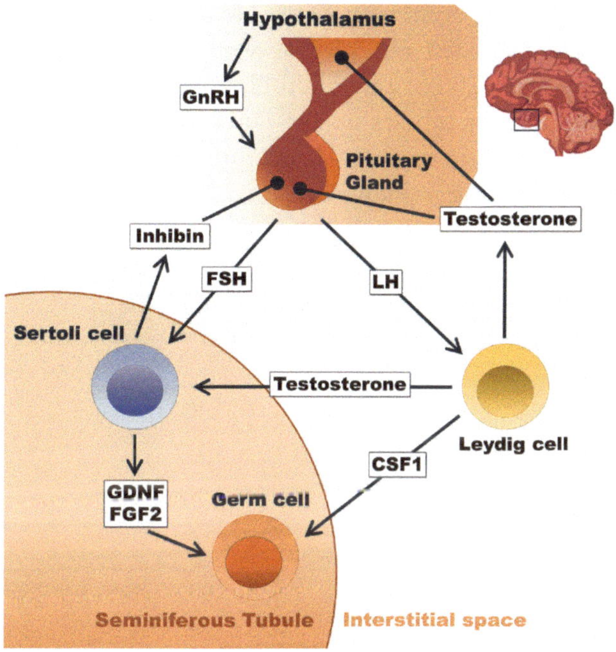

Fig. 2 The hypothalamic-pituitary axis. Gonadotropin-releasing hormone (GnRH) is a peptide hormone cyclically released by the neurons of the hypothalamus that stimulates the production of follicle-stimulating hormone (FSH) and luteinizing hormone (LH) by the anterior pituitary gland. These hormones are released into the blood stream and reach the gonads. In the testes, LH causes the production of testosterone by Leydig cells, and FSH plays a major role in Sertoli cell growth and maturation. Leydig cell-derived testosterone and Sertoli cell-produced growth factors, such as GDNF and bFGF, are key for regulating the SSC niche and spermatogenesis, since they drive SSC self-renewal and differentiation. Testosterone and inhibin form a negative feedback loop that reduce the hormone secretion in both hypothalamus and pituitary gland.

in this compartment, being Leydig cells the most frequent one, besides nerve cells, and endothelial cells that form capillaries. In histological sections, Leydig cells can be seen as clusters of polygonal cells with round-shaped nuclei and a granular eosinophilic cytoplasm. As response to LH, Leydig cells express testosterone, a hormone necessary for Sertoli cells to carry out their function as spermatogenesis supporters (Zhou et al., 2019). Colony stimulating factor 1 (CSF1) is another Leydig cell-derived factor with an important influence on the SSC niche, since it has been identified as an extrinsic stimulator of SSC self-renewal in a Sertoli cell-independent fashion (Oatley, Oatley, Avarbock, Tobias, & Brinster, 2009). Interestingly, CSF1 expression has been observed as well in peritubular myoid cells (PMCs) (Oatley et al., 2009), a myofibroblast-like cell population that surrounds the seminiferous tubules on the outer side of the basement membrane (Fig. 3A-C) (Silván et al., 2015). In mice and other rodents, PMC are organized by forming a unique layer, whereas in human testis, several

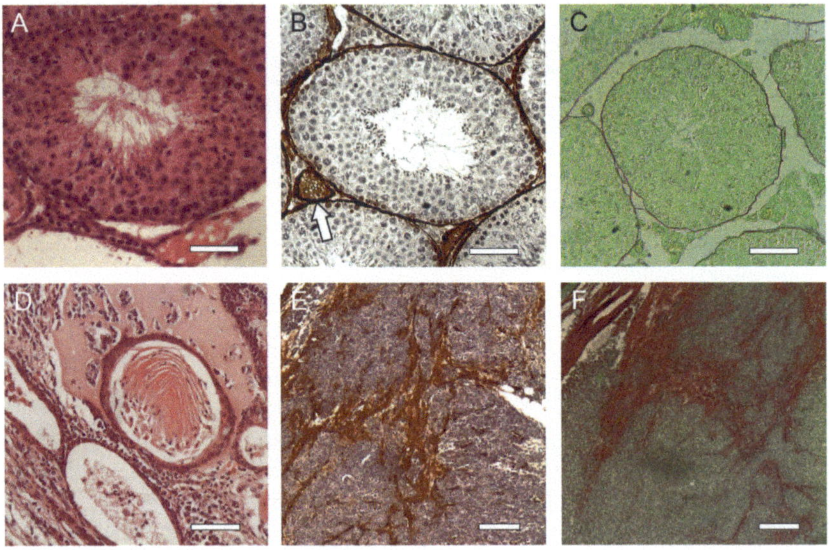

Fig. 3 Histological organization of the testis and TGCTs. (A) Hematoxylin-eosin stained section showing a murine seminiferous tubule, with Leydig cell aggregates and blood capillaries in the interstitial space. (B) Anti-SMA histochemistry reveals the presence of myofibroblasts surrounding the seminiferous tubules and in perivascular cells of blood capillaries (white arrow). (C) In the peritubular region the presence of collagen in the basement membrane is evidenced by red staining using Picrosirius red. (D) Experimental TGCTs display a chaotic histological organization that contains many different tissue types. (E) These tumors contain large populations of SMA-positive myofibroblasts forming thick barriers that compartmentalize the neoplastic tissue. (F) Similarly, collagen deposition appears increased in regions occupied by myofibroblasts, as seen in serial sections. Scale bars represent 50 μm in A-D and 100 μm in E and F. *Panel (B) is reproduced with permission from Silván, U., Díez-Torre, A., Bonilla, Z., Moreno, P., Díaz-Núñez, M., Aréchaga, J. (2015). Vasculogenesis and angiogenesis in nonseminomatous testicular germ cell tumors.* Urolologic Oncology, 33*(6), 268.e17–28. doi: 10.1016/j.urolonc.2015.01.005. Panel (D) is reproduced with permission from Silván, U., Arlucea, J., Andrade, R., Díez-Torre, A., Silió, M., Konerding, M. A., et al. (2008). Angiogenesis and vascular network of teratocarcinoma from embryonic stem cell transplant into seminiferous tubules.* British Journal of Cancer, 101*(1), 64–70. doi: 10.1038/sj.bjc.6605125.*

layers of PMCs can be found around each seminiferous tubule (Maekawa, Kamimura, & Nagano, 1996). PMCs are characterized by their cytoplasmic system of α-smooth muscle actin (α-SMA) and myosin filament clusters distributed in two independent layers, orthogonal to each other (Fig. 3B) (Losinno, Morales, Fernández, & Lopez, 2012). This cytoskeletal machinery is used by the PMCs for the propulsion of the tubular fluid that contains newly formed spermatozoa towards the rete testis. In addition to this contractile activity, a number of studies have shown that these cells are as well the source of signals with a direct effect on the SSC niche, and thus, on the regulation of spermatogenesis (Zhou et al., 2019).

The basement membrane of the seminiferous tubules is the outer limit of the SSC and is produced by Sertoli and peritubular myoid cells. It is mostly composed of type IV collagen, laminins (α1, α2, β1, β2, γ1), heparan sulfate proteoglycan 2 (Perlecan), and nidogen (Fig. 3C) (Park et al., 2016). Changes in the composition of the basement membrane have been related with the regulation of Sertoli cell differentiation and germ cell development (Hadley, Byers, Suárez Quian, Kleinman, & Dym, 1985), and are behind some male infertility cases (Häger, Gawlik, Nyström, Sasaki, & Durbeej, 2005; Wan, Mruk, Wong, & Cheng, 2013). More recently, it has been proven that the non-collagenous C-terminal domain (NC1) derived from the collagen α3(IV) chain, one of the main components of the basal membrane of seminiferous tubules, regulates cell-cell adhesion dynamics, specially the tight junctions between Sertoli cells in the BTB, and the apical ectoplasmic specializations, a testis-specific adherent junction that sustains the association between Sertoli cells and spermatids (Chen, Wang, et al., 2017; Chen, Zhang, et al., 2017). In another recent work, different effects of other components of the basement membrane on SSCs have been described and show that while laminin seems necessary for the maintenance of the undifferentiated state of the SSCs, signals derived from collagen type IV contribute to the onset of spermatogenesis (Park et al., 2016).

As mentioned before, each seminiferous tubule is surrounded by a network of nerves and blood capillaries. Leydig cells are usually associated with both nerves and vessels, in order to fulfill their endocrine functions. Undifferentiated spermatogonia, meanwhile, show a preferential position in those areas of the seminiferous tubules that are closer to the blood vessels (Chiarini-Garcia, Raymer, & Russell, 2003; Yoshida, Sukeno, & Nabeshima, 2007), suggesting that systemic factors could also be central for the establishment of SSC niche. In fact, other cellular components of interstitial tissue, such as infiltrating macrophages and mesenchymal cells, have as well been identified as active agents in the regulation of the niche of SSCs, either directly or indirectly (Heinrich & DeFalco, 2020; Meinhardt, Wang, Schulz, & Bhushan, 2018).

All these observations reveal a complex scenario with many actors involved, moreover, the contribution of all these agents is likely affected by external stimuli in order to adapt the SSC niche to different developmental stages, physiological situations or pathological conditions. Deciphering the complex network of interaction among all the SSC niche components might be essential to fully understand the origin and development of TGTCs.

4. Developmental origin of germ cell tumors

As previously mentioned, GCTs constitute an heterogeneous group of neoplasms that may differ in location, histological features and behavior. Based on these peculiarities, it has been proposed that the different types of GCTs may originate from different cells along the germline hierarchy and that the onset and development of each tumor type may have a different molecular basis.

As previously mentioned, it is broadly accepted that the common precursor of Type II GCTs is the so-called germ cell neoplasia *in situ* (GCNIS) that can progress to either seminomas or non-seminomatous germ cell tumors (NSGCTs). While seminomas show similar molecular and histological features to GCNIS, PGCs and gonocytes, NSGCTs display greater variability. In order to explain this observation, it has been proposed that NSGCTs may arise from intermediate malignant precursors, the so-called EC (Embryonal Carcinoma) cells. EC cells are considered the malignant counterparts of Embryonic Stem cells (ES cells) from the inner mass of the blastocyst. Like ES cells, EC cells show naive pluripotency, which gives them the ability to differentiate into all somatic and germ cell lines. Based on this plasticity, EC cells are considered the cancer stem cell (CSC) component from which all the NSGCT cells differentiate and contribute to the formation of the different subtypes (Fukawa & Kanayama, 2018; Nettersheim et al., 2015; Sada & Tumbar, 2013).

In late PGCs, gonocytes and GCNIS, *SOX2* expression declines and in turn, they activate the expression of *SOX17*, a crucial transcription factor for germ cell line specification. Although not expressing *SOX2*, late PGCs, gonocytes and GCNIS do express other pluripotency markers such as *OCT4*, *NANOG* and *KIT*. Thus, it has been proposed that late PGCs, gonocytes and GCNIS show rather a dormant pluripotency and they are able to give rise only to germ line-derived cells (De Jong et al., 2008; Fukawa & Kanayama, 2018). This would explain why seminomas, which may derive directly from GCNIS, show histological features that remind to the germ cell phenotype.

Among the GCNIS non-related TGCTs, Type I neoplasms include the infantile varieties that can develop as teratomas or yolk sac tumors, none of which contain cells that display germ cell characteristics. In contrast, yolk sac tumors show typical yolk sac histology and teratomas may contain many types of somatic tissue types. Based on these observations, it has been proposed that Type I GCTs originate from cells at the top of the germline hierarchy. In turn, prepubertal yolk sac tumors may arise from pluripotent ES cells that are influenced by certain factors that drive differentiation into extraembryonic tissues. As for prepubertal teratomas, which can contain cell types from the three germ layers, it has been proposed that they arise from early migrating PGCs that upon a tumorigenic stimulus or genomic aberration activate their 'dormant' pluripotency and subsequently differentiate into various somatic cell lines (Almstrup et al., 2006).

Both Type I and Type II GCTs can be found in gonadal and extragonadal locations. However, there is an interesting finding regarding the distribution of the extragonadal tumors. Type II extragonadal GCTs are usually found along the midline of the body, following the migration path of PGCs, while, Type I extragonadal GCTs are not restricted to that region. This fact supports the hypothesis of Type I GCTs arising from more undifferentiated PGCs (functionally closer to ES cells) that are more capable than late PGCs and gonocytes to survive and proliferate outside their niche (Oosterhuis et al., 2007).

Lastly, type III GCTs or spermatocytic seminomas usually develop in adulthood and are restricted to the gonads. Spermatocytic seminomas are histologically distinct to all other GCTs and they are thought to arise from adult stem cells located at the bottom level of the hierarchy of adult germline stem cells, that is, spermatogonia. Recent studies have proposed that a first spontaneous mutation may confer some adult spermatogonia growth and survival advantages that, when combined with secondary mutagenic events, may lead to the development of spermatocytic seminomas (Giannoulatou et al., 2017).

5. Tools to study the SSC niche

Germ cell tumorigenesis and early embryogenesis share many aspects, and, as previously mentioned, the undifferentiated stem cells of erminal tumors, EC cells, resemble ES cells in many aspects (Aréchaga & Damjanov, 2012; Dixon & Moore, 1953; Pierce, 1993). For instance, the gene expression pattern is similar in both cell types, and is characterized by the high expression levels of pluripotency genes (Henderson et al., 2002). Murine ES (mES) and EC (mEC) cells both express high levels of the transcription factor Oct4 and the surface marker SSEA1 (stage-specific embryonic antigen 1), both characteristic of embryonic development (Andrews et al., 2005; Henderson et al., 2002). Similarly, cDNA microarray analysis revealed that human EC (hEC) and ES (hES) cells express high levels of the transcription factor *OCT4*, together with the surface markers SSEA3 and SSEA4, TRA-1-60, TRA-1-81 and GCTM2, Thy1, MHC class 1 (Draper, Pigott, Thomson, & Andrews, 2002; Henderson et al., 2002).

mEC cell lines have been classically isolated from teratocarcinomas generated by the implantation of early mouse embryos or embryonic genital ridges in ectopic sites of acceptor mice (Damjanov & Wewer-Albrechtsen, 2013; Stevens, 1973). In turn, hEC cell lines are generally isolated from primary tumors, like for example, the hEC cell lines Tera-1 and Tera-2, which were established from primary teratocarcinomas of the testis. Due to their low culture requirements and differentiation potential, EC cells of human and murine origin have been widely used in the research of developmental processes, particularly neurogenesis (Marikawa, Tamashiro, Fujita, & Alarcón, 2009). Regarding the research field of testicular cancer, EC cell lines have been used to study the crosstalk of the cancer stem cell component of TGCTs and other cells present in their microenvironment. For instance, using indirect co-culture, it was described that the human EC cell line NCCIT can affect the miRNA and mRNA expression patterns of stromal cells isolated from healthy tissue to resemble that of stromal cells isolated from solid tumors (Vêncio et al., 2011). Similarly, co-culture of the murine F9 EC cell line with testicular PMCs revealed an upregulation in MMP-9 expression and increased migration of the PMCs (Díez-Torre, Silván, Moreno, Gumucio, & Aréchaga, 2011; Moreno-Ruiz et al., 2016). Despite the many benefits of these two-dimensional co-culture systems to study cell-cell interactions, solid tumors naturally grow in three-dimensional arrangements that affect how the malignant cells behave and interact with their environment. Importantly, these differences include changes in metabolic pathways and are also known to have an impact on the chemoresistance of tumor cells (Breslin & O'Driscoll, 2016). In this regard, three-dimensional cell aggregates, named embryoid bodies (EBs) when composed of pluripotent cells, replicate the *in vivo* conditions in a more faithful fashion. EBs can display a complex histological pattern including different tissues and stem-like cells, both of which are as well characteristic of spontaneous TGCTs. These EBs composed of EC cells have been used to compare the invasive potential of different cell lines using chick heart fragments. In these experiments murine EBs were confronted with spherical fragments of embryonic chick heart and incubated in suspension under shaking conditions. Histological analysis of the tissues after several days in culture revealed a higher invasive potential of an embryonic germ cell line (EG1) over the EC cell lines P19 and F9 (Díez-Torre et al., 2004).

In 1973, Stevens reported a new inbred subline of mice (named 129-terSv) that displayed a high incidence of spontaneous testicular teratomas (Stevens, 1973). Nevertheless, the generally fully differentiated nature of these tumors renders them non-malignant and therefore inadequate for the study of non-seminomatous TGCTs, which contain a population of pluripotent ECs. Years later, transplantation of pluripotent ES cells into the seminiferous tubules was shown to generate malignant testicular tumors with the histological aspect of spontaneous TGCTs (Fig. 3D and E) (Payne, 2013; Silván et al., 2009). In these experiments a sharp and thin glass capillary connected to a microinjection device is used to inject a suspension of cells in the lumen of the seminiferous tubules (Silván et al., 2011). Few hours after the injection, some of the transplanted cells migrate to the basal region of the tubules and over a period of weeks the development of TGCTs is evident. In addition to the potential of these approaches to study the early stage of TGCT development, it allows the genetic modification of the cells prior their transplantation, for instance for tracing them after being transplanted. As we will see below, this experimental model allowed us to describe the architecture of the vascular plexus of TGCTs and to establish the malignant origin of some of the endothelial cells that form it (Silván et al., 2009; Silván et al., 2011).

6. Vascular system and oxygenation

Testes receive their blood supply from the testicular arteries (also named male gonadal arteries) that emerge from the abdominal aorta and reach the testis through the sperminal cord. In their last section testicular arteries are surrounded by the pampiniform plexus, a venous network, which in addition to its vascular role, it is known to work as heat exchanger to regulate temperature of the arterial blood entering the testicles (Fig. 1D). Crossing the tunica albuginea testicular arteries are successively branched into smaller vessels that ultimately form a network composed of two differently oriented capillaries: longitudinal capillaries that run parallel to the seminiferous tubules, and transverse capillaries that connect the longitudinal ones every few hundred micrometers, forming a very regular structure, as seen in corrosion casting samples (Fig. 1E). As previously mentioned, SSCs are preferentially located in the basal compartment of the seminiferous tubules adjacent to the blood vessels and close to the blood vessel branch points, and move to other regions as they differentiate (Yoshida et al., 2007). Although the precise significance of this preferential spatial arrangement is not completely clear yet, it evidences the importance of the architectural organization of the vascular bed in the definition of the SSC niche. As previously discussed, GCNIS as well first appears in the basal compartment of the seminiferous tubules, however to date no spatial preference in relation with the capillary organization has been described.

In contrast with the structural stability of the vascular network of the testes, testicular endothelial cells (TECs) display higher proliferation rates than those endothelial cells found in other adult organs, such as liver and brain (Lissbrant, Lissbrant, Persson, Damber, & Bergh, 2003). Given that in adult testes no signs of vascular remodeling have been found, the physiological relevance of the high TEC turnover rate remains elusive. However, it is known that many cell types present in the testes produce vascularization-related factors. For instance, Sertoli cells secrete vascular endothelial growth factor (VEGF), TIMP-1, monocyte chemoattractant protein-1 (MCP-1) and bFGF (Mullaney & Skinner, 1992), and their presence is required for the maintenance of the capillary architecture *in vivo* (Rebourcet et al., 2016). In turn, *in vitro* experiments have revealed that medium conditioned by Sertoli cells significantly enhances the formation of capillary-like structures in an aortic explant model (Golat & Cameron, 2008). Other cell types, such as Leydig cells, PMCs and germ cells also secrete different factors, including members of the VEGF family (Caires, de Avila, & McLean, 2009), which are likely relevant for the vascular homeostasis of the tissue, and might explain the high proliferation rate of TECs.

Tumor vascularization can occur *via* two mechanisms. Firstly, as in adult organs the most frequent neovascularization process in tumoral tissues is angiogenesis, that is the formation of new capillaries from pre-existing ones by sprouting of endothelial cells or by vessel intussusception. Secondly, new vessels can arise by vasculogenesis, which consists in the differentiation of endothelial precursor cells (EPCs) into endothelial cells and the *de novo* formation of a primitive vascular network that at a later stage connects to the vascular system (Carmeliet, 2000). Because the same factors regulate both neovascularization mechanisms and considering that vasculogenic events are largely limited to the embryonic development of visceral organs, cancer research mostly refers to angiogenesis. However, given the embryonic nature of teratocarcinomas, including NSGCTs, it is not surprising that the histological analysis of human samples revealed, in addition to mature capillaries, a significant presence of embryonal-type vessels (Nacov, 1990). In certain tumors, such as glioblastoma and breast cancer, a third vasculogenic process, in which malignant cells give rise to endothelial cells, has been described (Ricci-Vitiani et al., 2010). Using the ES-transplantation TGCT model we later confirmed that in experimental TGCTs besides host-derived vasculature, a small portion of the newly formed vessels did originate from the transplanted cells (Silván et al., 2009), suggesting the existence of tumor-derived vasculogenesis in TGCTs. Analysis of the vascular bed of experimental TGCTs using corrosion casting revealed a highly disorganized network that displays immaturity features, such as loss of hierarchical branching, sudden changes in capillary diameter and frequent blind endings (Fig. 1F), all of which make it incompatible with correct tissue oxygenation. This leads to the formation of hypoxic pockets in the tissue and the consequent activation of Hypoxia-inducible factors (HIFs) and downstream pathways. In testicular tumors expression of VEGF, a key factor upregulated by HIFs, has been shown to correlate with microvessel density (MVD) and predict for metastatic dissemination (Viglietto et al., 1996). Nevertheless, other studies have reported higher levels of VEGF in non-malignant TGCTs over malignant ones (Jones et al., 2000), evidencing the complexity of the process, which is exacerbated by the numerous factors and cell types involved.

As previously mentioned, testicular function is tightly regulated by follicle-stimulating hormone (FSH) and luteinizing hormone (LH) that are produced in the pituitary gland upon Gonadotropin-releasing hormone (GnRH) stimulation. Because these hormones reach the gonads *via* blood circulation, and their function is therefore linked to the correct vascular perfusion of the tissue, the tumor-driven vascular remodeling can dramatically affect their distribution and availability across the solid malignancy.

Although gene expression analysis has discarded the presence of FSH and LH receptors in TGCTs, Leuprolide, a Gonadotropin-releasing hormone (GnRH) inhibitor, significantly reduces the growth of xenotransplanted TGCT tissues (Douglas, Richardson, & Nicol, 2006).

Taken together, the neovascularization process in TGCTs shares many similarities with that of other tumor types, such as a higher vascular density than the healthy organ in which they originate, the formation of hypoxic regions and the upregulation of the expression of vascularization-related factors. Nevertheless, their germinal origin and the presence of fully differentiated tissues, together with a pluripotent EC population, are unique to TGCTs and have a major impact on the remodeling of the vascular system during the course of the disease. A better understanding of this process, and of the cell types and factors influencing it, is required to improve the treatment of this disease while reducing the risk of metastatic spreading.

7. Immune system

At the beginning of puberty germ cells enter meiosis and activate the expression of a number of new antigens, which potentially could trigger the immune response of the host. However, mammalian testes contain a special immunological environment that protects germ cells from autoimmune aggressions. The mechanisms that enable the immune privilege of the testis include its structural organization as well as the local expression of immune-suppressive factors.

Regarding the tissue organization, the blood-testis barrier (BTB) has been proposed as an important feature linked to the immune privilege of the testis. As previously mentioned, BTB is a cellular structure formed by tight junctions, basal ectoplasmic specializations, desmosome-like junctions and gap junctions between Sertoli cells that divides the seminiferous tubules into two compartments: the basal and the adluminal one. In the basal compartment, the less differentiated germ cells are found, and the adluminal one contains germ cells at later steps of spermatogenesis (Luca et al., 2018). BTB plays an essential role in limiting the migration of immune factors through both compartments. Thus, most of the auto-antigens of late germ cells remain retained within the adluminal compartment conferring the testis with the above-mentioned protection against autoimmune reactions. In addition, the BTB limits the access of systemic immune contents to the adluminal compartment contributing to the immune-privileged state of the testis (Zhao, Zhu, Xue, & Han, 2014). The immune privilege of the testis is not only regulated by the physical barrier constituted by the BTB, but also by the local tissue expression of certain factors such as Fas ligand (FasL), a transmembrane protein of the TNF family. FasL binds to its receptor, cell surface Fas (Cd95), and activates a cascade of caspases that results in apoptosis (Nagata, 1997). In rodent and human testis, FasL forms a layer located at the periphery of the tubules, on the foot processes of the Sertoli cells. FasL expressing testis cells may trigger apoptosis in Fas-expressing inflammatory cells, mainly activated T lymphocytes, preventing their invasion into the seminiferous tubules (Li, Rosen, Sondel, & Berke, 2002). Interestingly, under physiological conditions, most of the cells that define the testis microenvironment, such as Sertoli cells, Leydig cells, peritubular myoid cells, and resident macrophages are known to express many cytokines involved in the regulation of inflammatory responses, including IL-1β, IL-6, IL-10, TNF-α and TGF-β, and all of them play an important role in the regulation of testis homeostasis and spermatogenesis (Li et al., 2002; Loveland et al., 2017).

Despite its immune privileged situation, testicular tissue is not fully exempt of specific and innate immune responses under certain clinical conditions. In some cases of infertility, for example, it is common to find infiltrating T lymphocytes in the interstitial region next to seminiferous tubules that show dysfunctional spermatogenesis (Fijak et al., 2018). Infiltrating immune cells are also typical features of certain testicular neoplasms, as they have been associated with GCNIS and seminomas (Yakirevich et al., 2002). T cells constitute the main component of these immune active populations, although B cell and macrophage recruitment are also commonly observed. Nevertheless, the function and biological significance of these events are not yet well understood.

The presence of infiltrating lymphocytes in testicular tumors was first reported in 1964 (Marshall & Dayan, 1964). Since then, many studies have shown an abundant presence of infiltrating immune cells in testicular tumors, especially in seminomas. Among these cell populations, T lymphocytes, both CD4+ and CD8+, stand out, while B lymphocytes and NK cells are less abundant (Hvarness et al., 2013; Lobo et al., 2019). In seminiferous tubules containing GCNIS, the highly proliferative neoplastic cells displace the Sertoli cells towards central areas and, as a consequence, the FasL layer becomes discontinuous and allows the infiltration of T lymphocytes. In the case of invasive seminomas, the FasL layer completely loses its function, resulting in an abundant lymphocyte infiltration, especially CD8+ cells. However, since neither GCNIS nor invasive seminomas cells express MHC class I or ICAM-1 molecules, necessary for the response of CD8+ T lymphocytes, their role in these tumors remains unclear (Bols, Jensen, Jensen, & Braendstrup, 2000). The low activity of infiltrating immune cells within testicular tumors suggests that either no lymphocytes with specificity towards tumor cells or Sertoli cells are present or, if present, the lymphocytes are not able to recognize the target cells (Bols et al., 2000).

The similarities, in terms of immune cell populations present, of testicular tissue independently of the presence of a malignancy (Hvarness et al., 2013) has classically supported the notion that immune surveillance is absent in these tumors. Nevertheless, in a recent study, some clear differences were found between the populations of immune cells present in testicular malignancies and those present in other types of inflammatory lesions associated with hypo-spermatogenesis (Klein et al., 2016). According to this study, the presence of B lymphocytes and dendritic cells could be a unique characteristic of the immune response associated with tumors in the testis since these populations seem to be almost absent in other types of inflammation-related lesions. However, it is possible that these B cells lack activity against tumor cells, as no potential antibody-producing plasma cells were detected in the neoplastic samples. Therefore, the presence of B cells could be a side effect of the proinflammatory microenvironment of the TGCTs, characterized by the presence of IL-6 and CXCL-13, two potent B cell chemoattractants.

The pro-inflammatory environment has been shown to be favorable for malignant progression in a number of tumors (Crusz & Balkwill, 2015), and the analysis of cytokines present in testicular tumors has shown that these neoplasms are rich in pro-inflammatory factors. Compared to healthy testis, biopsies of testis neoplastic tissue display higher levels of pro-inflammatory cytokines, such as IL-1β, IL-6 and TNF-α, whereas in the tissues of infertile men such changes are less evident (Klein et al., 2016). IL-6 overexpression has been observed in most types of tumors, where it promotes tumorigenesis by regulating many processes involved in neoplastic progression, such as apoptosis, survival, proliferation, angiogenesis, invasiveness and metastasis (Kumari, Dwarakanath, Das, & Bhatt, 2016). The major involvement of IL-6 in testis cancer, and the role of the neoplastic cells as one of the main sources of this and other pro-inflammatory cytokines, have been highlighted both by the aforementioned *ex vivo* observations and by a subsequent *in vitro* study. In this study, a co-culture model of TCam-2 seminoma cell line and peripheral blood mononuclear cells (PBMCs) demonstrated a potent cell-contact dependent up-regulation of IL-1β, TGF-β, TNF-α and CCL5. Seminoma cells were also identified as the main source of IL-6 in these co-cultures (Klein et al., 2017).

Interestingly, PMCs of the testis have also been identified as a source of IL-6 and some other proinflammatory interleukins, like Cox-2 and monocyte chemoattractant protein-1 (MCP-1). Moreover, this expression has been proven to be regulated through TNF-α signaling (Schell et al., 2008), a proinflammatory cytokine produced by cells of the immune system present within the testicular mesenchyme, mainly by monocytes, macrophages and mast cells. Both IL-6 and MCP-1 are likely involved in the recruitment of testis tumor-associated macrophages (TAMs), since the last has a chemoattractant effect on circulating monocytes and IL-6 induces the differentiation of monocytes into macrophages (Duluc et al., 2007).

In healthy testis macrophages and mast cells are found in the interstitial space (Mossadegh-Keller & Sieweke, 2018), but the number of macrophages and neutrophils is significantly higher in neoplastic microenvironments and these cells have been shown to play a key role in the progression of a number of tumors. Both tumor-associated neutrophils and macrophages have been identified as an important source of matrix metalloproteinases (MMPs), a family of proteolytic enzymes strongly associated with tumor progression, due to their involvement in processes such as invasion, apoptosis resistance, angiogenesis, and lymphangiogenesis (Gonzalez-Avila, Sommer, García-Hernández, & Ramos, 2020). Neutrophil derived MMP-9, for example, has been identified as a potent stimulator of angiogenesis on assays using chorioallantoic membrane chicken embryos (Ardi, Kupriyanova, Deryugina, & Quigley, 2007) and in colorectal cancer, where MMP-9 is associated with the higher availability of active vascular endothelial growth factor (VEGF) derived from the proteolytic release of its ECM-anchored form (Hawinkels et al., 2008). Besides the secretion of MMP-9, neutrophils have been shown to promote colorectal cancer invasion and angiogenesis through the direct production of VEGF, and hepatocyte growth factor (HGF) in the primary and metastatic sites (Mizuno et al., 2019). Tumor associated macrophages have been proven to promote the metastasis of gastric and breast cancer cells by the expression of Chitinase 3-like protein 1 and the subsequent activation of the MAPK signaling pathway (Chen, Zhang, et al., 2017).

Micro-RNAs (miRNAs) are short non-coding RNAs that modulate gene expression post-transcriptionally, and have both pro and anti-tumoral activity in different tumor types, in which they regulate tumor progression processes including proliferation, apoptosis, and invasion (Kim, Kasinski, & Slack, 2011). miR-125b is known to act as a tumor suppressor mRNA in several malignancies, and in a recent work, it has been identified as an important mediator in the crosstalk between neoplastic cells and infiltrating macrophages in TGCTs (Batool, Wang, Hao, Chen, & Liu, 2018). According to this study, low miR-125b expression in tumor cells increases the production of tumor-derived chemokine CSF1 and CX3CL1, which act as macrophage chemoattractants and, thus, contribute to tumor progression, whereas overexpression of miR-125b did inhibit TGCT growth *in vivo* through the reduction of pro-tumorigenic macrophage recruitment.

PD-1, PD-L1 and CTLA4 proteins are known to mediate immune check-point inhibition in several types of malignancies (Kim & Eder, 2014). PD-1 is constitutively expressed by a subset of T cells, it is upregulated on activated NK, T and B cells, monocytes and dendritic cells, and it is specially highly expressed on CD4+ follicular helper T cells. One of its ligands, PD-L1, seems to inhibit effector T-cell function and may be involved in protecting immune-privileged sites, such as the testis, from T-cell immune responses (Fife & Pauken, 2011). PD-L1 expression level on tumor infiltrating lymphocytes has been recently identified as a novel prognostic marker in TGCTs. The presence of PD-L1 positive lymphocytes is significantly higher in seminomas and embryonal carcinomas compared to yolk sac tumors, choriocarcinomas or teratomas. Also, a high infiltration of PD-L1 positive lymphocytes has been correlated with a better survival index in TGCT patients (Chovanec et al., 2017).

In summary, the cells of the immune system constitute an essential element of the active stroma in tumors and, although their original function is to attack neoplastic cells, their proinflammatory activity is closely associated with tumor progression and a poor prognosis. The role of infiltrating immune cells, especially macrophages and lymphocytes, in processes such as proliferation, invasion or angiogenesis in tumors has been an important object of study in different types of malignancies for decades. However, in the case of TGCTs, it was not until recently that these elements have received attention. The identification of the miR-125b/CSF-1/CX3CL1 and PD1/PD-L1 axes as potential therapeutic targets in TGCTs are good examples of the importance that this topic may have in future research in the field.

8. Concluding remarks

Testicular germ cell tumors represent a complex type of malignancy characterized by the presence of different tissues, in some cases accompanied by a pluripotent cell population. The unique features of TGCTs, such as the presence of diverse cell types and tissues at different states of differentiation, the high concentration of growth factors, and their location in an immune privileged tissue, invalidate many of the observations done in other tumor types and make the study of their development and progression a challenging task. Despite these difficulties, research done in this area has uncovered many of the mechanisms driving the initiation, development and metastatic spreading of TGCTs, and will certainly contribute to the development of new treatment strategies for these malignancies.

References

Almstrup, K., Sonne, S. B., Hoei-Hansen, C. E., Ottesen, A. M., Nielsen, J. E., Skakkebaek, N. E., et al. (2006). From embryonic stem cells to testicular germ cell cancer—Should we be concerned? *International Journal of Andrology*, *29*(1), 211–218. https://doi.org/10.1111/j.1365-2605.2005.00643.x.

Andrews, P. W., Matin, M. M., Bahrami, A. R., Damjanov, I., Gokhale, P., & Draper, J. S. (2005). Embryonic stem (ES) cells and embryonal carcinoma (EC) cells: Opposite sides of the same coin. *Biochemical Society Transactions*, *33*(Pt 6), 1526–1530. https://doi.org/10.1042/BST20051526.

Ardi, V. C., Kupriyanova, T. A., Deryugina, E. I., & Quigley, J. P. (2007). Human neutrophils uniquely release TIMP-free MMP-9 to provide a potent catalytic stimulator of angiogenesis. *Proceedings of the National Academy of Sciences of the United States of America*, *104*(51), 20262–20267. https://doi.org/10.1073/pnas.0706438104.

Aréchaga, J., & Damjanov, I. (2012). Above the borderland between normal and neoplastic development. *The International Journal of Developmental Biology*, *56*(10–12), 939–948. https://doi.org/10.1387/ijdb.120255ja.

Baroni, T., Arato, I., Mancuso, F., Calafiore, R., & Luca, G. (2019). On the origin of testicular germ cell tumors: From gonocytes to testicular cancer. *Frontiers in Endocrinology*, *10*(JUN). Frontiers Media S.A. https://doi.org/10.3389/fendo.2019.00343.

Batool, A., Wang, Y.-Q., Hao, X.-X., Chen, S.-R., & Liu, Y.-X. (2018). A miR-125b/CSF1-CX3CL1/tumor-associated macrophage recruitment axis controls testicular germ cell tumor growth. *Cell Death & Disease*, *9*(10), 962. https://doi.org/10.1038/s41419-018-1021-z.

Bols, B., Jensen, L., Jensen, A., & Braendstrup, O. (2000). Immunopathology of in situ seminoma. *International Journal of Experimental Pathology*, *81*(3), 211–217. https://doi.org/10.1046/j.1365-2613.2000.00151.x.

Bray, F., Ferlay, J., Soerjomataram, I., Siegel, R., Torre, L., & Jemal, A. (2018). *Global Cancer statistics 2018: GLOBOCAN estimates of incidence and mortality worldwide for 36 cancers in 185 countries: Global Cancer statistics 2018* (p. 68). CA: A Cancer Journal for Clinicians. https://doi.org/10.3322/caac.21492.

Brennan, J., Tilmann, C., & Capel, B. (2003). Pdgfr-alpha mediates testis cord organization and fetal Leydig cell development in the XY gonad. *Genes & Development, 17*(6), 800–810. https://doi.org/10.1101/gad.1052503.

Breslin, S., & O'Driscoll, L. (2016). The relevance of using 3D cell cultures, in addition to 2D monolayer cultures, when evaluating breast cancer drug sensitivity and resistance. *Oncotarget*, 7(29), 45745–45756. https://doi.org/10.18632/oncotarget.9935.

Caires, K. C., de Avila, J., & McLean, D. J. (2009). Vascular endothelial growth factor regulates germ cell survival during establishment of spermatogenesis in the bovine testis. *Reproduction (Cambridge, England), 138*(4), 667–677. https://doi.org/10.1530/REP-09-0020.

Carmeliet, P. (2000). Mechanisms of angiogenesis and arteriogenesis. *Nature Medicine, 6*(4), 389–395. https://doi.org/10.1038/74651.

Chen, H., Wang, Y., Ge, R., & Zirkin, B. R. (2017). Leydig cell stem cells: Identification, proliferation and differentiation. *Molecular and Cellular Endocrinology, 445*, 65–73. https://doi.org/10.1016/j.mce.2016.10.010.

Chen, Y., Zhang, S., Wang, Q., & Zhang, X. (2017). Tumor-recruited M2 macrophages promote gastric and breast cancer metastasis via M2 macrophage-secreted CHI3L1 protein. *Journal of Hematology & Oncology, 10*(1), 36. https://doi.org/10.1186/s13045-017-0408-0.

Chiarini-Garcia, H., Raymer, A. M., & Russell, L. D. (2003). Non-random distribution of spermatogonia in rats: Evidence of niches in the seminiferous tubules. *Reproduction (Cambridge, England), 126*(5), 669–680. https://doi.org/10.1530/rep.0.1260669.

Chovanec, M., Cierna, Z., Miskovska, V., Machalekova, K., Svetlovska, D., Kalavska, K., et al. (2017). Prognostic role of programmed-death ligand 1 (PD-L1) expressing tumor infiltrating lymphocytes in testicular germ cell tumors. *Oncotarget, 8*(13), 21794–21805. https://doi.org/10.18632/oncotarget.15585.

Cool, J., DeFalco, T. J., & Capel, B. (2011). Vascular-mesenchymal cross-talk through Vegf and Pdgf drives organ patterning. *Proceedings of the National Academy of Sciences of the United States of America, 108*(1), 167–172. https://doi.org/10.1073/pnas.1010299108.

Crusz, S. M., & Balkwill, F. R. (2015). Inflammation and cancer: Advances and new agents. *Nature Reviews. Clinical Oncology, 12*(10), 584–596. https://doi.org/10.1038/nrclinonc.2015.105.

Damjanov, I., & Wewer-Albrechtsen, N. (2013). Testicular germ cell tumors and related research from a historical point of view. *The International Journal of Developmental Biology, 57*(2–4), 197–200. https://doi.org/10.1387/ijdb.130143id.

De Jong, J., Stoop, H., Gillis, A. J. M., Van Gurp, R. J. H. L. M., Van De Geijn, G. J. M., De Boer, M., et al. (2008). Differential expression of SOX17 and SOX2 in germ cells and stem cells has biological and clinical implications. *Journal of Pathology, 215*(1), 21–30. https://doi.org/10.1002/path.2332.

Dieckmann, K. P., & Skakkebaek, N. E. (1999). Carcinoma in situ of the testis: Review of biological and clinical features. *International Journal of Cancer, 83*(6), 815–822. https://doi.org/10.1002/(sici)1097-0215(19991210)83:6<815::aid-ijc21>3.0.co;2-z.

Díez-Torre, A., Díaz-Núñez, M., Eguizábal, C., Silván, U., & Aréchaga, J. (2013). Evidence for a role of matrix metalloproteinases and their inhibitors in primordial germ cell migration. *Andrology, 1*(5), 779–786. https://doi.org/10.1111/j.2047-2927.2013.00109.x.

Díez-Torre, A., Silván, U., De Wever, O., Bruyneel, E., Mareel, M., & Aréchaga, J. (2004). Germinal tumor invasion and the role of the testicular stroma. *The International Journal of Developmental Biology, 48*(5–6), 545–557. https://doi.org/10.1387/ijdb.041897ad.

Díez-Torre, A., Silván, U., Moreno, P., Gumucio, J., & Aréchaga, J. (2011). Peritubular myoid cell-derived factors and its potential role in the progression of testicular germ cell tumours. *International Journal of Andrology, 34*(4 Pt 2), e252–e264. discussion e264-5. https://doi.org/10.1111/j.1365-2605.2011.01168.x.

Dixon, F. J., & Moore, R. A. (1953). Testicular tumors; a clinicopathological study. *Cancer, 6*(3), 427–454. https://doi.org/10.1002/1097-0142(195305)6:3<427::aid-cncr2820060302>3.0.co;2-u.

Douglas, M. L., Richardson, M. M., & Nicol, D. L. (2006). Testicular germ cell tumors exhibit evidence of hormone dependence. *International Journal of Cancer, 118*(1), 98–102. https://doi.org/10.1002/ijc.21330.

Draper, J. S., Pigott, C., Thomson, J. A., & Andrews, P. W. (2002). Surface antigens of human embryonic stem cells: Changes upon differentiation in culture. *Journal of Anatomy, 200*(Pt 3), 249–258. https://doi.org/10.1046/j.1469-7580.2002.00030.x.

Duluc, D., Delneste, Y., Tan, F., Moles, M.-P., Grimaud, L., Lenoir, J., et al. (2007). Tumor-associated leukemia inhibitory factor and IL-6 skew monocyte differentiation into tumor-associated macrophage-like cells. *Blood, 110*(13), 4319–4330. https://doi.org/10.1182/blood-2007-02-072587.

Eble, J. N., Sauter, G., Epstein, J. I., & Sesterhenn, I. A. (2004). *Pathology and genetics of tumours of the urinary system and male genital organs.*

Fife, B. T., & Pauken, K. E. (2011). The role of the PD-1 pathway in autoimmunity and peripheral tolerance. *Annals of the New York Academy of Sciences, 1217*, 45–59. https://doi.org/10.1111/j.1749-6632.2010.05919.x.

Fijak, M., Pilatz, A., Hedger, M. P., Nicolas, N., Bhushan, S., Michel, V., et al. (2018). Infectious, inflammatory and "autoimmune" male factor infertility: How do rodent models inform clinical practice? *Human Reproduction Update, 24*(4), 416–441. https://doi.org/10.1093/humupd/dmy009.

Fujimoto, T., Yoshinaga, K., & Kono, I. (1985). Distribution of fibronectin on the migratory pathway of primordial germ cells in mice. *The Anatomical Record, 211*(3), 271–278. https://doi.org/10.1002/ar.1092110307.

Fukawa, T., & Kanayama, H. O. (2018). Current knowledge of risk factors for testicular germ cell tumors. *International Journal of Urology, 25*(4), 337–344. Blackwell Publishing. https://doi.org/10.1111/iju.13519.

Ge, R.-S., Dong, Q., Sottas, C. M., Papadopoulos, V., Zirkin, B. R., & Hardy, M. P. (2006). In search of rat stem Leydig cells: Identification, isolation, and lineage-specific development. *Proceedings of the National Academy of Sciences of the United States of America, 103*(8), 2719–2724. https://doi.org/10.1073/pnas.0507692103.

Giannoulatou, E., Maher, G. J., Ding, Z., Gillis, A. J. M., Dorssers, L. C. J., Hoischen, A., et al. (2017). Whole-genome sequencing of spermatocytic tumors provides insights into the mutational processes operating in the male germline. *PLoS One, 12*(5), e0178169. https://doi.org/10.1371/journal.pone.0178169.

Golat, B. T., & Cameron, D. F. (2008). Sertoli cells enhance formation of capillary-like structures in vitro. *Cell Transplantation, 17*(10−11), 1135–1144. https://doi.org/10.3727/096368908787236512.

Gonzalez-Avila, G., Sommer, B., García-Hernández, A. A., & Ramos, C. (2020). Matrix Metalloproteinases' role in tumor microenvironment. *Advances in Experimental Medicine and Biology, 1245*, 97–131. https://doi.org/10.1007/978-3-030-40146-7_5.

Griswold, M. D. (2016). Spermatogenesis: The commitment to meiosis. *Physiological Reviews, 96*(1), 1–17. https://doi.org/10.1152/physrev.00013.2015.

Griswold, S. L., & Behringer, R. R. (2009). Fetal Leydig cell origin and development. *Sexual Development: Genetics, Molecular Biology, Evolution, Endocrinology, Embryology, and Pathology of Sex Determination and Differentiation, 3*(1), 1–15. https://doi.org/10.1159/000200077.

Hadley, M. A., Byers, S. W., Suárez-Quian, C. A., Kleinman, H. K., & Dym, M. (1985). Extracellular matrix regulates Sertoli cell differentiation, testicular cord formation, and germ cell development in vitro. *The Journal of Cell Biology, 101*(4), 1511–1522. https://doi.org/10.1083/jcb.101.4.1511.

Häger, M., Gawlik, K., Nyström, A., Sasaki, T., & Durbeej, M. (2005). Laminin {alpha} 1 chain corrects male infertility caused by absence of laminin {alpha}2 chain. *The American Journal of Pathology*, *167*(3), 823–833. https://doi.org/10.1016/s0002-9440(10)62054-8.

Harpelunde Poulsen, K., Nielsen, J. E., Grønkær Toft, B., Joensen, U. N., Rasmussen, L. J., Blomberg Jensen, M., et al. (2020). Influence of nodal signalling on pluripotency factor expression, tumour cell proliferation and cisplatin-sensitivity in testicular germ cell tumours. *BMC Cancer*, *20*(1), 349. https://doi.org/10.1186/s12885-020-06820-6.

Hawinkels, L. J. A. C., Zuidwijk, K., Verspaget, H. W., de Jonge-Muller, E. S. M., van Duijn, W., Ferreira, V., et al. (2008). VEGF release by MMP-9 mediated heparan sulphate cleavage induces colorectal cancer angiogenesis. *European Journal of Cancer (Oxford, England: 1990)*, *44*(13), 1904–1913. https://doi.org/10.1016/j.ejca.2008.06.031.

Heinrich, A., & DeFalco, T. (2020). Essential roles of interstitial cells in testicular development and function. *Andrology*, *8*(4), 903–914. https://doi.org/10.1111/andr.12703.

Henderson, J. K., Draper, J. S., Baillie, H. S., Fishel, S., Thomson, J. A., Moore, H., et al. (2002). Preimplantation human embryos and embryonic stem cells show comparable expression of stage-specific embryonic antigens. *Stem Cells (Dayton, Ohio)*, *20*(4), 329–337. https://doi.org/10.1634/stemcells.20-4-329.

Hermann, B. P., Sukhwani, M., Hansel, M. C., & Orwig, K. E. (2010). Spermatogonial stem cells in higher primates: Are there differences from those in rodents? *Reproduction (Cambridge, England)*, *139*(3), 479–493. https://doi.org/10.1530/REP-09-0255.

Hvarness, T., Nielsen, J. E., Almstrup, K., Skakkebaek, N. E., Rajpert-De Meyts, E., & Claesson, M. H. (2013). Phenotypic characterisation of immune cell infiltrates in testicular germ cell neoplasia. *Journal of Reproductive Immunology*, *100*(2), 135–145. https://doi.org/10.1016/j.jri.2013.10.005.

Jones, A., Fujiyama, C., Turner, K., Fuggle, S., Cranston, D., Turley, H., et al. (2000). Angiogenesis and lymphangiogenesis in stage 1 germ cell tumours of the testis. *BJU International*, *86*(1), 80–86. https://doi.org/10.1046/j.1464-410x.2000.00660.x.

Jørgensen, A., Macdonald, J., Nielsen, J. E., Kilcoyne, K. R., Perlman, S., Lundvall, L., et al. (2018). Nodal signaling regulates germ cell development and establishment of seminiferous cords in the human fetal testis. *Cell Reports*, *25*(7), 1924–1937.e4. https://doi.org/10.1016/j.celrep.2018.10.064.

Kaleva, M., & Toppari, J. (2003). Genetics and hormones in testicular descent. *Hormones (Athens, Greece)*, *2*(4), 211–216. https://doi.org/10.14310/horm.2002.11102.

Kim, J. W., & Eder, J. P. (2014). Prospects for targeting PD-1 and PD-L1 in various tumor types. *Oncology (Williston Park, N.Y.)*, *28*(Suppl 3), 15–28.

Kim, M., Kasinski, A. L., & Slack, F. J. (2011). MicroRNA therapeutics in preclinical cancer models. *The Lancet. Oncology*, *12*(4), 319–321. https://doi.org/10.1016/S1470-2045(11)70067-5.

Klein, B., Haggeney, T., Fietz, D., Indumathy, S., Loveland, K. L., Hedger, M., et al. (2016). Specific immune cell and cytokine characteristics of human testicular germ cell neoplasia. *Human Reproduction (Oxford, England)*, *31*(10), 2192–2202. https://doi.org/10.1093/humrep/dew211.

Klein, B., Schuppe, H.-C., Bergmann, M., Hedger, M. P., Loveland, B. E., & Loveland, K. L. (2017). An in vitro model demonstrates the potential of neoplastic human germ cells to influence the tumour microenvironment. *Andrology*, *5*(4), 763–770. https://doi.org/10.1111/andr.12365.

Kumari, N., Dwarakanath, B. S., Das, A., & Bhatt, A. N. (2016). Role of interleukin-6 in cancer progression and therapeutic resistance. *Tumour Biology: The Journal of the International Society for Oncodevelopmental Biology and Medicine*, *37*(9), 11553–11572. https://doi.org/10.1007/s13277-016-5098-7.

Laird, D. J., Altshuler-Keylin, S., Kissner, M. D., Zhou, X., & Anderson, K. V. (2011). Ror2 enhances polarity and directional migration of primordial germ cells. *PLoS Genetics*, 7(12), e1002428. https://doi.org/10.1371/journal.pgen.1002428.

Li, J.-H., Rosen, D., Sondel, P., & Berke, G. (2002). Immune privilege and FasL: Two ways to inactivate effector cytotoxic T lymphocytes by FasL-expressing cells. *Immunology*, *105*(3), 267–277. https://doi.org/10.1046/j.1365-2567.2002.01380.x.

Lissbrant, I. F., Lissbrant, E., Persson, A., Damber, J.-E., & Bergh, A. (2003). Endothelial cell proliferation in male reproductive organs of adult rat is high and regulated by testicular factors. *Biology of Reproduction*, *68*(4), 1107–1111. https://doi.org/10.1095/biolreprod.102.008284.

Lobo, J., Rodrigues, Â., Guimarães, R., Cantante, M., Lopes, P., Maurício, J., et al. (2019). Detailed characterization of immune cell infiltrate and expression of immune checkpoint molecules PD-L1/CTLA-4 and MMR proteins in testicular germ cell tumors disclose novel disease biomarkers. *Cancers*, *11*(10). https://doi.org/10.3390/cancers11101535.

Losinno, A. D., Morales, A., Fernández, D., & Lopez, L. A. (2012). Peritubular myoid cells from rat seminiferous tubules contain actin and myosin filaments distributed in two independent layers. *Biology of Reproduction*, *86*(5), 1–8,150. https://doi.org/10.1095/biolreprod.111.095158.

Loveland, K. L., Klein, B., Pueschl, D., Indumathy, S., Bergmann, M., Loveland, B. E., et al. (2017). Cytokines in male fertility and reproductive pathologies: Immunoregulation and beyond. *Frontiers in Endocrinology*, *8*, 307. https://doi.org/10.3389/fendo.2017.00307.

Luca, G., Baroni, T., Arato, I., Hansen, B. C., Cameron, D. F., & Calafiore, R. (2018). Role of Sertoli cell proteins in immunomodulation. *Protein and Peptide Letters*, *25*(5), 440–445. https://doi.org/10.2174/0929866525666180412163151.

Maekawa, M., Kamimura, K., & Nagano, T. (1996). Peritubular myoid cells in the testis: Their structure and function. *Archives of Histology and Cytology*, *59*(1), 1–13. https://doi.org/10.1679/aohc.59.1.

Mäkelä, J.-A., Koskenniemi, J. J., Virtanen, H. E., & Toppari, J. (2019). Testis Development. *Endocrine Reviews*, *40*(4), 857–905. https://doi.org/10.1210/er.2018-00140.

Mamsen, L. S., Brøchner, C. B., Byskov, A. G., & Møllgard, K. (2012). The migration and loss of human primordial germ stem cells from the hind gut epithelium towards the gonadal ridge. *The International Journal of Developmental Biology*, *56*(10 – 12), 771–778. https://doi.org/10.1387/ijdb.120202lm.

Marikawa, Y., Tamashiro, D. A. A., Fujita, T. C., & Alarcón, V. B. (2009). Aggregated P19 mouse embryonal carcinoma cells as a simple in vitro model to study the molecular regulations of mesoderm formation and axial elongation morphogenesis. *Genesis (New York, N.Y. 2000)*, *47*(2), 93–106. https://doi.org/10.1002/dvg.20473.

Marshall, A. H., & Dayan, A. D. (1964). An immune reaction in man against seminomas, dygerminomas, pinealomas, and the mediastinal tumors of similar histological appearance? *Lancet (London, England)*, *2*(7369), 1102–1104. https://doi.org/10.1016/s0140-6736(64)92618-2.

Masaki, K., Sakai, M., Kuroki, S., Jo, J.-I., Hoshina, K., Fujimori, Y., et al. (2018). FGF2 has distinct molecular functions from GDNF in the mouse germline niche. *Stem Cell Reports*, *10*(6), 1782–1792. https://doi.org/10.1016/j.stemcr.2018.03.016.

Meinhardt, A., Wang, M., Schulz, C., & Bhushan, S. (2018). Microenvironmental signals govern the cellular identity of testicular macrophages. *Journal of Leukocyte Biology*, *104*(4), 757–766. https://doi.org/10.1002/JLB.3MR0318-086RR.

Mitchell, R. T., Cowan, G., Morris, K. D., Anderson, R. A., Fraser, H. M., Mckenzie, K. J., et al. (2008). Germ cell differentiation in the marmoset (Callithrix jacchus) during fetal and neonatal life closely parallels that in the human. *Human Reproduction (Oxford, England)*, *23*(12), 2755–2765. https://doi.org/10.1093/humrep/den295.

Mizuno, R., Kawada, K., Itatani, Y., Ogawa, R., Kiyasu, Y., & Sakai, Y. (2019). The role of tumor-associated neutrophils in colorectal Cancer. *International Journal of Molecular Sciences*, *20*(3). https://doi.org/10.3390/ijms20030529.

Moch, H., Cubilla, A. L., Humphrey, P. A., Reuter, V. E., & Ulbright, T. M. (2016). The 2016 WHO classification of Tumours of the urinary system and male genital organs—Part a: Renal, penile, and testicular Tumours. *European Urology*, *70*(1), 93–105. https://doi.org/10.1016/j.eururo.2016.02.029.

Moreno-Ruiz, P., Arluzea, J., Silván, U., Díez-Torre, A., Andrade, R., Bonilla, Z., et al. (2016). Testis peritubular myoid cells increase their motility and express matrix-metalloproteinase 9 (MMP-9) after interaction with embryonal carcinoma cells. *Andrology*, *4*(1), 111–120. https://doi.org/10.1111/andr.12140.

Mossadegh-Keller, N., & Sieweke, M. H. (2018). Testicular macrophages: Guardians of fertility. *Cellular Immunology*, *330*, 120–125. https://doi.org/10.1016/j.cellimm.2018.03.009.

Mullaney, B. P., & Skinner, M. K. (1992). Basic fibroblast growth factor (bFGF) gene expression and protein production during pubertal development of the seminiferous tubule: Follicle-stimulating hormone-induced Sertoli cell bFGF expression. *Endocrinology*, *131*(6), 2928–2934. https://doi.org/10.1210/endo.131.6.1446630.

Nacov, E. (1990). Tumor angiogenesis formation of vessels de novo at germ cell tumors. *Cancer*, *66*(5), 916–922. https://doi.org/10.1002/1097-0142(19900901)66:5<916::aid-cncr2820660517>3.0.co;2-m.

Nagata, S. (1997). Apoptosis by death factor. *Cell*, *88*(3), 355–365. https://doi.org/10.1016/s0092-8674(00)81874-7.

Nakata, H., Wakayama, T., Sonomura, T., Honma, S., Hatta, T., & Iseki, S. (2015). Three-dimensional structure of seminiferous tubules in the adult mouse. *Journal of Anatomy*, *227*(5), 686–694. https://doi.org/10.1111/joa.12375.

Nettersheim, D., Jostes, S., Sharma, R., Schneider, S., Hofmann, A., Ferreira, H. J., et al. (2015). BMP inhibition in seminomas initiates Acquisition of Pluripotency via NODAL signaling resulting in reprogramming to an embryonal carcinoma. *PLoS Genetics*, *11*(7). https://doi.org/10.1371/journal.pgen.1005415.

Oatley, J. M., Oatley, M. J., Avarbock, M. R., Tobias, J. W., & Brinster, R. L. (2009). Colony stimulating factor 1 is an extrinsic stimulator of mouse spermatogonial stem cell self-renewal. *Development (Cambridge, England)*, *136*(7), 1191–1199. https://doi.org/10.1242/dev.032243.

Oduwole, O. O., Peltoketo, H., Poliandri, A., Vengadabady, L., Chrusciel, M., Doroszko, M., et al. (2018). Constitutively active follicle-stimulating hormone receptor enables androgen-independent spermatogenesis. *The Journal of Clinical Investigation*, *128*(5), 1787–1792. https://doi.org/10.1172/JCI96794.

Oosterhuis, J. W., & Looijenga, L. H. J. (2005). Testicular germ-cell tumours in a broader perspective. *Nature Reviews Cancer*, *5*(3), 210–222. https://doi.org/10.1038/nrc1568.

Oosterhuis, J. W., Stoop, H., Honecker, F., & Looijenga, L. H. J. (2007). Why human extragonadal germ cell tumours occur in the midline of the body: Old concepts, new perspectives. *International Journal of Andrology*, *30*(4), 256–264. https://doi.org/10.1111/j.1365-2605.2007.00793.x.

Ostrer, H., Huang, H. Y., Masch, R. J., & Shapiro, E. (2007). A cellular study of human testis development. *Sexual Development: Genetics, Molecular Biology, Evolution, Endocrinology Embryology, and Pathology of Sex Determination and Differentiation*, *1*(5), 286–292. https://doi.org/10.1159/000108930.

Park, M. H., Park, J. E., Kim, M. S., Lee, K. Y., Hwang, J. Y., Yun, J. I., et al. (2016). Effects of extracellular matrix protein-derived signaling on the maintenance of the undifferentiated state of Spermatogonial stem cells from porcine neonatal testis. *Asian-Australasian Journal of Animal Sciences*, *29*(10), 1398–1406. https://doi.org/10.5713/ajas.15.0856.

Payne, C. J. (2013). Cycling to and from a stem cell niche: The temporal and spatial odyssey of mitotic male germ cells. *The International Journal of Developmental Biology*, *57*(2–4), 169–177. https://doi.org/10.1387/ijdb.130006cp.

Pierce, G. B. (1993). On the boundary between development and neoplasia. An interview with professor G. Barry Pierce. Interview by Juan Arechaga. *The International Journal of Developmental Biology*, *37*(1), 5–16.

Rajpert-De Meyts, E., McGlynn, K. A., Okamoto, K., Jewett, M. A. S., & Bokemeyer, C. (2016). Testicular germ cell tumours. *Lancet (London, England)*, *387*(10029), 1762–1774. https://doi.org/10.1016/S0140-6736(15)00991-5.

Rebourcet, D., Wu, J., Cruickshanks, L., Smith, S. E., Milne, L., Fernando, A., et al. (2016). Sertoli cells modulate testicular vascular network development, structure, and function to influence circulating testosterone concentrations in adult male mice. *Endocrinology*, *157*(6), 2479–2488. https://doi.org/10.1210/en.2016-1156.

Ricci-Vitiani, L., Pallini, R., Biffoni, M., Todaro, M., Invernici, G., Cenci, T., et al. (2010). Tumour vascularization via endothelial differentiation of glioblastoma stem-like cells. *Nature*, *468*(7325), 824–828. https://doi.org/10.1038/nature09557.

Rodprasert, W., Virtanen, H. E., Mäkelä, J.-A., & Toppari, J. (2019). *Hypogonadism and Cryptorchidism. Frontiers in Endocrinology*, *10*, 906. https://doi.org/10.3389/fendo.2019.00906.

Runyan, C., Schaible, K., Molyneaux, K., Wang, Z., Levin, L., & Wylie, C. (2006). Steel factor controls midline cell death of primordial germ cells and is essential for their normal proliferation and migration. *Development (Cambridge, England)*, *133*(24), 4861–4869. https://doi.org/10.1242/dev.02688.

Sada, A., & Tumbar, T. (2013). New insights into mechanisms of stem cell daughter fate determination in regenerative tissues. *International Review of Cell and Molecular Biology*, *300*, 1–50. https://doi.org/10.1016/B978-0-12-405210-9.00001-1.

Saitou, M., & Yamaji, M. (2012). Primordial germ cells in mice. *Cold Spring Harbor Perspectives in Biology*, *4*(11). https://doi.org/10.1101/cshperspect.a008375.

Schell, C., Albrecht, M., Mayer, C., Schwarzer, J. U., Frungieri, M. B., & Mayerhofer, A. (2008). Exploring human testicular peritubular cells: Identification of secretory products and regulation by tumor necrosis factor-alpha. *Endocrinology*, *149*(4), 1678–1686. https://doi.org/10.1210/en.2007-1064.

Sekido, R., & Lovell-Badge, R. (2013). Genetic control of testis development. *Sexual Development: Genetics, Molecular Biology, Evolution, Endocrinology, Embryology, and Pathology of Sex Determination and Differentiation*, 7(1–3), 21–32. https://doi.org/10.1159/000342221.

Silván, U., & Aréchaga, J. (2012). Anatomical basis for cell transplantation into mouse seminiferous tubules. *Reproduction (Cambridge, England)*, *144*(3), 385–392. https://doi.org/10.1530/REP-12-0043.

Silván, U., Díez-Torre, A., Andrade, R., Arluzea, J., Silió, M., & Aréchaga, J. (2011). Embryonic stem cell transplantation into seminiferous tubules: A model for the study of invasive germ cell tumors of the testis. *Cell Transplantation*, *20*(5), 637–642. https://doi.org/10.3727/096368910X536581.

Silván, U., Díez-Torre, A., Arluzea, J., Andrade, R., Silió, M., & Aréchaga, J. (2009). Hypoxia and pluripotency in embryonic and embryonal carcinoma stem cell biology. *Differentiation; Research in Biological Diversity*, *78*(2–3), 159–168. https://doi.org/10.1016/j.diff.2009.06.002.

Silván, U., Díez-Torre, A., Bonilla, Z., Moreno, P., Díaz-Núñez, M., & Aréchaga, J. (2015). Vasculogenesis and angiogenesis in nonseminomatous testicular germ cell tumors. *Urolologic Oncology*, *33*(6). https://doi.org/10.1016/j.urolonc.2015.01.005. 268.e17–28.

Sridharan, S., Simon, L., Meling, D. D., Cyr, D. G., Gutstein, D. E., Fishman, G. I., et al. (2007). Proliferation of adult sertoli cells following conditional knockout of the gap junctional protein GJA1 (connexin 43) in mice. *Biology of Reproduction*, *76*(5), 804–812. https://doi.org/10.1095/biolreprod.106.059212.

Stevens, L. C. (1973). A new inbred subline of mice (129-terSv) with a high incidence of spontaneous congenital testicular teratomas. *Journal of the National Cancer Institute, 50*(1), 235–242. https://doi.org/10.1093/jnci/50.1.235.

Takashima, S., Kanatsu-Shinohara, M., Tanaka, T., Morimoto, H., Inoue, K., Ogonuki, N., et al. (2015). Functional differences between GDNF-dependent and FGF2-dependent mouse spermatogonial stem cell self-renewal. *Stem Cell Reports, 4*(3), 489–502. https://doi.org/10.1016/j.stemcr.2015.01.010.

Vêncio, E. F., Pascal, L. E., Page, L. S., Denyer, G., Wang, A. J., Ruohola-Baker, H., et al. (2011). Embryonal carcinoma cell induction of miRNA and mRNA changes in co-cultured prostate stromal fibromuscular cells. *Journal of Cellular Physiology, 226*(6), 1479–1488. https://doi.org/10.1002/jcp.22464.

Viglietto, G., Romano, A., Maglione, D., Rambaldi, M., Paoletti, I., Lago, C. T., et al. (1996). Neovascularization in human germ cell tumors correlates with a marked increase in the expression of the vascular endothelial growth factor but not the placenta-derived growth factor. *Oncogene, 13*(3), 577–587.

von Kopylow, K., & Spiess, A.-N. (2017). Human spermatogonial markers. *Stem Cell Research, 25*, 300–309. https://doi.org/10.1016/j.scr.2017.11.011.

Wan, H.-T., Mruk, D. D., Wong, C. K. C., & Cheng, C. Y. (2013). The apical ES-BTB-BM functional axis is an emerging target for toxicant-induced infertility. *Trends in Molecular Medicine, 19*(7), 396–405. https://doi.org/10.1016/j.molmed.2013.03.006.

Wu, J., Jeong, H. K., Bulin, S. E., Kwon, S. W., Park, J. H., & Bezprozvanny, I. (2009). Ginsenosides protect striatal neurons in a cellular model of Huntington's disease. *Journal of Neuroscience Research, 87*(8), 1904–1912. https://doi.org/10.1002/jnr.22017.

Yakirevich, E., Lefel, O., Sova, Y., Stein, A., Cohen, O., Izhak, O. B., et al. (2002). Activated status of tumour-infiltrating lymphocytes and apoptosis in testicular seminoma. *The Journal of Pathology, 196*(1), 67–75. https://doi.org/10.1002/path.996.

Yao, H. H.-C., Whoriskey, W., & Capel, B. (2002). Desert hedgehog/patched 1 signaling specifies fetal Leydig cell fate in testis organogenesis. *Genes & Development, 16*(11), 1433–1440. https://doi.org/10.1101/gad.981202.

Yoshida, S. (2010). Stem cells in mammalian spermatogenesis. *Development, Growth & Differentiation, 52*(3), 311–317. https://doi.org/10.1111/j.1440-169X.2010.01174.x.

Yoshida, S., Sukeno, M., & Nabeshima, Y.-I. (2007). *A vasculature-associated niche for undifferentiated spermatogonia in the mouse testis. Science (New York, N.Y.), 317(5845), 1722–1726.* https://doi.org/10.1126/science.1144885.

Yoshida, S., Sukeno, M., Nakagawa, T., Ohbo, K., Nagamatsu, G., Suda, T., et al. (2006). The first round of mouse spermatogenesis is a distinctive program that lacks the self-renewing spermatogonia stage. *Development (Cambridge, England), 133*(8), 1495–1505. https://doi.org/10.1242/dev.02316.

Zhao, S., Zhu, W., Xue, S., & Han, D. (2014). Testicular defense systems: Immune privilege and innate immunity. *Cellular & Molecular Immunology, 11*(5), 428–437. https://doi.org/10.1038/cmi.2014.38.

Zhou, R., Wu, J., Liu, B., Jiang, Y., Chen, W., Li, J., et al. (2019). The roles and mechanisms of Leydig cells and myoid cells in regulating spermatogenesis. *Cellular and Molecular Life Sciences : CMLS, 76*(14), 2681–2695. https://doi.org/10.1007/s00018-019-03101-9.

FSC
www.fsc.org
MIX
Papier | Fördert gute Waldnutzung
FSC® C083411

Kolibri 360 GmbH
Ferdinand-Jühlke-Straße 7
99095 Erfurt, Deutschland
produktsicherheit@kolibri360.de